Introduction to Tunnel Construction

Applied Geotechnics

Titles currently in this series:

Geotechnical Modelling
David Muir Wood
Hardback ISBN 978–0–415–34304–6
Paperback ISBN 978–0–419–23730–3

Sprayed Concrete Lined Tunnels
Alun Thomas
Hardback ISBN 978–0–415–36864–3

Forthcoming:

Practical Engineering Geology
Steve Hencher
Hardback ISBN 978–0–415–46908–1
Paperback ISBN 978–0–415–46909–8

Landfill Engineering
Geoff Card
Hardback ISBN 978–0–415–37006–6

Particulate Discrete Element Modelling
Catherine O'Sullivan
Hardback ISBN 978–0–415–49036–8

Advanced Soil Mechanics Laboratory Testing
Richard Jardine et al.
Hardback ISBN 978–0–415–46483–3

Introduction to Tunnel Construction

David Chapman, Nicole Metje and Alfred Stärk

LONDON AND NEW YORK

First published 2010
by Spon Press
2 Park Square, Milton Park, Abingdon, Oxon OX14 4RN

Simultaneously published in the USA and Canada
by Spon Press
270 Madison Avenue, New York, NY 10016, USA

Spon Press is an imprint of the Taylor & Francis Group, an informa business

Typeset in Sabon by
Florence Production Ltd, Stoodleigh, Devon
Printed and bound in Great Britain by
CPI Antony Rowe, Chippenham, Wiltshire

This publication presents material of a broad scope and applicability. Despite stringent efforts by all concerned in the publishing process, some typographical or editorial errors may occur, and readers are encouraged to bring these to our attention where they represent errors of substance. The publisher and author disclaim any liability, in whole or in part, arising from information contained in this publication. The reader is urged to consult with an appropriate licensed professional prior to taking any action or making any interpretation that is within the realm of a licensed professional practice.

The authors have gone to every effort to seek permission from and acknowledge the sources of images which appear in this publication, that have been previously published elsewhere. Nevertheless, should there be any cases where copyright holders have not been correctly identified and suitably acknowledged, the authors and the publisher welcome advice from such copyright-holders and will endeavour to amend the text accordingly on future prints.

British Library Cataloguing in Publication Data
A catalogue record for this book is available from the British Library

Library of Congress Cataloging-in-Publication Data
Chapman, David N.
Introduction to tunnel construction/David N. Chapman, Nicole Metje, and Alfred Stärk.
p. cm. – (Applied geotechnics)
Includes bibliographical references and index.
1. Tunneling. 2. Tunnels – Design and construction.
I. Metje, Nicole. II. Stärk, Alfred. III. Title.
TA805.C45 2010
624.1′93 – dc22 2009044487

ISBN10: 0–415–46841–8 (hbk)
ISBN10: 0–415–46842–6 (pbk)
ISBN10: 0–203–89515–0 (ebk)

ISBN13: 978–0–415–46841–1 (hbk)
ISBN13: 978–0–415–46842–8 (pbk)
ISBN13: 978–0–203–89515–3 (ebk)

Dedicated to Professor Reinhard Rokahr who provided the inspiration and first introduced some of us to the eldorado of tunnelling.
Also dedicated to our families.

Contents

Preface xv
Acknowledgements and permissions xvii
Abbreviations xxi
Symbols xxiii

1 Introduction 1

1.1 Philosophy of tunnelling 1
1.2 Scope of this book 3
1.3 Historical context 3
1.4 The nature of the ground 6
1.5 Tunnel cross section terminology 7
1.6 Content and layout of this book 7

2 Site investigation 9

2.1 Introduction 9
2.2 Site investigation during a project 10
2.2.1 Introduction 10
2.2.2 Desk study 11
2.2.3 Site reconnaissance 11
2.2.4 Ground investigation (overview) 12
2.3 Ground investigation 13
2.3.1 Introduction 13
2.3.2 Field investigations 13
2.3.2.1 Non-intrusive methods 13
2.3.2.2 Intrusive exploration 18
2.3.3 Laboratory tests 31
2.4 Ground characteristics/parameters 41
2.4.1 Influence of layering on Young's modulus 44
2.4.2 Squeezing and swelling ground 45
2.4.3 Typical ground parameters for tunnel design 46

2.4.4 Ground (rock mass) classification 49
2.4.4.1 Rock Quality Designation 49
2.4.4.2 Rock Mass Rating 53
2.4.4.3 Rock Mass Quality Rating (Q-method) 54
2.4.4.4 A few comments on the rock mass classification systems 58

2.5 Site investigation reports 60
2.5.1 Types of site investigation report 60
2.5.2 Key information for tunnel design 61

3 Preliminary analyses for the tunnel 64

3.1 Introduction 64

3.2 Preliminary stress pattern in the ground 64

3.3 Stability of soft ground 66
3.3.1 Stability of fine grained soils 67
3.3.2 Stability of coarse grained soils 69

3.4 The coefficient of lateral earth pressure (K_0) 70

3.5 Preliminary analytical methods 73
3.5.1 Introduction 73
3.5.2 Bedded-beam spring method 74
3.5.3 Continuum method 74
3.5.4 Tunnel support resistance method 76

3.6 Preliminary numerical modelling 78
3.6.1 Introduction 78
3.6.2 Modelling the tunnel construction in 2-D 79
3.6.3 Modelling the tunnel construction in 3-D 81
3.6.4 Choice of ground and lining constitutive models 82

4 Ground improvement techniques and lining systems 84

4.1 Introduction 84

4.2 Ground improvement and stabilization techniques 84
4.2.1 Ground freezing 85
4.2.2 Lowering of the groundwater table 89
4.2.3 Grouting 90
4.2.4 Ground reinforcement 95
4.2.5 Forepoling 98
4.2.6 Face dowels 100
4.2.7 Roof pipe umbrella 101
4.2.8 Compensation grouting 102
4.2.9 Pressurized tunnelling (compressed air) 105

4.3 Tunnel lining systems 108
4.3.1 Lining design requirements 108
4.3.2 Sprayed concrete (shotcrete) 109
4.3.3 Ribbed systems 114
4.3.4 Segmental linings 115
4.3.5 In situ *concrete linings 123*
4.3.6 Fire resistance of concrete linings 125

5 Tunnel construction techniques 127

5.1 Introduction 127
5.2 Open face construction without a shield 128
5.2.1 Timber heading 128
5.2.2 Open face tunnelling with alternative linings 128
5.3 Partial face boring machine (roadheader) 129
5.4 Tunnelling shields 132
5.5 Tunnel boring machines 138
5.5.1 Introduction 138
5.5.2 Tunnel boring machines in hard rock 140
5.5.2.1 Gripper tunnel boring machine 140
5.5.2.2 Shield tunnel boring machines 145
5.5.2.3 General observations for hard rock tunnel boring machines 147
5.5.3 Tunnel boring machines in soft ground 150
5.5.3.1 Introduction 150
5.5.3.2 Slurry tunnelling machines 153
5.5.3.3 Earth pressure balance machines 158
5.5.3.4 Multi-mode tunnel boring machines 161
5.5.3.5 Choice of slurry or earth pressure balance tunnel boring machine 163
5.6 Drill and blast tunnelling 164
5.6.1 Introduction 164
5.6.2 Drilling 165
5.6.3 Charging 168
5.6.4 Stemming 169
5.6.5 Detonating 169
5.6.5.1 Detonating effect 169
5.6.5.2 Types of explosive 170
5.6.5.3 Detonators 172
5.6.5.4 Cut types 174
5.6.5.5 Explosive material requirements 180
5.6.6 Ventilation 180
5.6.7 Mucking and support 182

5.7 New Austrian Tunnelling Method and sprayed concrete lining 183
5.7.1 New Austrian Tunnelling Method 183
5.7.2 Sprayed concrete lining 187
5.7.3 LaserShell™ technique 192

5.8 Cut-and-cover tunnels 193
5.8.1 Introduction 193
5.8.2 Construction methods 193
5.8.3 Design issues 195
5.8.4 Excavation support methods (shoring systems) for the sides of the excavation 196

5.9 Immersed tube tunnels 201
5.9.1 Introduction 201
5.9.2 Stages of construction for immersed tube tunnels 203
5.9.3 Types of immersed tube tunnel 206
5.9.3.1 Steel shell 206
5.9.3.2 Concrete 206
5.9.4 Immersed tube tunnel foundations and settlements 209
5.9.5 Joints between tube elements 209
5.9.6 Analysis and design 211
5.9.7 Examples of immersed tube tunnels 213

5.10 Jacked box tunnelling 216
5.10.1 Introduction 216
5.10.2 Outline of the method and description of key components 216
5.10.3 Examples of jacked box tunnels 221
5.10.3.1 Vehicular under-bridge, M1 motorway, J15A, Northamptonshire, UK 221
5.10.3.2 I-90 Highway Extension, Boston, Massachusetts, USA 226

5.11 Pipe jacking and microtunnelling 230
5.11.1 Introduction 230
5.11.2 The pipe jacking construction process 231
5.11.3 Maximum drive length for pipe jacking and microtunnelling 235

5.12 Horizontal directional drilling 235

6 Health and safety, and risk management in tunnelling 244

6.1 The health and safety hazards of tunnel construction 244
6.1.1 Introduction 244
6.1.2 Hazards in tunnelling 245
6.1.3 Techniques for risk management 245

6.1.4 Legislation, accidents and ill health statistics 246
6.1.5 Role of the client, designer and contractors 247
6.1.6 Ground risk 248
6.1.7 Excavation and lining methods 249
6.1.8 Tunnel boring machines 249
6.1.9 Tunnel transport 250
6.1.10 Tunnel atmosphere and ventilation 250
6.1.11 Explosives 251
6.1.12 Fire, flood rescue and escape 251
6.1.13 Occupational health 252
6.1.14 Welfare and first aid 253
6.1.15 Work in compressed air 253
6.1.16 Education, training and competence 254
6.1.17 Concluding remarks 255

6.2 Risk management in tunnelling projects 255
6.2.1 Introduction 255
6.2.2 Risk identification 258
6.2.3 Analyzing risks 258
6.2.4 Evaluating risks 259
6.2.5 Risk monitoring and reviewing 259

7 Ground movements and monitoring 262

7.1 Ground deformation in soft ground 262
7.1.1 Surface settlement profiles 263
7.1.1.1 Estimating the trough width parameter, i 266
7.1.1.2 Volume loss 268
7.1.2 Horizontal displacements 269
7.1.3 Long-term settlements 270
7.1.4 Multiple tunnels 271

7.2 Effects of tunnelling on surface and subsurface structures 271
7.2.1 Effect of tunnelling on existing tunnels, buried utilities and piled foundations 272
7.2.2 Design methodology 276

7.3 Monitoring 280
7.3.1 Challenges and purpose 280
7.3.2 Trigger values 282
7.3.3 Observational method 283
7.3.4 In-tunnel monitoring during New Austrian Tunnelling Method tunnelling operations 285
7.3.4.1 Measurements 285
7.3.4.2 General development of displacements 287

7.3.4.3 Interpretation of the measurements: displacements 289
7.3.4.4 Interpretation of the measurements: comparative observation 291
7.3.4.5 Interpretation of the measurements: deformation 293
7.3.4.6 Interpretation of the measurements: stress-intensity-index 296
7.3.4.7 Measuring frequency and duration 298
7.3.4.8 Contingency measures 298
7.3.5 Instrumentation for in-tunnel and ground monitoring 304
7.3.6 Instrumentation for monitoring existing structures 307

8 Case studies 311

8.1 Eggetunnel, Germany 311
8.1.1 Project overview 311
8.1.2 Invert failure of the total cross section in the Eggetunnel 312
8.1.3 Sprayed concrete invert – its purpose and monitoring 314

8.2 London Heathrow T5, UK: construction of the Piccadilly Line Extension Junction 319
8.2.1 Project overview 319
8.2.2 The 'Box' 319
8.2.3 Construction of the sprayed concrete lining tunnels 321
8.2.4 Ground conditions 321
8.2.5 The LaserShell™ method 322
8.2.6 TunnelBeamer™ 323
8.2.7 Monitoring 325
8.2.7.1 Existing Piccadilly Tunnel Eastside 325
8.2.7.2 Existing Piccadilly Tunnel Westside 325

8.3 Lainzer Tunnel LT31, Vienna, Austria 330
8.3.1 Project overview 330
8.3.2 Geology 333
8.3.3 Starting construction from the shafts 333
8.3.4 Side wall drift section: excavation sequence and cross section 334
8.3.5 Monitoring of the sprayed concrete lining of the side wall drift section 339
8.3.6 Cracks in the sprayed concrete lining 339

Appendix A: Further information on rock mass classification systems **345**

A.1 Rock Mass Rating 345

A.2 Rock Mass Quality Rating (Q) 350
A.2.1 Use of the Q-method for predicting TBM performance 354

Appendix B: Analytical calculation of a sprayed concrete lining using the continuum method **356**

B.1 Introduction 356

B.2 Analytical model using Ahrens *et al.* (1982) 357

B.3 Required equations and calculation process 358

B.4 Example for a tunnel at King's Cross Station, London 361

References and bibliography 368
Index 385

Preface

This book seeks to provide an introduction to tunnel construction for people who have little experience of the subject. Tunnelling is an exciting subject and is unlike any other form of construction, as the ground surrounding the tunnel is an integral part of the final structure and plays a pivotal role in its stability. The 'art' of tunnelling cannot be learnt purely from books and a lot of essential decisions are based on engineering judgement, experience and even emotion. There is often no single answer to any question: often the response has to be 'it depends'.

So how can this book help the reader to understand tunnelling? The aim of the book is to provide the reader with background information so that he or she can either make an informed decision and/or consult more specialist references on a specific topic. It will hopefully give the reader the tools needed to critically assess tunnel construction techniques and to realize that not all can be learnt from textbooks. In addition, the book hopes to demonstrate the breadth of the subject and that to become a tunnelling expert, many years of experience are required. At the same time, the book hopes to show the reader the excitement associated with tunnelling and the fact that many unknowns exist which require engineering judgement.

Disclaimer

While every effort has been made to check the integrity and quality of the contents, no liability is accepted by either the publisher or the authors for any damages incurred as the result of the application of information contained in this book. Where values for parameters have been stated, these should be treated as indicative only. Readers should independently verify the properties of materials they are dealing with as they may differ substantially from those referred to in this book.

This publication presents material of a broad scope and applicability. Despite stringent efforts by all concerned in the publishing process, some typographical or editorial errors may occur. Readers are encouraged to bring these to our attention where they represent errors of substance. The publisher and authors disclaim any liability, in whole or in part, arising from information contained in this publication. Readers are urged to consult with an appropriate licensed professional prior to taking any action or making any interpretation that is within the realm of a licensed professional practice.

Acknowledgements and permissions

The authors would like to express their deep gratitude to their colleagues at the Institute of Tunnelling and Underground Construction (IUB), especially Professor Reinhard Rokahr without whose support and encouragement this book would not have materialised. Special thanks also go to Dr Donald Lamont who contributed to the health and safety section of this book, Dr Alexander Royal for his contribution to the sections on pipe jacking and horizontal directional drilling, Graham Chapman for reading through some of the manuscript and Qiang Liu for producing some of the figures. The authors would also like to thank all those people who reviewed the book critically before it went to print and thus making the book better for it, especially Dr Douglas Allenby (BAM Nuttall Ltd), Martin Caudell (Soil Mechanics), Dr Michael Cooper, Colin Eddie (Underground Professional Services Ltd), Robert Essler (RD Geotech Ltd), Dr Dexter Hunt (University of Birmingham), Christian Neumann (ALPINE BeMo Tunnelling GmbH Innsbruck), Dr Barry New (Geotechnical Consulting Group), Casper Paludan-Müller (Cowi A/S), Roy Slocombe (Herrenknecht UK), Dr Alun Thomas (Mott MacDonald) and Dr-Ing. Rudolf Zachow (IUB, Hanover University).

The authors would like to acknowledge the following people and organizations who have assisted and/or kindly granted permission for certain figures, tables and photographs to be reproduced in this book:

Companies and persons who gave permission to use photographs, figures and tables (acknowledged in the text):

- Aker Wirth GmbH
- ALPINE BeMo Tunnelling, GmbH Innsbruck
- Atlas Copco
- Bachy Soletanche Ltd
- BAM Nuttall Ltd. and John Ropkins Ltd
- John Bartlett
- Dr Nick Barton
- Dr John Billam
- British Drilling & Freezing Co. Ltd
- David Caiden
- Professor E.J. Cording
- COWI A/S
- Dosco Overseas Engineering Ltd
- Don Deere
- Dyno Nobel Inc.
- Geopoint Systems BV

- Herrenknecht GmbH
- Peter Jewell
- Dr Ron Jones
- Mike King
- London Underground Ltd
- Lovat
- Professor Robert Mair
- Massachusetts Turnpike Authority
- Mitsubishi Heavy Industries Mechatronics Systems Ltd
- NoDig Media Services
- Prime Drilling HDD-Technology
- The Robbins Company
- Professor Dr-Ing. habil. Reinhard B. Rokahr
- Rowa Tunnelling Logistics
- Dr Alexander Royal
- Alex Sala
- Soil Mechanics
- Wilde FEA Ltd

Those who granted us permission to use figures and tables (in addition to those acknowledged in the main text):

- Figures 2.6, 2.12, 2.13, 2.16, 2.17, 2.18, 2.19, 2.20, 3.1, 3.5, 3.6, 3.7, 4.17, 5.3, 5.12, 5.20, 5.38, 5.39, 5.42, 5.43, 5.44, 5.45, 5.46, 5.48, 7.11, 7.12, 7.13, 7.23 and Table 1.1: Institute of Tunnelling and Underground Construction, Hanover University, Germany
- Figures 2.2a, 2.2b and 2.3: from *Transportation Research Circular E-C130: Geophysical Methods Commonly Employed for Geotechnical Site Characterization*, Transportation Research Board of the National Academies, Washington, DC, 2008, Figures 2a and b (p. 6); Figure 3a (p. 7); and Figure 4a (p. 8). Reproduced with permission from the Transportation Research Board and Dr N.L. Anderson.
- Figure 2.10a: from *NCHRP Synthesis 368: Cone Penetration Testing*, Transportation Research Board, P.W. Mayne, National Research Council, Washington, DC, 2007, Figure 1 (p. 6). Reproduced with permission from the Transportation Research Board.
- Figure 2.11: reproduced with permission of CIRIA from B2 – *Cone Penetration Testing: Methods and Interpretation*, CIRIA, London, 1987, Figure 10 (p. 20).
- Figure 2.15: from DIN 18196 *Earthworks and Foundations: Soil Classification for Civil Engineering Purposes*. Reproduced by permission of DIN Deutsches Institut für Normung e.V. The definitive version for the implementation of this standard is the edition bearing the most recent date of issue, obtainable from Beuth Verlag GmbH, 10772 Berlin, Germany.
- Figures 2.22, A.1 and Tables 2.17, A.1: reproduced with permission from John Wiley & Sons, Inc., from *Engineering Rock Mass Classifications*, Z.T. Bieniawski, 1989, Figure 4.1 (p. 61); Charts A–D (pp. 56–7); Table 4.4 (p. 62); and Table 4.1 (p. 54).
- Figures 2.24, 2.25: reprinted from 'Use and misuse of rock mass classification systems with particular reference to the Q-system', *Tunnelling and Underground Space Technology*, 21(6), A. Palmström and E. Broch, 2006, Figure 7 (p. 584) and Figure 10 (p. 588), with permission from Elsevier; Figure 7 (p. 584), with additional permission from N. Barton.

- Tables 2.2, 2.4 and 2.6: permission to reproduce extracts from BS EN 1997-2:2007, BS EN 1997-2:2007 and BS EN ISO 14688–2: 2004, respectively, is granted by BSI. British Standards can be obtained in PDF or hard copy formats from the BSI online shop: www.bsigroup.com/Shop or by contacting BSI Customer Services for hard copies only: Tel: +44 (0)20 8996 9001, Email: cservices@bsigroup.com.
- Table 2.15: reprinted with permission from *Applied Sedimentation*, 1950 by the National Academy of Sciences, courtesy of the National Academies Press, Washington, DC, and also with kind permission from Springer Science & Business Media and James Thomson, *Pipejacking and Microtunnelling*, Table 9.1, original copyright Chapman and Hall, 1993.
- Table 2.16: used with kind permission from the American Institute of Mining, Metallurgical, and Petroleum Engineers, New York from 'Failure and breakage of rock', *Proceedings of the 8th US Symposium on Rock Mechanics (USRMS)*, C. Fairhurst (ed.), 1967, 'Design of surface and near-surface construction in rock', D.U. Deere, A.J. Hendron Jr., F.D. Patton and E.J. Cording, Figure 5b (p. 250).
- Table 2.18: with kind permission from Springer Science & Business Media: 'Engineering classification of rock masses for the design of tunnel support', *Rock Mechanics and Rock Engineering*, 6, N. Barton, R. Lien and J. Lunde, 1974, Table 3, permission also obtained from N. Barton.
- Figures 3.2, 4.11 and 4.12: reprinted from 'Settlements induced by tunneling in soft ground', *Tunnelling and Underground Space Technology*, 22(2), International Tunnelling Association, 2007, Figure 8 (p. 122); Figure 19 (p. 140) and Figure 18 (p. 140), with permission from Elsevier.
- Figure 4.2: reproduced with kind permission from Pearson Education from F.C. Harris, *Exploring Modern Construction & Ground Engineering Equipment & Methods*, 1994.
- Figures 4.4a, 4.5a and b, 4.7, 4.8, 5.56, 5.58a and b: reproduced with kind permission from Taylor & Francis from *An Introduction to Geotechnical Processes*, J. Woodward, 2005, pages 36, 97, 96, 96, 53 and 56, respectively.
- Figure 4.15: reproduced with permission of CIRIA from SP200 – *Building Response to Tunnelling: Case Studies from Construction of the Jubilee Line Extension, London. Volume 1: The Project*, CIRIA, London, 2002, Figure 11.3 (p. 141).
- Figures 4.22, 5.24 and 5.29: reproduced with kind permission from Maney Publishing (www.maney.co.uk) from B.N. Whittaker and R.C. Frith, *Tunnelling: Design, Stability and Construction*, 1990, The Institution of Mining and Metallurgy, London, Figure 4.6 (p. 83); Figure 4.7 (p. 83) and Figure 14.2 (p. 334).
- Figures 5.52, 5.53 and 5.58c: reproduced with kind permission from Springer Publishers from *Tunnel Engineering Handbook*, Second Edition, T.R. Kuesel and E.H. King (eds), 1996, Figure 17.6 (p. 325) and Figures 17.9b and e (p. 330).
- Figures 5.61 and 5.62: reprinted from 'State of the art report in immersed and floating tunnels', *Tunnelling and Underground Space Technology*, 12(2), International Tunnelling Association, 1997, Figure 3.1 (p. 97) and Figure 3.2 (p. 98), with permission from Elsevier.
- Figures 5.73, 5.74, 5.75 and 5.76: used with kind permission from Thomas Telford Ltd and Dr Douglas Allenby, from 'The use of jacked-box tunnelling under a live motorway', *Proceedings of the Institution of Civil Engineers, Geotechnical Engineering*, D. Allenby and J.W.T. Ropkins, 2004, Figure 3 (p. 232); Figure 5 (p. 234); Figure 9 (p. 237) and Figure 10 (p. 237).

- Figures 7.3 and 7.5: reproduced with kind permission from Taylor and Francis, from 'Theme lecture: bored tunnelling in the urban environment', *Proceedings of the 14th International Conference on Soil Mechanics and Foundation Engineering*, R.J. Mair and R.N. Taylor, 1997, Figure 22 (p. 2362) and Figure 25 (p. 2363).
- Figure 7.4: reproduced with kind permission from John Wiley and Sons Ltd, from 'The response of buried pipelines to ground movements caused by tunnelling in soil', *Ground Movements and Structures*, J.D. Geddes (ed.), J. Yeates, 1985, Figure 1 (p. 131), original copyright Pentech Press, UK.
- Figure 7.6: reproduced with kind permission from Taylor & Francis from *Soil Movements Induced by Tunneling and their Effects on Pipelines and Structures*, P.B. Attewell, J. Yeates and A.R. Selby, 1986, Figure 4.35 (p. 283), permission also obtained from Dr A.R. Selby.
- Figure 7.7: reproduced with kind permission from Taylor and Francis, from 'The response of full-scale piles to tunnelling', *Proceedings of the 5th International Conference of TC28 of the ISSMGE, Geotechnical Aspects of Underground Construction in Soft Ground*, D. Selemetas, J.R. Standing and R.J. Mair, 2006, Figure 10 (p. 768).
- Figure 7.8: from *Geotechnical Instrumentation for Monitoring Field Performance*, J. Dunnicliff and G.E. Green, copyright 1988, 1993 John Wiley & Sons, Inc., Figure 7.1 (p. 76). This material is reproduced with permission of John Wiley & Sons, Inc.
- Figure 7.26: reproduced with kind permission from Taylor & Francis, from 'The measurement of ground movements due to tunnelling at two control sites along the Jubilee Line Extension', *Proceedings of the International Conference of TC28 of the ISSMGE, Geotechnical Aspects of Underground Construction in Soft Ground*, J.R. Standing, R.J. Nyren, J.B. Burland and T.I. Longworth, 1996, Figures 2 and 3 (p. 752).
- Table 7.3: with kind permission from Springer Science & Business Media, 'Behaviour of foundations and structures', *Proceedings of the Ninth International Conference on Soil Mechanics and Foundation Engineering*, J.B. Burland, B.B. Broms and V.F.B. de Mello 1977, Table 1 (p. 500), permission also obtained from Professor J.B. Burland.
- Figure A.2: reprinted from 'TBM performance estimation using rock mass classification', *International Journal of Rock Mechanics and Mining Sciences*, 39, M. Sapigni, M. Berti, E. Bethaz, A. Busillo and G. Cardone, 2002, pp. 771–88, with permission from Elsevier.
- Tables A.2–A.7: reprinted from 'Some new Q-value correlations to assist in site characterization and tunnel design', *International Journal of Rock Mechanics and Mining Sciences*, 39(2), N. Barton, pp. 185–216, 2002, with permission from Elsevier.
- Quoted material used with permission from Thomas Telford Ltd and Springer Publishers.

Abbreviations

2-D	two-dimensional
3-D	three-dimensional
ADS	anti-drag system
BSI	British Standards Institute
BTS	British Tunnelling Society
CDM	cement deep mixing
CPT	cone penetration test
CTRL	Channel Tunnel Rail Link
EPBM	earth pressure balance machine
EPDM	ethylene-propylene-diene monomer
ESR	excavation support ratio
FSTT	French Society for Trenchless Technology
GBR	Geotechnical Baseline Report
GFR	Geotechnical Factual Report
GIR	Geotechnical Interpretive Report
GSL	ground surface level
GWL	groundwater level (table)
HDD	horizontal directional drilling
HDPE	high density polyethylene
HME	Hypothetical Modulus of Elasticity model
HSE	Health and Safety Executive, UK
ICE	Institution of Civil Engineers, UK
ISRM	International Society for Rock Mechanics
ITA	International Tunnelling Association
ITIG	International Tunnelling Insurance Group
LF	load factor (= N/N_c)
LHS	left-hand side
LVDT	linear variable differential transformer
NATM	New Austrian Tunnelling Method
ÖBV	Österreichischer Beton Verein
PFA	pulverized fuel ash
PiccEx	Piccadilly Line Extension
PJA	Pipe Jacking Association
RHS	right-hand side
RMR	Rock Mass Rating
RQD	Rock Quality Designation

SCL sprayed concrete lining
SCR solid core recovery
SGI spheroidal graphite (cast) iron
SISG Site Investigation Steering Group, ICE, UK
SPT standard penetration test
SRF stress reduction factor
SSP seismic soft-ground probing
STM slurry tunnelling machine
SWOT Storm Water Outfall Tunnel
TAM tube-a-manchette
TBM tunnel boring machine
TCR total core recovery
TSG tail shield grease
UK United Kingdom
VSP vertical seismic profiling

Symbols

γ	(bulk) unit weight of ground (kN/m^3)
γ_d	(bulk) unit weight for ground above the groundwater table (kN/m^3)
γ_{sat}	(bulk) unit weight for ground below the groundwater table (kN/m^3)
γ_w	unit weight of water (kN/m^3)
$\Delta\varepsilon$	change in strain
$\Delta\sigma$	change in stress (MN/m^2)
$\Delta\sigma_m$	average normal stress on the load plates (MN/m^2)
$\Delta S_{Z,R}$	average settlements of the centre and the edge of the load plate (mm)
ΔV	potential difference
$\dot{\varepsilon}$	strain rate
ε_u	ultimate strain at failure
ε_{horiz}	horizontal strain
ε_{pl}	plastic strain
ε_{vert}	vertical strain
η	stress-intensity-index
λ	parameter to describe the proportion of unloading in the convergence–confinement method
λ_d	predetermined value of the parameter λ
μ	Poisson's ratio
σ	total stress (kN/m^2)
σ'	effective stress (kN/m^2)
σ_1, σ_2, (σ_3)	principal stresses (kN/m^2)
σ_3, (σ_2)	confining stress for triaxial test (kN/m^2)
σ_v	total vertical stress (kN/m^2)
σ_v'	effective vertical stress (kN/m^2)
σ_h	total horizontal stress (kN/m^2)
σ_h'	effective horizontal stress (kN/m^2)
σ_s	surcharge acting on the ground surface (kN/m^2)
σ_T	tunnel face support pressure (kN/m^2)
σ_u	ultimate stress at failure (MN/m^2)
$\sigma_{u,adj}$	adjusted σ_u for uniaxial test (MN/m^2)
ϕ	internal friction angle (°)
ϕ'	effective internal friction angle = angle of shearing resistance (°)

ϕ_u	undrained internal friction angle (°)
ω	constant for the type of loading plate
c	apparent cohesion (kN/m^2)
c'	effective apparent cohesion (kN/m^2)
C	overburden to tunnel crown (or cover depth) (m)
c_u, s_u	undrained shear strength (kN/m^2)
c_v	coefficient of consolidation (mm^2/min)
d	sample diameter for uniaxial test and point load index test (mm)
D	diameter of tunnel (m)
D_e	equivalent dimension of the excavation (m)
D_r	relative density of coarse grained soils
E	Young's modulus (kN/m^2)
E'	drained deformation modulus (kN/m^2)
E_d	deformation modulus (kN/m^2)
E_s	stiffness modulus (kN/m^2)
E_v'	vertical drained deformation modulus from oedometer test (kN/m^2)
f_1	factor to allow for the plasticity index
f_s	sleeve friction for CPT (MN/m^2)
G_{max}	shear stiffness/modulus (kN/m^2)
Gs	specific gravity (kN/m^3)
H	depth from the ground surface to tunnel axis (C+D/2) (m)
h	sample height for a uniaxial test (mm)
h	horizontal displacement of footing (mm)
I	current (A)
i	trough width parameter (m)
I_c	consistency index
I_L	liquidity index
I_P	plasticity index
I_S	point load index strength (MN/m^2)
J_a	joint alteration number for Q-method
J_n	joint set number for Q-method
J_r	joint roughness number for Q-method
J_v	sum of the number of joints per unit length for the RQD index
J_w	joint water reduction factor for Q-method
k	permeability (m/s)
K	trough width factor
K_a	active coefficient of lateral earth pressure
K_p	passive coefficient of lateral earth pressure
K_0	coefficient of lateral earth pressure at rest
L	failure load in point load index test (MN)
L_1	interface between two strata
m_v	coefficient of volume compressibility (m^2/MN)
N_{SPT}	standard penetration test blow count
N	stability ratio
N_c	critical stability ratio or stability ratio at collapse
P	length of unsupported tunnel ahead of tunnel shield or lining (m)
P_T	support resistance (kN/m^2)
p_h	horizontal pressure (kN/m^2)

p_v	vertical pressure (kN/m^2)
Q	Q-value for Rock Mass Quality Rating method
q_c	cone tip resistance for CPT (MN/m^2)
Q_c	normalized Q-value
Q_{TBM}	Q-value for TBM tunnelling
r	radius of the load plate (m)
R_f	friction ratio for CPT (%)
S	surface settlement (mm)
S_h	horizontal ground displacement (mm)
S_{max}	maximum surface settlement directly above the tunnel centreline (mm)
S_v	vertical ground displacement (mm)
T_γ	tunnel stability number for the soil load
T_s	tunnel stability number for surface surcharge
u	pore pressure (kN/m^2)
UCS, q_u	unconfined compressive strength (MN/m^2)
V_l	volume loss per metre length of tunnel (m^3/m)
V_o	excavated volume of the tunnel per metre length of tunnel (m^3/m)
V_p	seismic velocity (m/s)
V_S	volume of the surface settlement trough per metre length of tunnel (m^3/m)
V_t	estimated volume loss per metre length of tunnel (m^3/m)
v	vertical displacement of footing (mm)
w	water (moisture) content (%)
w	settlement of the tunnel crown (mm)
w_{crit}	critical settlement of the tunnel crown (mm)
w_L	liquid limit (%)
w_P	plastic limit (%)
x, y, z	co-ordinate axes
y	transverse horizontal distance from tunnel centreline (m)
z	depth from the ground surface (m)
z_w	depth below groundwater table (m)

Note that all logarithmic terms are $\log_{10}$ in this book.

1 Introduction

1.1 Philosophy of tunnelling

Tunnels are unlike any other civil engineering structures. In buildings or bridges the building materials have defined and testable properties, whereas this is not the case in tunnelling. Table 1.1 illustrates some of the issues associated with tunnel design when compared to above ground construction projects.

Although a tunnel structure often needs support systems made up of concrete and steel, it is the ground that is the major part of the structure, and this can have both a supporting and a loading role. The key to successful tunnel construction is therefore to understand this material, in particular

Table 1.1 Comparison between tunnels and above ground construction projects

	Above ground construction	*Tunnel construction*
Construction material	The defined properties of the construction materials are guaranteed by the quality control procedures during the production process, including control testing.	The ground, with all its uncertainty, and the general inability to influence its properties (notwithstanding ground improvement techniques) is the construction material.
Loads	The loads for which the structural analysis is carried out are mostly known.	Only by making assumptions is it possible to estimate the loads possible, which means that the magnitude of the load is based on assumption and is thus basically unknown.
Safety	Because the properties of the construction materials and the loads are known, the safety factor relative to failure can be determined.	Because of the number of uncertainties related to the loads and material properties it is not possible to calculate a quantitative factor regarding the safety of the tunnel construction.

its strength and stability characteristics. No matter how much of the ground we test in preliminary site investigations, how many borehole cores we take for testing in the laboratory, we can only ever test a small fraction of the total ground to be affected by the tunnel construction. Therefore, it is up to the engineer to determine the relevant ground conditions and its associated properties. But as only a small fraction of the material can be tested and with limited knowledge of, for example, the effects of layering, fissures and discontinuities, much of this assessment is based on judgement and experience. One might even suggest that emotions are involved. So how can this then be used as the basis for tunnel design? It is up to engineering judgement to interpret the site investigation report and suggest suitable design and construction techniques.

Often, the assumption is that the ground acts as a continuum and allows three-dimensional stress redistribution around the tunnel void, thus taking some of the load, so that not the full overburden acts as the load on the tunnel. But how can anyone determine the percentage of this load-bearing capacity? Again, this comes down to engineering judgement. If a tunnel is lined using sprayed concrete, how can the residual stress-intensity-index be determined for the lining? If the displacement of discrete points is measured, how do we know that the maximum displacement has not been exceeded and the tunnel is not in danger of collapse? When is a crack in the tunnel lining significant and a sign of worse to come? Often it simply comes down to engineering judgement and experience. Many of these questions do not have a single answer, but depend on the individual case. No new tunnel construction is the same as a previous one. During the construction of a tunnel it is important to listen to the miners who have worked in many tunnels and use their experience to respond to different behaviours of the ground when excavating the tunnel. The key is to understand that tunnelling is not a discrete science with definite answers. There are many unknowns and the answer to most of the above questions is ‘it depends’.

Experience and engineering judgement help to make a considered and informed decision, but continuous measurements during construction are essential to compare actual behaviour with those predicted. This book does not propose to give the reader all the answers related to any tunnel construction. Rather, its aim is to provide the reader with background information so that he or she can either make an informed decision and/or consult more specialist references on a specific topic. It will hopefully give the reader the tools needed to critically assess tunnel constructions and to realize that not all can be learnt from textbooks but that, to become a tunnelling expert, many years of experience are required. At the same time, this book hopes to demonstrate to the reader the excitement associated with tunnelling and to make it clear that there are many unknowns that require engineering judgement. Solving these issues is the challenge the civil engineer faces. If the reader takes away one message from this book, it should be that the answer to a lot of questions regarding tunnelling design and construction is 'it depends' and sometimes using emotions is essential to overcome the challenges posed by tunnelling.

1.2 Scope of this book

Tunnelling is an extensive topic and so the objective of this book is to provide a general knowledge base and guidance for further reading. It not only concentrates on different tunnel construction techniques but also brings in associated relevant topics such as site investigation, which have a large impact on the final tunnel design and its subsequent construction. It is important to note that tunnels in the context of this book include all types of tunnels not only the larger-scale metro, road and rail tunnels, but also utility tunnels for water, sewerage and cables.

This textbook aims to provide a comprehensive introduction to tunnel construction. It is aimed at undergraduate and postgraduate students with little or no previous experience and knowledge of tunnel construction, as well as recently graduated engineers who find themselves working in this exciting field of civil engineering.

1.3 Historical context

There has been considerable development in tunnel construction techniques in the last 200 years, especially since Marc Brunel's famous first use of a tunnelling shield when constructing the first tunnel under the River Thames in London in 1825. Nevertheless, if Marc or his son Isambard Kingdom Brunel were to look at today's tunnelling methods they would see certain similarities with the techniques used in their day, particularly drill and blast and even tunnel boring machines (TBMs). The primary purpose of a TBM is to provide stability to the face and the surrounding ground, thus improving health and safety for the tunnellers, just as Brunel's own Thames Tunnel shield did. Although they would also notice great advances in technology, it would probably be the extent to which tunnelling has been used around the world and the sheer scale of many of these tunnels in terms of diameter, length and difficult construction conditions that would amaze them the most.

There are a number of detailed histories of the engineering art that is tunnelling, and this history is not reproduced here. The reader is directed to Sandström (1963), Beaver (1973), Megaw and Bartlett (1981), West (1988) and Muir Wood (2000) for further information. However, some tunnel constructions that marked key developments for 'modern' (from 1666) tunnelling are as follows.

- The first recorded use of gunpowder as a construction tool was for a pioneering tunnel of the canal age. This was constructed on the Canal du Midi, a canal built across France in the years 1666–81 connecting the Atlantic Ocean to the Mediterranean Sea. The main tunnel on this route was 157 m long with a rectangular cross section of 6.5 m by 8 m, and was built during the years 1679–81.

- Civil engineering as a profession was largely created in the UK by the development of the canal system, which itself was part of the industrial revolution of the eighteenth century. Two significant tunnels of this era included the 2090 m Harecastle Tunnel, constructed using gunpowder as part of the Grand Trunk canal during the 1770s, and the 5000 m long tunnel at Standedge, constructed through millstone grit. This latter took 17 years to complete and opened in 1811.
- The first tunnel underneath a navigable waterway was a tunnel under the River Thames in London, between Rotherhithe and Wapping. This involved using a tunnelling shield known as 'Brunel's Shield', designed by Marc Brunel. Construction of this brick-lined tunnel started in 1825 and it finally opened in 1842. The key function of this shield was to support the face and provide safety for the miners. The shield was made from cast iron (81.3 tonnes/80 tons), was 11.6 m (38 ft) wide and 6.8 m (22 ft. 6 in) tall and was made up of 12 parallel frames, each 0.9 m (3 ft) wide (Figure 1.1). In addition, there was a movable working platform on which the miners threw the spoil, and which was also used by the masons erecting the brick lining (Sandström 1963).
- Considerable amounts of tunnelling took place in the UK as a result of the coming of the railways, which started with the Liverpool to Manchester Railway opening in 1830. Water was a major problem for many of the tunnel projects. Between 1830 and 1890 over 50 railway tunnels exceeding one mile (1.61 km) in length were completed. I.K. Brunel was appointed Engineer of the Great Western Railway in 1833, at the age of 26, and planned the route from Bristol to London.

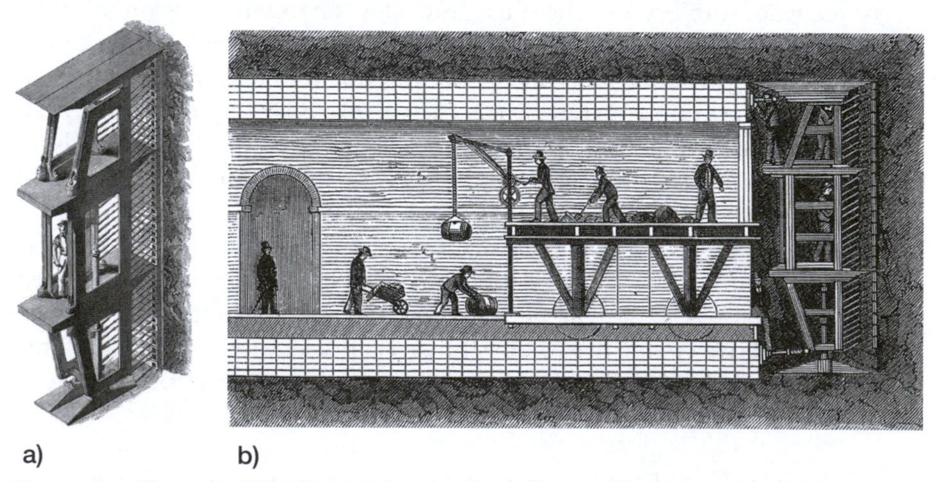

Figure 1.1 'Brunel's Shield' used for the first Thames Tunnel, a) one of the twelve frames making up the shield, and b) a cross section through this tunnel during construction (after Beamish 1862)

A major tunnel on this route was the Box Tunnel with a length of 2937 m. Water was a major problem on several sections of this tunnel, but it opened successfully in 1841.

- 1857 saw the start of construction on the first major tunnel in the Alpine regions of Europe. The Fréjus Tunnel involved construction between two portals, one at 1344 m above sea level at Bardonnéche and the other at 1202 m at Fourneaux, with the distance between portals being 12,221 m. Rock drills were used extensively on the project and drill carriages mounting four to eight drills were introduced in 1863 and used until the completion of the project in 1870.
- At about the same time as the first Alpine tunnels were being constructed, the Hoosac Tunnel in Massachussetts, USA was started (1855–76). This became known as 'the Great Bore'. It was 7.44 km long (4.62 miles) and was constructed mainly through schist and gneiss. The rate of construction was very slow at 0.32 m per day in 1865, but this improved with the introduction of compressed air rock drills to about 1.65 m per day in 1873.
- 1869 was an important year for subaqueous tunnelling as it marked the successful completion of the Tower subway in London using a shield (designed by J. H. Greathead) and cast iron lining. The shield used is the ancestor of almost all subsequent tunnelling shields (it was circular as compared to Brunel's rectangular shield used on the earlier Thames Tunnel). It even incorporated grouting behind the cast iron lining to fill the void. The system was very efficient and allowed progress of 3 m per day. The tunnel was 2.18 m in diameter and 402 m long.
- Greathead made a number of further developments in shield technology, including a closed face shield with the ground being broken up with jets and the spoil being removed as a slurry, i.e. the forerunner of the slurry shield. (A slurry shield was first used in 1971 at New Cross in London, UK.)
- The first use of hydraulic jacks to propel a shield forward was designed by Beach in 1869 and used under Broadway in New York, USA.
- There were a number of developments in rotary tunnelling machines as part of the various attempts at the Channel Tunnel, UK in the 1880s.
- Compressed air was used as a means of preventing water inflow into the tunnel during the construction of the Hudson river tunnel in New York, completed in 1910. This project also introduced the 'medical lock' for treatment of caisson disease. At about the same time the first (old) Elbtunnel under the Elbe river in Hamburg also used compressed air during construction between 1907 and 1911. It suffered a blowout in 1909 with an 8 m high water fountain being observed. It should be noted that a patent for working in compressed air had been taken out in 1830 by Lord Cochrane in the UK.
- The first use of a combination of a shield and compressed air (together with cast iron segmental lining) was on the City and South London

Railway completed in 1890 (now part of the Northern Line on the London Underground system). The tunnels were twin tubes with a diameter of 3.1–3.2 m and constructed through mainly London Clay, but with occasional water bearing gravel. (Most of the original London Tube lines were constructed by the cut-and-cover technique.)

- Probably the first highway tunnel to use the submerged tube method of construction was the Posey Tunnel in California, USA, opened in 1928. It used 62 m lengths of steel circular shells encased in concrete and lowered into a dredged trench on the river bed.
- The Liverpool to Birkenhead Tunnel under the river Mersey, UK constructed between 1925 and 1933, was at the time the largest underwater tunnel ever built, with a length of 2 miles and 230 yards (3.49 km) and wide enough for four lanes of traffic.

Since these modest beginnings there has been an explosion of tunnelling all over the world and we can now probably claim on a technical level to be able to build tunnels anywhere, through any ground.

Looking to the future, the importance of tunnelling to the sustainability of megacities (defined as metropolitan areas with a total population in excess of 10 million people, or a minimum population density of 2000 persons per km^2) cannot be underestimated as it is vital for the development of the underground space.

1.4 The nature of the ground

There is a tendency for tunnelling projects to be classified as either 'soft ground' or 'hard ground (rock)' tunnels, and in this book the authors have adopted this terminology. However, it must be remembered that there is a transition between these terms and tunnelling projects often have to deal with much more complicated ground conditions, often with mixed ground components. This book uses the broad description of these 'categories' adopted by the British Tunnelling Society and the Institution of Civil Engineers in the UK (BTS/ICE 2004) in their tunnel lining design guide, which suggests that all types of soil and weak rocks would normally fall into the category of 'soft ground' (weak rocks include poorer grade chalk, weak mudstones and weakly cemented and/or highly fractured sandstone). 'Hard ground' would generally comprise all other forms of rock. In this book the word 'ground' is used as a generic term when referring to the material surrounding a tunnel and includes the rock material and, for example, any discontinuities and faults. An alternative term used in the literature is 'rock mass'.

There are many options available these days for the construction of tunnels. The selection of which tunnelling technique to use must be made on the basis of the known and suspected ground conditions, in combination with other aspects such as access, possibly local tunnelling traditions and skills, as well as costs. Adaptability of the technique to variability of the ground could also be an important factor.

One of the key aspects of any civil engineering construction project, particularly relevant for tunnelling, is the observational nature of the process. The ground in particular is not man-made and is infinitely variable. We therefore must treat it with respect. Based on observations, either from previous projects in the area or from the current project as it is progressing, engineering judgement based on performance is essential to inform the design and construction processes. This point is discussed further in section 7.3.3.

1.5 Tunnel cross section terminology

Some useful terminology related to a tunnel cross section is shown in Figure 1.2

Other terminology is explained throughout the book, i.e. when the term first appears in the text. The index can be used to find these explanations.

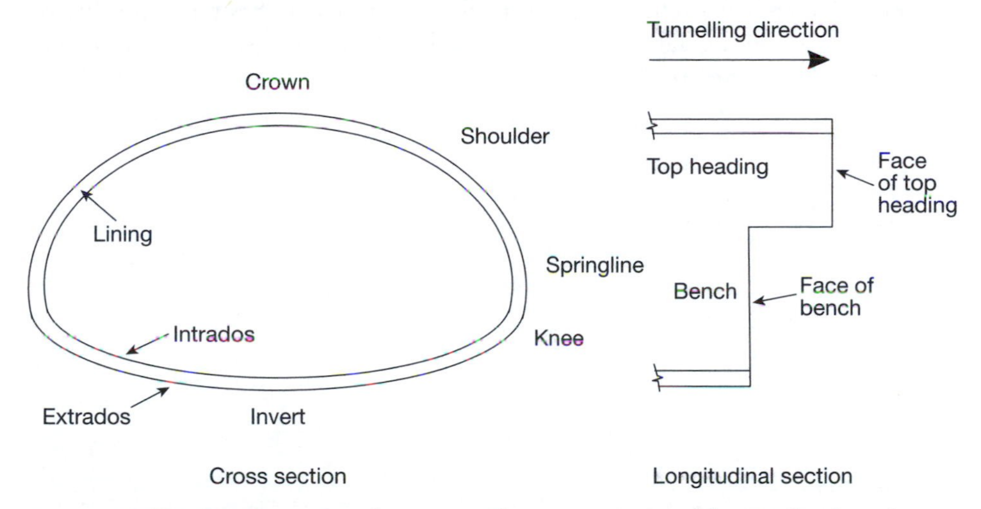

Figure 1.2 Terminology related to a tunnel cross section and longitudinal section

1.6 Content and layout of this book

The book consists of eight chapters (including this chapter) containing the following:

Chapter 2: Introduces the subject of site investigation and the issues of geological properties and classification, including laboratory and field testing.

Chapter 3: Covers preliminary analysis issues, such as calculation of primary stresses, stability in soft ground, preliminary analysis methods and numerical modelling.

Chapter 4: Covers methods of improving the stability of the ground prior to or during tunnel construction to ensure that the tunnel can be constructed safely. This chapter also describes the various methods available for lining a tunnel.

Chapter 5: Covers the main techniques of constructing tunnels.

Chapter 6: Introduces health and safety in tunnelling projects and the concept of risk management.

Chapter 7: Covers additional important issues associated with the construction of tunnels, including aspects related to tunnelling in soft ground such as ground movements and the effect of these ground movements on adjacent structures and services. This chapter also describes the observational method and monitoring and instrumentation related to tunnelling projects.

Chapter 8: Describes three case histories that are used to put into context some of the techniques and issues related to the construction of tunnels as experienced in practice.

There are extensive references within the text in each chapter and a list of these references is given at the end of the book. In addition, a bibliography list suggests further reading material.

2 Site investigation

2.1 Introduction

Tunnel construction is governed by the ground and hence site investigation is vital to obtain ground characteristics and geotechnical parameters. Knowledge of the ground conditions plays a key role in the choice of construction technique, and hence the success of a tunnel project. It is important to realize that the ability to influence the project outcome (in terms of cost and schedule) is easier earlier on in the project programme and much more difficult at a later stage, and the site investigation results can be a key influence on the early decisions. In many respects the site investigation for tunnelling projects is similar to other civil engineering projects and thus general textbooks and standards should be consulted (for example SISG 1993a, b and c, Attewell 1995, Clayton *et al.* 1995, BSI 1999, Simons *et al.* 2002, BSI 2007). However, more specific information related to tunnelling can be found in Dumpleton and West (1976) and BTS/ICE (2005). The new ICE Specification for Site Investigation to be published imminently will have a Tunnelling Addendum (current reference SISG 1993c).

Site investigation is defined in this book as the overall investigation of a site(s) associated with a tunnel construction, including the above and the below ground surface investigations. Ground investigation is defined as a sub-section of the site investigation and is associated specifically with defining the subsurface conditions. The aim of the site investigation is to produce a full three-dimensional model of the site, both above and below ground, and to highlight the associated impact (risks) of the tunnelling works on this environment and also the possible risks to the tunnelling works themselves. These risks can then be assessed and mitigated using appropriate construction techniques (risk management is discussed further in section 6.2). The site investigation comprises a number of key elements as shown in Figure 2.1. This chapter of the book describes these key elements in detail and highlights the investigation necessary for each step. The site investigation culminates in the site investigation report(s) described in section 2.5. It is important to realize that the site investigation information is not fixed at the start of the project and that the ground model develops and evolves with the project.

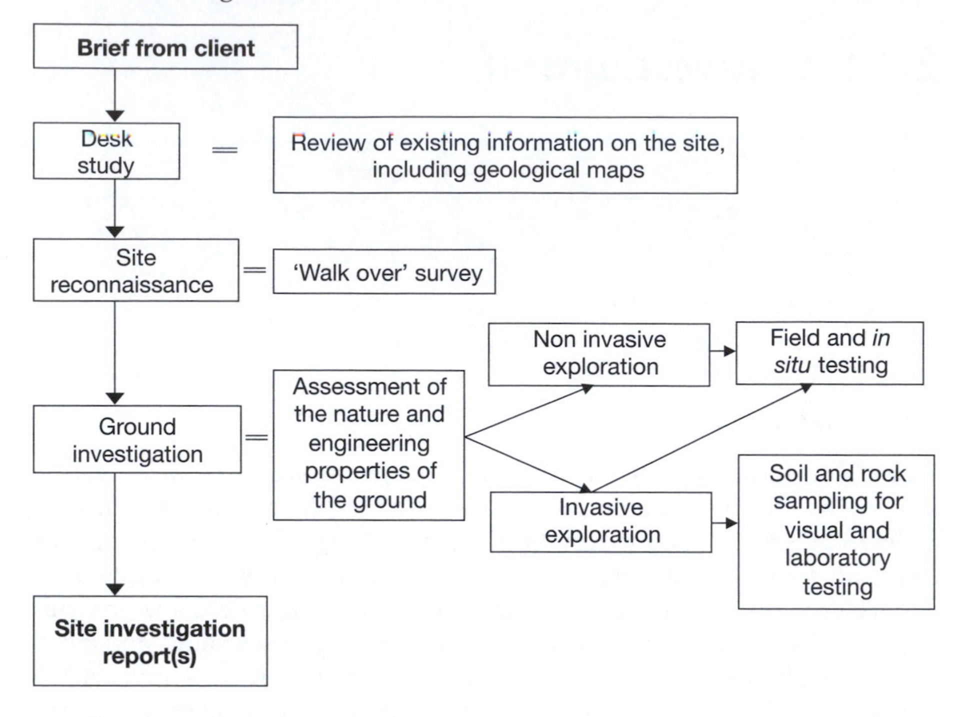

Figure 2.1 Elements of a typical site investigation

The money available to spend on site investigation is usually between 1 to 3% of the total tunnelling project costs. It is therefore important to use this money wisely in order to minimize the subsequent risks during construction. The traditional view is that the more one pays for site investigation the more likely one is to reduce additional costs resulting from unforeseen circumstances. There is some evidence to support this, although it is important to make informed decisions on how this money is spent. However, it is unlikely that more money will be spent on site investigation for tunnelling projects so it is important to use a risk-based approach to maximize the impact of the money available and minimize the risk of overlooking something important.

2.2 Site investigation during a project

2.2.1 Introduction

For any given project there are a number of different types of site investigation, namely: preliminary investigation, design investigation and control investigation (BSI 2007). These may be carried out during different stages of the project and have varying objectives. The main focus of the preliminary investigation is to assess the general suitability of the site and compare different alignments, with due consideration to third parties. The main aim of the design investigations is to provide information required for the design of the tunnel, including the construction method. In addition, control investigations may be required during the construction or execution of the project, and include, for example, checking ground characteristics and groundwater conditions.

A typical site investigation comprises four key elements: the desk study, site reconnaissance, ground investigation and the production of the site investigation report(s) (Figure 2.1). However, when designing a site investigation it is important to be objective and make sure that what is done can be clearly justified and that the desired outcome of the investigation is clear. The desk study and site reconnaissance can help design the subsequent ground investigation. It is essential that the specified sampling and testing are appropriate for the materials and parameters required for the subsequent design.

2.2.2 Desk study

The desk study is a very important stage of any site investigation, which, if done well, can save considerable time, and hence money, later on in the investigation process. The aim of the desk study exercise is to assess the conceptual model developed for the tunnel scheme using all the available records of the area where the proposed scheme is to take place. Desk studies cover all aspects of the site, including current usage, overlying and adjacent structures, historical usage and geology.

It is important that the desk study highlights any issues that could affect the health and safety of personnel during the subsequent site investigation and also the construction of the project. It should also provide as much information as possible to aid the planning of the subsequent stages of the site investigation, which in the case of tunnelling projects is usually the location, depth and type of boreholes.

In most countries there are numerous sources of information available that can aid a desk study, for example, geological maps, geological memoirs, old and new topographic maps (for example Ordnance Survey maps in the UK), aerial photographs, utility company records, site investigation databases (the British Geological Survey in the UK) and local councils.

It is also important to use site investigation companies that are familiar with the local area, as previous experience can be invaluable.

2.2.3 Site reconnaissance

Site reconnaissance (sometimes termed 'walkover survey') is the first site-specific work. With tunnelling projects it is rarely possible to walk along the entire length of the tunnel alignment, but this should be attempted as it can provide excellent detailed site knowledge for future planning. This is particularly important when planning any intrusive ground investigation and for the location of shafts. The objectives of a site reconnaissance include but are not limited to (after Allen 2006):

- location/confirmation of buried services;
- assessment of structures, particularly historic structures likely to be affected by the tunnelling works;
- identification of access restrictions;
- identification of any evidence of existing geology (e.g. exposed cut faces);
- identification of any evidence of existing structural or geotechnical problems, cracks and settlements of structures;
- identification of any new construction works (not shown on current maps);
- identification of any unexpected hazards.

It is important to record site details via photographs, sketches and notes. The information is checked against the desk study findings and further desk studies and/or further site visits undertaken as appropriate. As with the desk study this stage is relatively low cost compared to the later stages of a site investigation and can produce valuable qualitative information (Allen 2006).

2.2.4 Ground investigation (overview)

The ground investigation element of the site investigation should be planned based on the findings of the desk study and the site reconnaissance.

The ground investigation should give information about the stratigraphy of the ground. This is the genesis of the underground strata, i.e. the layering and the types of layers. It is important to conduct description and classification and testing, to determine information on the properties and parameters of the ground. Key general parameters for tunnel design include the strength of the ground to assess stability and the loading on the lining, modulus values such as Young's modulus, E, to assess how much the ground will deform with changes in stress, and the water conditions and permeability of the ground, as water can influence stability and make tunnel construction difficult. Water (the hydraulic regime) is extremely important when conducting underground construction and any investigation should include determining the groundwater level(s), water pressure, confined aquifers and water chemistry (with respect to how aggressive it is towards concrete). Other aspects include determining the swelling properties of clays, cavities (karsts) and abrasiveness characteristics.

This may sound straightforward. However, the ground is generally highly variable and its parameters can change over relatively short distances. It is therefore often a challenge to develop a model of the ground and associated risks along the route of a potential tunnel, and establishing the necessary design parameters is rarely 'straightforward' with considerable engineering judgement being required.

Due to the complexity of this aspect of the site investigation, the ground investigation is covered in detail in a separate section below (2.3).

2.3 Ground investigation

2.3.1 Introduction

A tunnel is commonly a composite structure made up of the tunnel lining and the surrounding material, although there are bare rock tunnels which do not need a lining. The surrounding material not only has a loading function, but is also the medium in which a void is created with the help of the supporting role of the surrounding material. Without this supporting role of the ground an economical tunnel design would not be possible, i.e. the ground is an integral part of the tunnel construction.

To make a judgement about the stability of the tunnel, as with any civil engineering design, the characteristics of the 'building materials' must be known: this includes both the tunnel lining and the surrounding ground material. There is difficulty in determining the ground parameters particularly when there are faults, inhomogeneity and weathering, all of which make it difficult to assign simple statements about the ground behaviour. Laboratory and field experiments can be carried out to give an indication of the soil and rock stability, which can be used to give some, albeit limited, idea of the ground stability.

In this book the ground investigation comprises of field investigations and laboratory experiments to obtain information about the subsurface and its properties. Table 2.1 provides a list of potential parameters required from a site investigation in order to aid the design of a tunnelling project. Many of these parameters are determined during the ground investigation, although some may have been obtained from past investigations as highlighted by the desk study.

The decision as to which techniques should be used during the ground investigation must be considered carefully and in relation to the budget and goals required. It is important to identify the investigation goals in order to avoid wasting time and, consequently, money.

2.3.2 Field investigations

A variety of investigation techniques can be employed as part of the ground investigation. These include intrusive and non-intrusive methods. A combination of various methods is usually the best approach.

2.3.2.1 Non-intrusive methods

Although intrusive methods allow the inspection and testing of the ground itself, they are normally restricted to discrete locations. Non-intrusive

Table 2.1 Ground parameters for the design of tunnel projects (adapted from Morgan 2006, after Jewell 2002, used with permission from Peter Jewell)

Geotechnical design parameter	*Symbol*	*Units*	*Application to tunnel design and construction*
Soil description from light cable percussion boring			Defines type of ground
Soil and/or rock description from rotary coring			Defines type of ground
Grade of rock			Extent of ground support
Percentage core recovery and core condition (total core recovery, solid core recovery and rock quality designation respectively)	TCR, SCR, RQD		State of weak or hard rock
Unit weight	γ	kN/m^3	Overburden pressure
Relative density of coarse grained soils	D_r		State of natural compaction of cohesionless soft ground
Moisture content	w	percent	Profiling of property changes with depth
Specific gravity	Gs	kN/m^3	Type of ground
Plasticity and Liquidity Indices (liquid limit, plastic limit, plasticity index, liquidity index, consistency index respectively)	w_L, w_P, I_P, I_L, I_c	percent	Type and strength of cohesive soft ground
Particle size distribution			Composition of soft ground
Unconfined (or Uniaxial) compressive strength, UCS	q_u	MPa or MN/m^2	Intact strength of hard rock

Table 2.1 (continued)

Geotechnical design parameter	*Symbol*	*Units*	*Application to tunnel design and construction*
Point load index strength	I_S	MPa or MN/m^2	Intact strength of hard rock
Undrained shear strength	c_u, s_u	kPa or kN/m^2	Shear strength of soft ground (short-term strength of fine grained 'cohesive' soils)
Effective shear strength	c'	kPa or kN/m^2	Long-term apparent 'cohesion' of soft ground (fine grained soils)
Angle of internal shearing resistance	ϕ'	degrees	Long-term shear strength of 'cohesive' soft ground (fine grained soils) Short and long-term shear strength of 'cohesionless' soft ground (coarse grained soils)
Ultimate stress at failure	σ_u	MPa or MN/m^2	Characterizing rock
Ultimate strain at failure	ε_u		Characterizing rock
Modulus of elasticity (Young's modulus)	E	MPa or MN/m^2	Stress increment per strain increment, i.e. directly related to strength
Drained deformation modulus	E'	MPa or MN/m^2	Long-term stiffness
Poisson's ratio	μ		Influences stiffness values
Coefficient of lateral earth pressure (at rest, active and passive values respectively)	K_0, K_a, K_p		Ratio between horizontal and vertical effective stresses
Permeability	k	m/s	Characteristic ground permeabilities and variations, waterproofing
pH, sulphate and chloride content	pH, SO_3, Cl		Concrete and steel durability
Chemical contamination			Extent of ground contamination
Abrasion			Rate of cutter tool wear

methods can be used for determining additional information about the ground and include geophysical methods. Geophysical methods can be used to obtain information over a relatively large area of the subsurface ground, and hence can be used to help locate boreholes, provide information about the nature and variability of the subsurface between existing boreholes, or can be used where access for intrusive methods is not possible. It should be noted that interpretation of the output from these methods is not easy and usually requires a borehole(s) to correlate results. Some of the more appropriate geophysical methods for tunnel projects are briefly described below.

SEISMIC METHODS

Seismic techniques are based on the generation of seismic waves on the ground surface at a source, S, and the measurement of the time taken by the waves to travel from the source, through the ground to a series of receivers, R. They utilize the fact that elastic waves travel with different velocities in different rocks. The seismic wave can be generated using a drop hammer or a 3 kg sledgehammer to give a penetration depth of up to 20 m. For deeper penetration depths falling weight devices or even explosives can be used (Waltham 2002). Geophones are commonly used as receivers. Two main travel paths for the seismic wave are possible. The wave can travel along the interface between two rock types (L1), i.e. it is refracted (Figure 2.2a), or it can be reflected off this interface (Figure 2.2b).

Knowing the distance between the source and the receiver and the travel time it is possible to determine the shear velocity, and hence the depth to the refracting/reflecting interface. If the density is known (calibrated from borehole information) the shear stiffness, G_{max}, of the material can be inferred from the surface waves resulting from seismic surveys. Seismic reflection and refraction can be useful for determining depth to bedrock and depth to groundwater table, but reflection can give better resolution and can also identify multiple layers and faults.

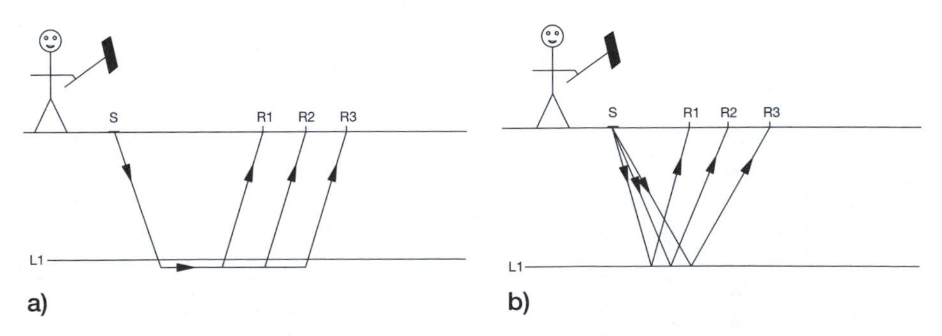

Figure 2.2 a) Seismic refraction and b) seismic reflection (after Anderson *et al.* 2008)

RESISTIVITY/CONDUCTIVITY

The results from these methods are particularly useful when combined with seismic refraction. They are especially useful for determining the soil/water interface, soil profiles and also for characterizing contaminated groundwater plumes.

Figure 2.3 shows the principle of these techniques. In the electrical resistivity technique a current (I) is induced between paired electrodes (C1, C2). The potential difference (ΔV) between paired voltmeter electrodes P1 and P2 is measured. Apparent resistivity is then calculated (based on I, ΔV, and the electrode spacings). If the electrode spacing is expanded about a central location, a resistivity–depth sounding can be generated. If the array is expanded and moved along the surface, 2-D or 3-D resistivity–depth models can be created (after Anderson *et al.* 2008).

BOREHOLE GEOPHYSICAL LOGGING

A wide variety of in-hole methods are available adapted from the petroleum industry. Borehole geophysical logging can be useful for special circumstances and includes sonic and electrical resistivity methods. It is good for determining the properties of the ground at depth such as density, but must be used selectively in order to be cost-effective.

CROSS-HOLE SEISMIC TECHNIQUES

This technique provides improved definition of geology at depth when compared to surface seismics. It can be good for the characterization of underground caverns. This method, however, can be expensive due to the need for closely spaced boreholes. It typically involves high-frequency acoustic pulses generated at predetermined source locations at different levels in the source borehole. The amplitude and arrival time of direct arrivals

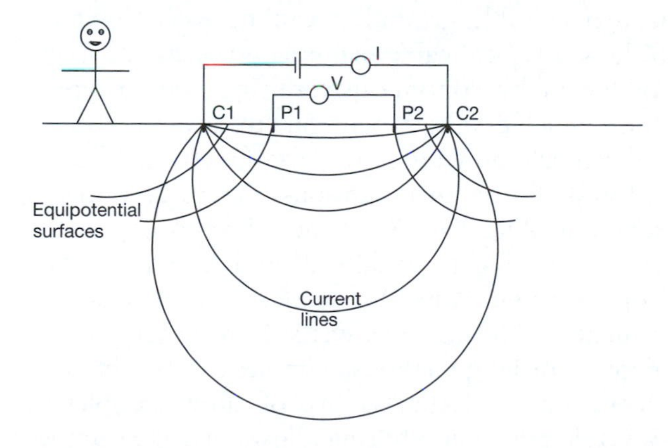

Figure 2.3 Resistivity/conductivity (after Anderson *et al.* 2008)

(and others) is recorded at predetermined receiver locations at different levels in the receiver borehole. The recorded travel time–amplitude data are statistically analysed and used to generate a velocity–attenuation cross sectional model of the area between the source and receiver boreholes. Attenuation is defined as the reduction in signal strength as a result of it passing through a medium, in this case the ground (after Anderson *et al.* 2008).

Other geophysical techniques include magnetic methods (good for locating buried foundations, mineshafts and ferrous utilities and obstructions), gravitational methods (good for cavity detection) and electromagnetic methods (good for locating utilities, ground and pollution mapping). For further information on geophysical methods commonly employed for site characterization, the reader is referred to McDowell (2002) and the Transportation Research Circular E-C130 (Anderson *et al.* 2008).

2.3.2.2 Intrusive exploration

Intrusive exploration is used for obtaining samples/cores of the ground for visual examinations and laboratory testing, and also for conducting *in situ* testing to determine the ground characteristics and primary stress conditions.

IN SITU SAMPLING

The principle methods for obtaining samples/cores include trial pit excavations, percussive drilling, rotary drilling techniques and even trial tunnels. Trial pit excavations are used for relatively shallow investigations to a few metres, but depending on available space can open up a relatively large area of the ground. Percussive boring (known as either cable percussion or shell and auger boring) is the most common technique in the UK for soft ground (soil/weak or weathered rock) as it is relatively cheap, simple, flexible and robust. Through suitable ground it can be used down to 60 m. Figures 2.4a and b show a typical cable percussion rig. As the name implies, the boring is conducted by continuously raising and dropping weighted hollow drilling tools which gradually penetrate the ground. Rotary drilling is used in rocks and can drill down to hundreds of metres, although smaller rigs are available for shallower investigations. Figure 2.4c shows an example of a small rotary drilling rig. The standard approach in the UK is to use cable percussion boring to rockhead, and if required the borehole is extended by rotary coring. However, some strata, for example weathered rock, overconsolidated clay and most chalk, may be sampled by either cable percussion or rotary drilling methods. Samples can be obtained during these intrusive operations and the quality of these samples is described in section 2.3.3. Cable percussion boring allows discrete samples to be obtained, for example using a U100 driven sample tube (100 mm

Figure 2.4 a) and b) Cable percussion rig and c) rotary drilling rig (courtesy of Soil Mechanics)

Figure 2.5 Example of continuous rock cores obtained from rotary drilling

Figure 2.6 Triple core barrel sampler (double core barrel with a plastic liner) used with rotary drilling

diameter, 450 mm long sample). BSI (1999) suggests that in soft ground, samples should be obtained at the top of each new stratum and thereafter every 1.0 to 1.5 m, and standard penetration tests (see *in situ* testing) conducted immediately afterwards. Figure 2.5 shows an example set of rock cores obtained from a rotary drilling operation, which allows continuous samples to be obtained.

Figure 2.6 shows a triple core barrel sampler used with rotary drilling, which incorporates an outer drill tube that rotates and has the cutter attached to the end, an inner steel tube (core barrel) that does not rotate (the gap between these two tubes is used to pass drilling fluid to the drill bit), and a plastic liner to help preserve the sample during retrieval from the core barrel and transportation to the laboratory for logging and sampling.

In soft ground it is also usual to obtain disturbed samples from cable percussion techniques. These are samples where there is no attempt made to preserve the shape or the fabric of the soil, but if sealed correctly (bagged) can give useful information on particle size and water content, and if the soil is clay, information on plasticity indices (see section 2.3.2).

A recent development in intrusive investigation is the use of horizontal directional drilling that allows core drilling in practically any direction, for example along the alignment of the tunnel. Horizontal directional drilling is discussed in detail in section 5.12.

The transportation, storage and labelling of samples needs to done carefully and a satisfactory procedure adopted to ensure they can be readily identified. If not done adequately it can lead to sample deterioration and hence influence subsequent laboratory test results.

It is important to consider the position of the boreholes carefully. Although it is essential to get a representative sample of the ground, accessibility to the location for the drilling rig might be a limiting factor. The possibility of lateral realignment of the tunnel should be considered in the drilling plan. Even though boreholes should be filled in properly, it is not always done satisfactorily, and this can create problems during the later construction stages, for example water ingress and pressure losses. Therefore, these should not be drilled directly on the alignment of the tunnel. However, when creating large openings (caverns) it may be necessary to drill into the later void. It should be noted that creating a hole is expensive and it is therefore often worth utilizing the borehole to incorporate instrumentation used during the construction of the tunnel (monitoring and instrumentation is described in section 7.3).

For tunnels and shafts it is important to take the exploration to a generous depth below the proposed invert level of the tunnel because changes in design may result in a lowering of the level of the tunnel, and because the zone of influence of the tunnel may be extended by the nature of the ground at a greater depth (BSI 1999).

The number of boreholes associated with a particular tunnelling project depends on the ground conditions and extent of tunnelling works, and there are no fixed rules on the spacing or number. As a rule of thumb, however, for relatively long tunnels 300 m spacing would be sensible for the main tunnel and 30 m spacing at the portals.

From the borehole results and associated testing (see section 2.3.3) it is possible to obtain a geological section along the tunnel and in a plan showing the layering, i.e. a geological model along the route of the tunnel (Figure 2.7). It may also be useful to conduct some shorter angled drillings to help develop the stratigraphic section and produce a more complete

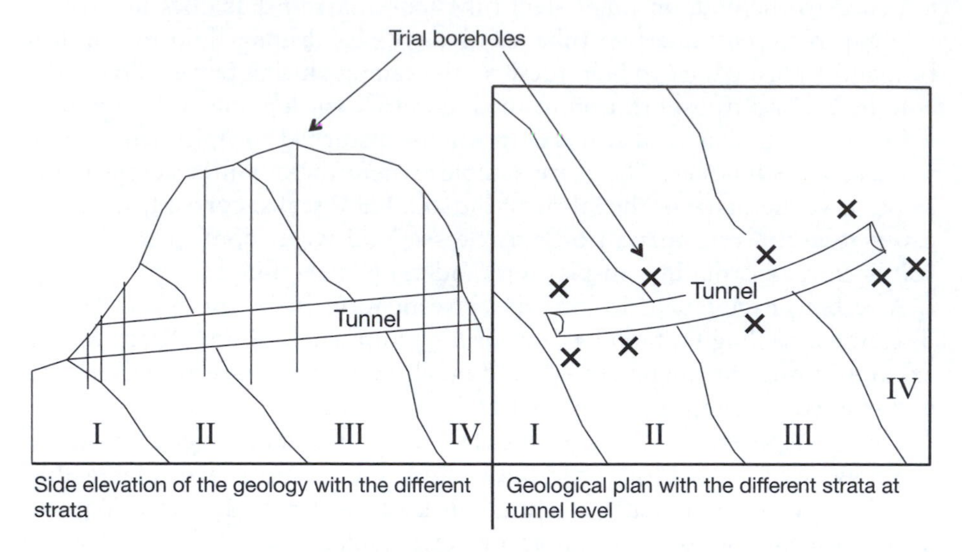

Figure 2.7 Possible borehole locations for a mountain tunnel (Note: exaggerated vertical scale)

picture. This is important if the strata are highly dipping and relatively thin, as the vertical boreholes could miss some of these strata.

The following points should also be noted related to the borehole investigation:

- The borehole positions should be shown accurately on the proposed plans for the tunnel and the ground level at each borehole position must be recorded.
- The majority of information from the investigatory boreholes is derived from cores from the whole depth. This allows undisturbed samples of the rock to be obtained and tested.
- If the boreholes are very deep, it is sometimes preferred to get cores only at specific sections, not for the whole depth.
- It is desirable to create a 3-D model of the geology associated with the project from the boreholes. However, it is also important, if possible, to get a personal observation of the site during the borehole drilling and hence obtain a good 'feel' for the ground conditions.
- Simply looking at photographs of the cores and reading the associated reports can give a false impression. Colours on photographs, for example, can be misleading and information could be missing from the report such as, for example, whether it was wet? What was the Rock Quality Designation (defined in section 2.4.4.1)? It is important to spend time studying the cores directly and noting any irregularities.
- Boreholes should also obtain information on the hydrology, groundwater level and layers holding water.

If it is not possible to determine aspects of the subsurface details using boreholes from the ground surface, trial excavations/tunnels can be used. The Gotthard Base Tunnel in Switzerland, for example, used a trial tunnel of a couple of kilometres to determine further details of the 'Piora fault'.

Once samples have been obtained, it is then possible to conduct laboratory tests on the material, including uniaxial, triaxial testing and basic index tests. Some of these tests are discussed in more detail in section 2.3.3.

IN SITU TESTING

It is also possible to conduct *in situ* testing (with or without associated sampling) and these can include standard penetration testing, cone penetration tests and pressuremeter or dilatometer testing. The reader is directed to the appropriate standards in their own country for further details, for example BSI (2007). It should be noted that most field tests only provide indirect measures of the ground properties and very often empirically derived relationships need to be applied to obtain design parameters. Several of the more common tests are briefly described below:

Standard penetration test In soft ground, cable percussion (or shell and auger) boring techniques are often employed during field investigations. As part of this boring process, *in situ* standard penetration tests (SPTs) can be conducted as the borehole is created. The SPT test is performed at the bottom of the borehole (the boring level reached at that time). It is carried out by driving a standard sampling tool (for example a split barrel sampler) through a prescribed distance (450 mm), with a known weight (63.5 kg). The weight is dropped from a fixed height of 760 mm onto an anvil placed on top of the drill rods and the number of blows required for it to penetrate 450 mm is recorded (generally in 75 mm intervals). Normally the number of blows recorded to penetrate the first 150 mm is discarded as the ground in this region is normally disturbed by the boring process. The number of blows recorded to drive the sampler the next 300 mm is taken as the blow count, N_{SPT}. Figure 2.8a shows a standard penetration test being conducted and Figure 2.8b shows a schematic of the test procedure. The split barrel sampler is shown in Figure 2.9a and although this is a poor quality sampler from the point of view of soil sampling (the dimensions are such that the soil is disturbed too much as it enters the sampler), it does provide a sample for visual inspection at the level at which the test was conducted (Figure 2.9b). A significant effect on the results may begin to occur when the borehole diameter is greater than 150 mm. The water level in the borehole casing (boreholes are often cased to prevent collapse) should be kept above the natural groundwater level to avoid instability at the base of the borehole due to water flowing towards the borehole.

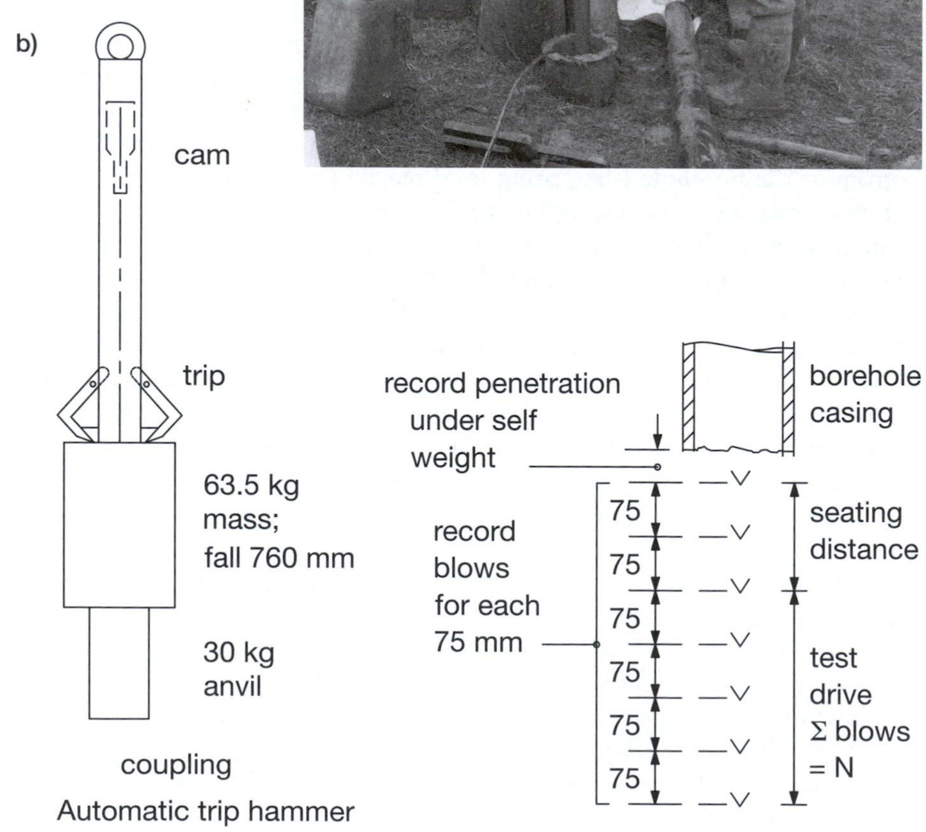

Figure 2.8
a) Standard penetration test carried out using a cable percussion rig (shell and auger) (courtesy of Soil Mechanics) and
b) schematic of the SPT automatic trip hammer arrangement and procedure (courtesy of J. Billam)

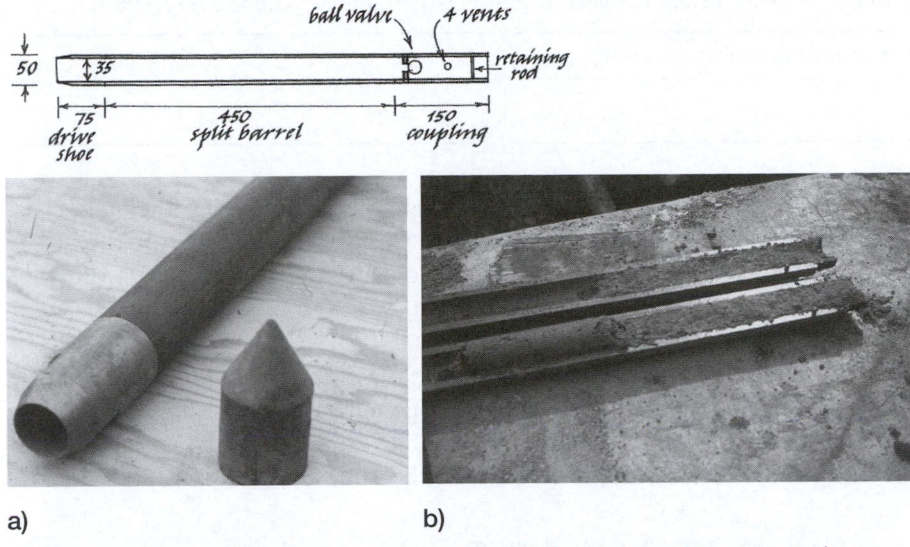

Figure 2.9 Split barrel sampler used for the standard penetration test: a) the open tube is used for fine grained soils and the cone for coarse grained soils and b) in fine grained soils a sample can be obtained for visual inspection (courtesy of Dr Ron Jones)

Table 2.2 Approximate relationship between blow count and density (BSI 2007)

Density	*Very loose*	*Loose*	*Medium*	*Dense*	*Very dense*
Normalized blow count (N_{SPT})[a]	0–3	3–8	8–25	25–42	42–58

Note: (a) Normalized blow count is adjusted for the energy transmitted down the rod and the effective overburden pressure. These descriptors should not be used for very coarse soils.

The SPT is popular because it is relatively simple and cheap to carry out. However, interpretation can be difficult as although the word 'standard' is used in the title of the test, there is a large variation in the equipment used. It was originally developed for coarse grained soils because of the difficulty in obtaining samples in these soils, and it is useful for giving an indication of density (Table 2.2).

Empirical relationships have been developed for other design parameters, for example the undrained shear strength, c_u, of overconsolidated clays can be approximated using equation 2.1.

$$c_u = f_1 N_{SPT} \tag{2.1}$$

where f_1 is a factor to allow for plasticity index, I_P, as shown in Table 2.3. (Plasticity index is defined in section 2.3.3.)

Table 2.3 Relationship between f_1 and plasticity index (after Stroud 1989)

Plasticity index (I_P)	*15*	*25*	*35 and over*
f_1	7.0	4.8	4.2

There are also correlations with the angle of shearing resistance (effective internal friction angle), ϕ', however, as this value depends on the stress level, care must be taken when determining this value from charts. Further information on SPT testing can be found in Clayton (1995).

Cone penetration test The cone penetration test (CPT) is used in soft ground and consists of pushing a cone (penetrometer tip) attached to the base of a series of rods into the ground. As the rods are pushed into the ground (at a constant rate of 20 mm/s), the cone tip resistance, q_c, and the

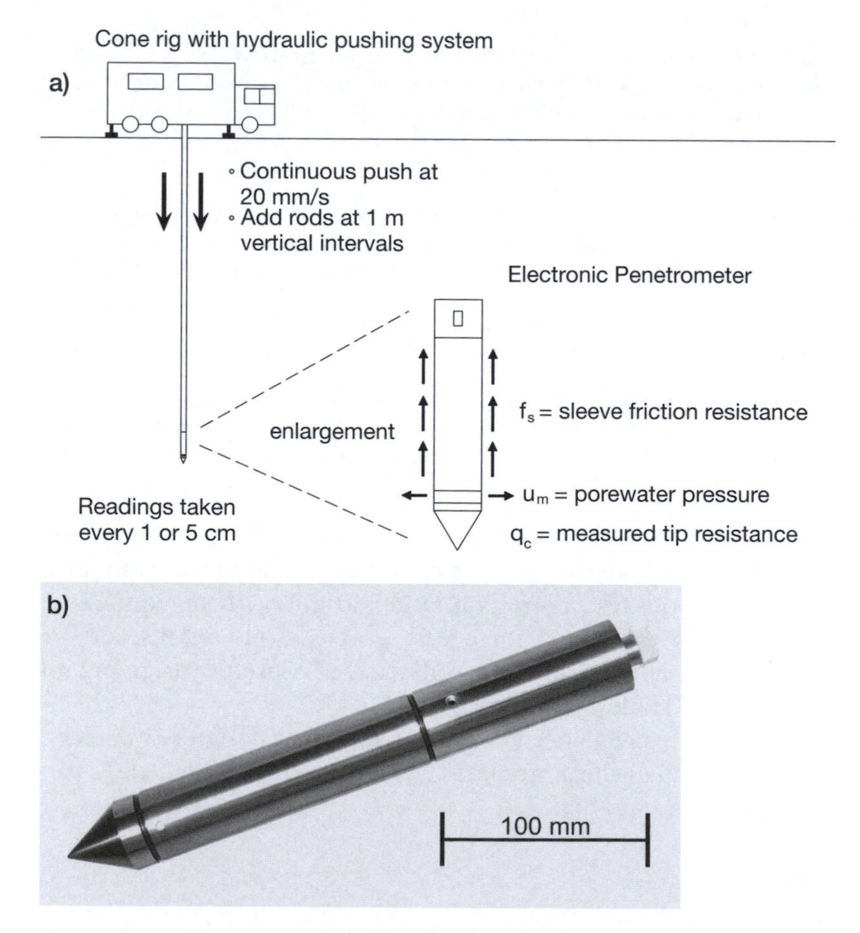

Figure 2.10 Cone penetration testing: a) test arrangement (after Mayne 2007) b) example of an electric penetrometer cone (courtesy of Geopoint Systems BV)

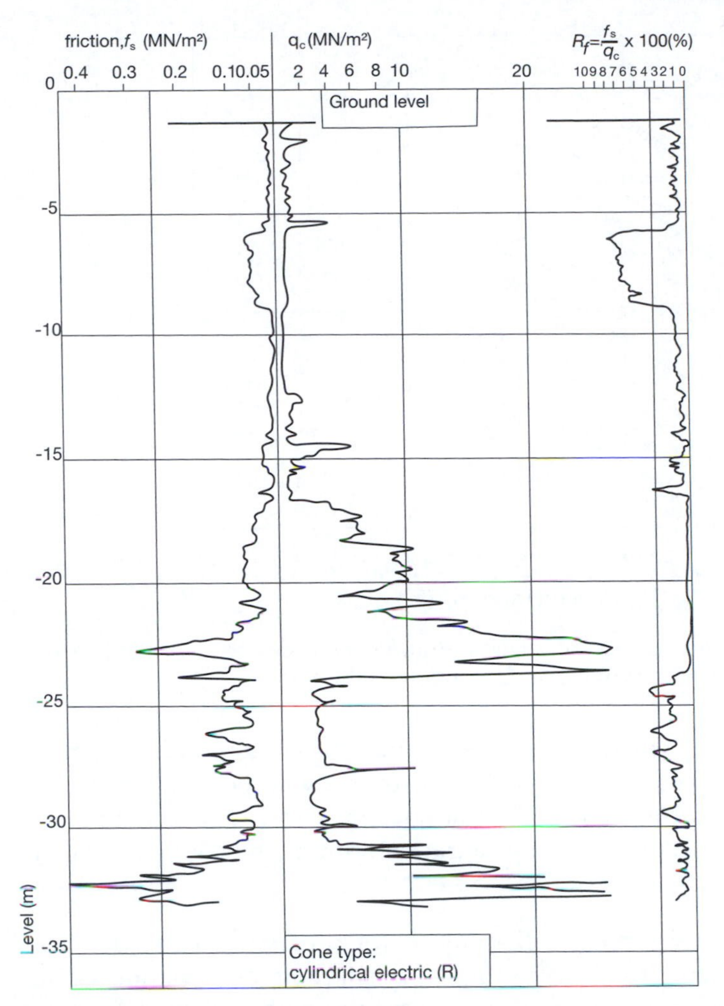

Figure 2.11 Typical plot from a cone penetration test (after Meigh 1987)

sleeve friction, f_s, are recorded by the penetrometer tip (Figures 2.10a&b). The cone angle on the penetrometer tip is 60° and the cross sectional area is 1000 mm². There are different types of equipment available on the market to conduct this test, but the most common involves a 20 tonne truck, to push the penetrometer tip into the ground.

A typical plot from a CPT is shown in Figure 2.11. The friction ratio, R_f is the ratio of the sleeve friction divided by the cone tip resistance, i.e. $R_f = f_s/q_c$. The zone in Figure 2.11 where the f_s and q_c values are higher indicate stiffer ground.

It has been found that by plotting the values of q_c against R_f an approximate description of the soil type can be obtained (Figure 2.12). Other relationships have been developed over the years between q_c and a number of other parameters, for example undrained shear strength, and angle of

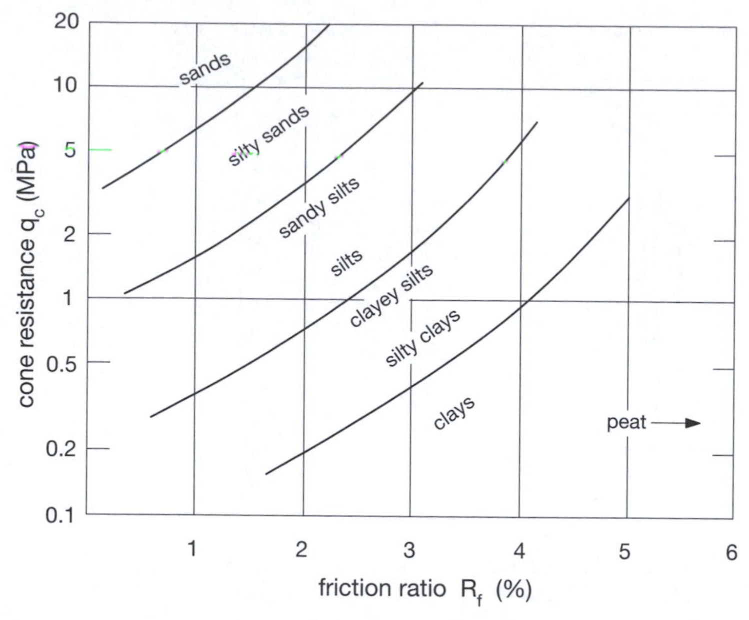

Figure 2.12 Simplified version of a friction ratio cone resistance plot used to obtain an approximate description of the ground

shearing resistance, but local experience and correlation with laboratory tests is essential. There are also penetrometer tips with pore water pressure measurement (piezocone) and geophysical testing (seismic cone and resistivity cone). Further details on cone penetration testing can be found in Mayne (2007).

Dilatometer/pressuremeter A dilatometer (or a pressuremeter, the terms seem interchangeable for rock testing) is a borehole deformation device. It is used as a rock/soil loading test in boreholes with a defined diameter. The aim of the dilatometer test is to determine the deformation modulus of the ground (see Figure 2.17, 'Definitions of different modulus values') and horizontal stress. The dilatometer consists of a cylindrical pressure cell containing strain arms within a cylindrical rubber membrane, which is pressed hydraulically against the borehole wall (Figure 2.13a). The borehole walls are loaded and then unloaded (cyclically) causing the borehole walls to deform (measured by the strain arms), thus allowing an estimate to be made of the deformation modulus of the material for an associated change in radii. By conducting the test in different directions within the ground, it is possible to determine the deformation modulus in different directions and hence obtain information on anisotropy within the ground. As with all *in situ* experiments the validity of the results are potentially limited because only small areas/sections of the ground are tested. Problems in validation can occur if the ground has many fractures, the borehole wall is not even or when the borehole is not stable and rock is collapsing into the bore.

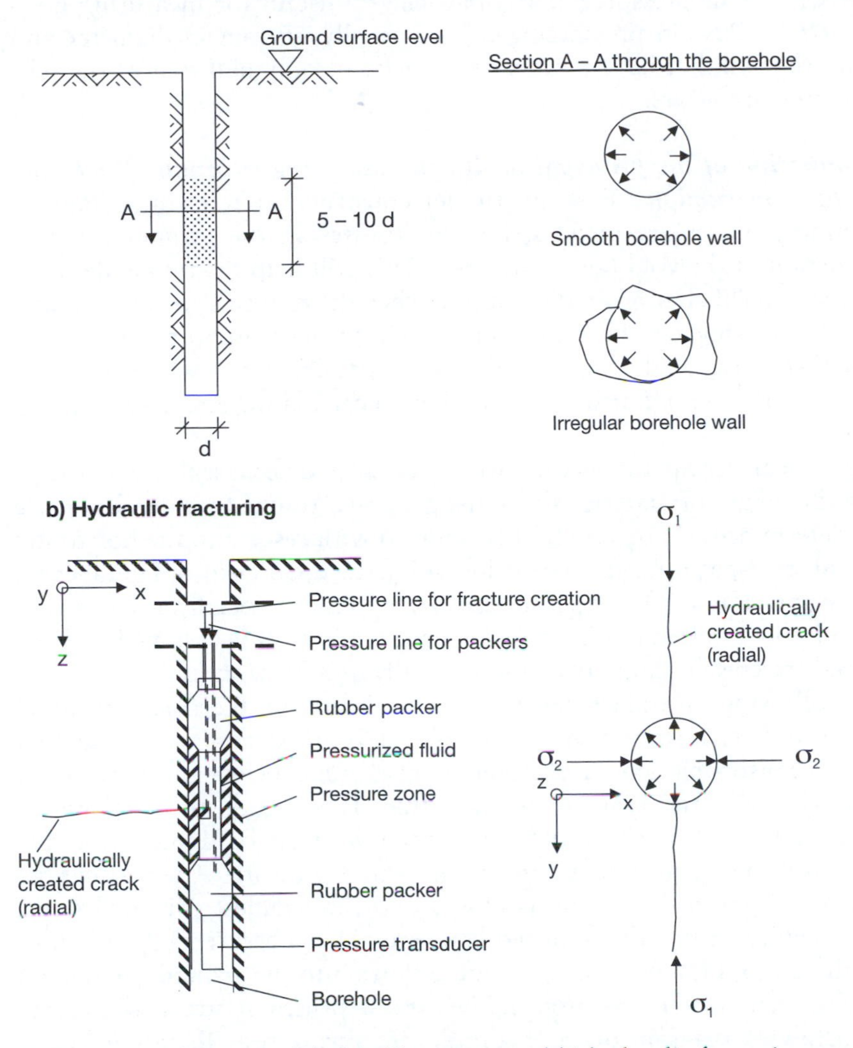

Figure 2.13 Schematic of the a) dilatometer and b) hydraulic fracturing tests

In these cases, it is necessary to secure the sides of the borehole or smooth out the contours to allow the experiment to be conducted. This is usually done with concrete or cement slurry resulting in an improvement of the ground at the borehole but can lead to possible false results for the test.

In soft ground (soils and weak rock), there are three types of pressuremeters available; a pre-bored pressuremeter (dilatometer type or Ménard type – Ménard invented the original pressuremeter), a self-boring pressuremeter and push-in pressuremeters. The pre-bored pressuremeters, typically 1 m long and 74 mm diameter, are lowered into a slightly oversized pre-bored hole. As the name implies, the self-boring pressuremeter bores itself into the ground, replacing the soil, and can operate up to pressures of approximately 4.5 MN/m^2. This pressuremeter is particularly useful for measuring horizontal stress. Push-in pressuremeters are usually 50 mm in diameter and displace the soil, but have a pressure capacity of only half that of the self-boring pressuremeters.

Determination of the principal* in situ *stresses using hydraulic fracturing prior to construction For any tunnel construction it is important to determine the primary stresses, i.e. the stresses in the ground prior to construction of the void (i.e. the tunnel). This will help the tunnel designer to estimate the likely stress redistribution when the tunnel is constructed and hence the loading on the tunnel lining. Of primary importance are the principal stresses, i.e. the largest and smallest possible stress where the shear stress is equal to zero. Further information on calculating stresses is given in section 3.2.

The vertical principal stress can easily be calculated because the unit weight, γ, and the height of material above the proposed tunnel axis depth, H, are easily determined, i.e. $\sigma_v = \gamma H$. This is not so with respect to the horizontal principal stress, $\sigma_h = K_0 \gamma H$. The value of K_0, the coefficient of lateral earth pressure at rest, is a difficult parameter to determine, especially as it can vary in magnitude in different directions (see section 3.4). It is assumed that the principal stresses are initially acting vertically and horizontally.

The following procedure can be used to determine the smallest lateral pressure and its direction in a borehole. It is important to ensure that there is a reasonable length of borehole above and below the location of the measurement and that this is crack free. This section of the borehole is then sealed with packers and pressurized with air or liquid (generally water) until there is a sudden drop in measured pressure. After noting the maximum pressure the system is closed and the smallest principal stress can be deduced from the adjusted pressure. The pressure drop develops when the ground fractures and the liquid flows into the ground. Two main fractures occur in the direction of the largest principal stress, σ_1. Figure 2.13b shows a diagram of the hydraulic fracturing test. In this figure, x and y are the principal stress directions and σ_1 and σ_2 are the principal stresses. In this example, the direction of the largest deformation gives the smallest principal stress direction and thus the smallest value of K_0. The largest value of K_0 in this example is found in the y-direction, but cannot be determined in this experiment. The value of K_0 determined from this experiment is still only an estimation and it is therefore advisable to do design calculations for a range of K_0 values.

Double load plate test With the double load plate test, load plates are pressed against the rock using a hydraulic jack (Figure 2.14). At the load plates the deformation of the ground is measured and is used to determine the deformation characteristics, and hence elasticity (Young's modulus, E).

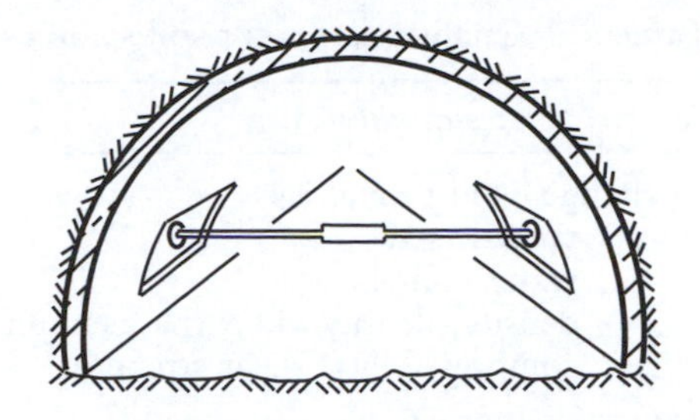

Figure 2.14
Diagram showing one method of conducting the double load plate test

The double load plate experiment is normally performed in test or trial tunnels where it is possible to use the opposite wall, or the crown or invert, as a reaction. Equation 2.2, based on Hooke's law, can be used to determine deformation modulus E_d:

$$E_d = \omega\left(1-\mu^2\right) r \frac{\Delta\sigma_m}{\Delta s_{Z,R}} \tag{2.2}$$

where μ is Poisson's ratio, ω is a constant for the type of load plate, flexible or rigid, r is the radius of the load plate, $\Delta\sigma_m$ is the difference between two load stages of the average normal stress on the plates, $\Delta s_{Z,R}$ is the difference between two load stages of the average settlements of the centre and edge of the plate.

2.3.3 Laboratory tests

After conducting *in situ* sampling it is possible to visually inspect these samples, describe the material in accordance with appropriate standards and carry out testing in the laboratory. It should be noted that the results from the samples tested only provide information on the sample itself and engineering judgement is essential to translate this information into ground characteristics (section 2.4).

It is important to visually inspect the samples collected in order to gain a preliminary profile with depth before conducting any laboratory experiments. For rock cores, total core recovery (percentage of core recovered, solid and pieces, relative to the overall length of the core interval), solid core recovery (total length of pieces of core recovered which have a full diameter, expressed as a percentage of the overall length of the core interval) and rock quality designation (RQD, see section 2.4.4.1) values should be obtained as these give an indication of the fracturing and fragmentation.

Depending on how the sampling has been carried out on site and the care taken in their subsequent handling, soil samples are classified into five quality classes with respect to the soil characteristics that remain unchanged (BSI 2007). These classes are described in Table 2.4, with quality class 1 being the best. The table shows the possible information, as indicated by the dots, which can be obtained for the five different quality samples. Also indicated on Table 2.4 are sample categories A, B and C (A being the best

Table 2.4 Quality classes for soil samples for laboratory testing (after BSI 2007)

<table>
<tr><th>Soil properties/quality class</th><th>1</th><th>2</th><th>3</th><th>4</th><th>5</th></tr>
<tr><td>Unchanged soil properties</td><td></td><td></td><td></td><td></td><td></td></tr>
<tr><td>particle size</td><td>•</td><td>•</td><td>•</td><td>•</td><td></td></tr>
<tr><td>water content</td><td>•</td><td>•</td><td>•</td><td></td><td></td></tr>
<tr><td>density, density index, permeability</td><td>•</td><td>•</td><td></td><td></td><td></td></tr>
<tr><td>compressibility, shear strength</td><td>•</td><td></td><td></td><td></td><td></td></tr>
<tr><td>Properties that can be determined</td><td></td><td></td><td></td><td></td><td></td></tr>
<tr><td>sequence of layers</td><td>•</td><td>•</td><td>•</td><td>•</td><td>•</td></tr>
<tr><td>boundaries of strata – broad</td><td>•</td><td>•</td><td>•</td><td>•</td><td></td></tr>
<tr><td>boundaries of strata – fine</td><td>•</td><td>•</td><td></td><td></td><td></td></tr>
<tr><td>Atterberg limits, particle density, organic content</td><td>•</td><td>•</td><td>•</td><td>•</td><td></td></tr>
<tr><td>water content</td><td>•</td><td>•</td><td>•</td><td></td><td></td></tr>
<tr><td>density, density index, porosity, permeability</td><td>•</td><td>•</td><td></td><td></td><td></td></tr>
<tr><td>compressibility, shear strength</td><td>•</td><td></td><td></td><td></td><td></td></tr>
<tr><td>Sampling category to be used</td><td colspan="5">A</td></tr>
<tr><td></td><td></td><td></td><td colspan="3">B</td></tr>
<tr><td></td><td></td><td></td><td></td><td></td><td>C</td></tr>
</table>

quality and C being the worst). These relate to the techniques used in the field for obtaining the samples. For example, drive sampling, in which a tube or a split-tube sampler having a sharp edge at its lower end is forced into the ground either by a static thrust (by pushing), by dynamic impact or by percussion are mostly category A or B sampling methods. Rotary core sampling methods, in which a tube with a cutter at its lower end is rotated into the ground, are usually category B. Auger sampling with hand or mechanical augers are usually category C sampling methods.

Although samples are often described as disturbed or 'undisturbed', there is no such thing as a truly undisturbed sample as the very act of retrieving the sample from the ground disturbs it, the stress conditions are changed for example. Hence the term 'undisturbed' is often written in parentheses to indicate this fact. The quality of the sampling technique also dictates how disturbed the sample is. For example a 'bagged' sample as described above is highly disturbed.

Table 2.5 Particle size ranges

Component	*Size range (mm)*
Clay	< 0.002
Silt (fine, medium and coarse)	0.002–0.006, 0.006–0.02, 0.02–0.06
Sand (fine, medium and coarse)	0.06–0.2, 0.2–0.6, 0.6–2.0
Gravel (fine, medium and coarse)	2.0–6.0, 6.0–20.0, 20.0–60.0
Cobbles	60.0–200.0
Boulders	> 200.0

For investigations involving soil samples in the laboratory, it is important to determine the water content, the particle size distribution, the mineralogy of clay soils, and the percentage of air voids associated with the material. Table 2.5 gives an indication of the particle size ranges associated with various soil components.

For soils containing clay-sized particles, plasticity information is particularly useful to gauge its behaviour, for example Atterberg limits (liquid limit (w_L) and plastic limit (w_P)), and plasticity indices (plasticity index ($I_P = w_L - w_P$) and liquidity index ($I_L = (w - I_P)/(w_L - w_P)$ where w is the water content). Figure 2.15 shows a plasticity chart showing the A-line, i.e. the distinction between soils with predominantly clay-sized particles (above the A-line) and those with predominantly silt sized particles (below the A-line). For example, a high plasticity clay, i.e. one with high I_P and w_L values, can be susceptible to large swelling and shrinkage behaviour when subjected to small changes in water content.

From the Atterburg limits and the water content, w, of the soil, the consistency index, I_c, can be determined as shown in equation 2.3.

$$I_c = \frac{w_L - w}{I_P} \tag{2.3}$$

Consistency index is a useful way of estimating the state or condition of silts and clays (Table 2.6).

Furthermore, it is important to determine the time dependent behaviour of the materials in the long-term. For example, for fine grained soils it is useful to do oedometer (one dimensional consolidation) tests as this can

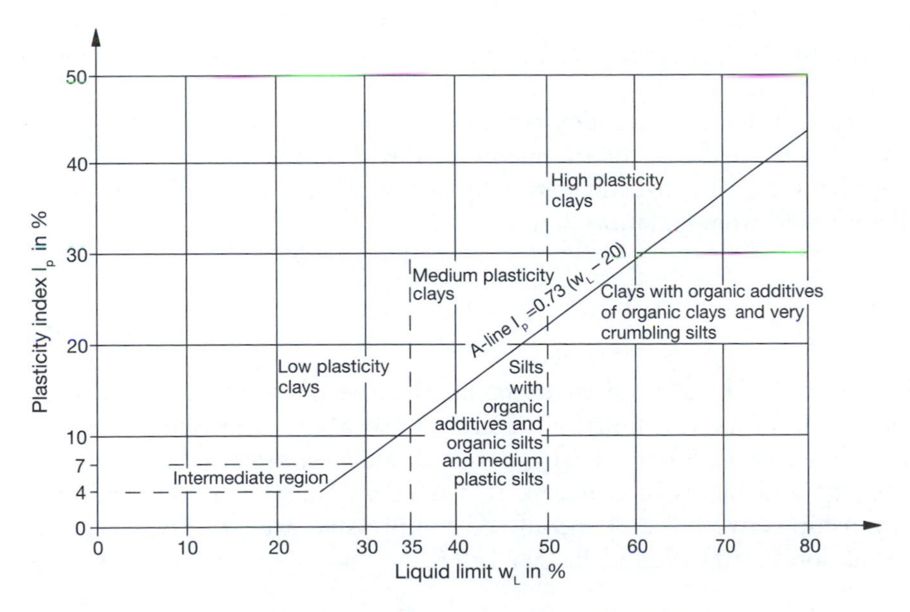

Figure 2.15 Plasticity (Casa grande) chart (after DIN 2006)

Table 2.6 Consistency index (BSI 2004c and Stein 2005)

Consistency of silts and clays	*Consistency index (I_c)*	*Description of soil*
Very soft	< 0.25	When pressing it in the fist, the soil squirts between the fingers.
Soft	0.25 to 0.50	Easily kneaded
Firm	0.50 to 0.75	Hard to knead, but can be rolled in the hand to threads of about 3 mm in diameter without tearing or crumbling.
Stiff	0.75 to 1.00	When trying to roll it into threads of about 3 mm in diameter, it crumbles or tears, but it remains moist enough to be able to reform it into a lump.
Very stiff	> 1.00	When dried its appearance (colour) is very light. It cannot be kneaded, but only broken. Reforming it into a lump is no longer possible.

give information on the characteristics of how it will deform with time and also its permeability can be estimated. The parameters obtained from this test are coefficient of volume compressibility, m_v (note, this parameter is stress dependent and so the value has to be calculated for the appropriate stress range), coefficient of consolidation, c_v, and vertical drained deformation modulus, E_v'.

In addition, for clayey and unsaturated soils, the swelling characteristics should be established as any increases in volume could induce large forces onto the tunnel lining that need to be included in any structural analysis of tunnel linings.

For rock, a point load index test can be conducted to obtain the point load index strength, I_S, for the material (ISRM 1985). This is conducted by applying a point load diametrically across the rock core. The I_S value is determined from equation 2.4.

$$I_S = \frac{L}{d^2} \quad (2.4)$$

where L is the load required to break the specimen and d is the core diameter. I_S varies as a function of size and so a size corrected value corresponding to a d = 50 mm, i.e. $I_{S(50)}$, is used. This test can also be conducted on blocks and lumps of material. If d is in millimetres, an approximate relationship between I_S and unconfined compressive strength, UCS, is given by equation 2.5 (Hoek and Brown 1997).

$$UCS = (14 + 0.175d)I_S \quad (2.5)$$

There are also laboratory tests for determining the abrasiveness of rocks in order to gauge the wear on cutting tools. One such test is the CERHAR Abrasiveness Test (CAI Test) developed at the Centre d'Étude et de Recherche du Charbon (Büchi *et al.* 1995). This test provides an index value that can be used as a gauge for the abrasiveness of different rock types. The index value ranges from 0.3 for very low abrasiveness to 6.0 for extremely abrasive. Using this test, basalt has an abrasiveness index of 2.7, gneiss 4.4 and granite 4.9.

Further details on laboratory tests can be obtained from standards, such as Eurocode 7: Geotechnical Design – Part 2: Ground investigation and testing (BSI 2007). This document includes guidance on both soil and rock testing. In addition, details on identification and description, and classification of soft ground can be obtained from BSI (2002a) and BSI (2004a) respectively. For the identification and classification of rocks the reader is referred to BSI (2003). Details of shear strength tests can be obtained from BSI (1990). In addition, Head (1997 and 2008) and Head and Keeton (2010) provide extensive descriptions of all the main soil laboratory tests.

In order to determine the strength parameters for rock and soil, and the modulus, E, uniaxial and triaxial tests can be conducted. These tests are briefly described below.

UNIAXIAL TEST

This is a standard experiment for rock cores in order to obtain failure parameters (unconfined compressive strength UCS, σ_u, ε_u), E and μ. For the uniaxial test the core sample is loaded in one direction. Laboratory samples are made up of cylindrical shaped cores with diameters of at least 90–100 mm. During the test, load is applied to the end of the sample (Figure 2.16a). Some considerations for sample preparation include:

- the end surfaces must be flat and even;
- the ends must be parallel and at right angles to the sample axis in order to avoid bending stresses being induced into the sample (which will give a reduced value of strength);
- during the test, friction is generated between the end surfaces of the sample and the end loading platens. This has the potential for increasing the failure load of the sample as it restricts the sample expansion. This is negligible if the height (h) to diameter (d) ratio is greater than or equal to 2 (h/d $\geq$ 2). If h/d is less than 2 (h/d $<$ 2), for example if there is not enough intact material in the core sample, then equation 2.6 can be used to adjust the stress, $\sigma_{u,adj}$.

$$\sigma_{u,adj} = \frac{8 \cdot \sigma_u}{7 + 2 \cdot \frac{d}{h}} \tag{2.6}$$

Care should be taken as the failure load can be reduced by up to 11% between stumpy and slender samples.

The rate of loading is also important. However, as the load increment is dependent on the deformation of the material, there are no set values. The strain calculated from the displacement measuring transducer attached to the test rig should increase at a rate of, $\dot{\varepsilon}$ = 0.05%/min, with the expectation that the test should last at least 5 minutes. This is the general guideline for a material that does not deform much. If a more ductile material is being tested and large strains are expected, the rate can be a lot higher, for example rock salt $\varepsilon_u \approx 10\%$, therefore the rate is increased to 0.25%/min. In comparison, for concrete $\varepsilon_u \approx$ 6 to 8‰, i.e. approximately 10 times smaller than for soft rock, and for granite $\varepsilon_u \approx$ 3 to 4‰.

Experiments in which either the strain or stress are regulated are called strain and stress controlled tests respectively. For strain controlled tests it is unavoidable that the sample is completely destroyed at ε_u and so there is no information on the after failure strain response. However, the demands put on the testing machine, i.e. the control techniques, are less than those for a stress controlled test.

One of the most important material parameters is obtained from Hooke's law, which is the ratio of stress over strain (where E is the elastic modulus in equation 2.7).

$$E = \frac{\Delta\sigma}{\Delta\varepsilon} \qquad (2.7)$$

The required values are taken from the stress/strain graph (Figures 2.16b, 2.18c). Depending on the section of the curve investigated, several different moduli can be identified as shown in Figure 2.17.

As a rule, to determine the value of E from the results of laboratory experiments, look at the $\sigma - \varepsilon$ graph where the sample behaves elastically and also where it is linear, i.e. where E is constant (equation 2.7 assumes linearity). When doing an analysis on hard rock, the most reliable results are obtained using a reloading modulus. As a rule the middle third of the reloading modulus should be used and additionally the initial load should not exceed 60% of the failure load as this avoids local overstressing because of inhomogeneities and microcracks within the sample. This means that E can be determined from the intact sample. The value of E is often dependent on the stress situation; generally the higher the isotropic stress, the higher E. It is therefore worth noting that in the case of the actual ground, there is a possibility that E will change with depth and as a consequence the value of E at the crown and invert of the tunnel could be different.

The value of E in rock can well exceed 100,000 MN/m^2 (this is an order of magnitude greater than the value of concrete). This means that E of the ground cannot be higher than that of the rock/soil itself and in reality, on

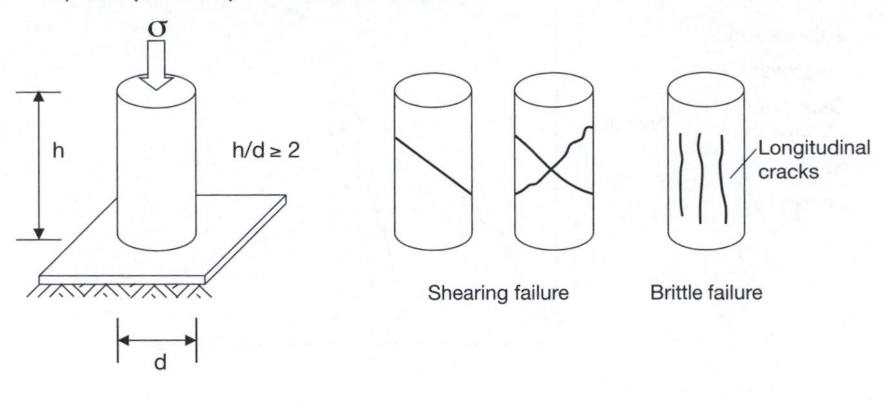

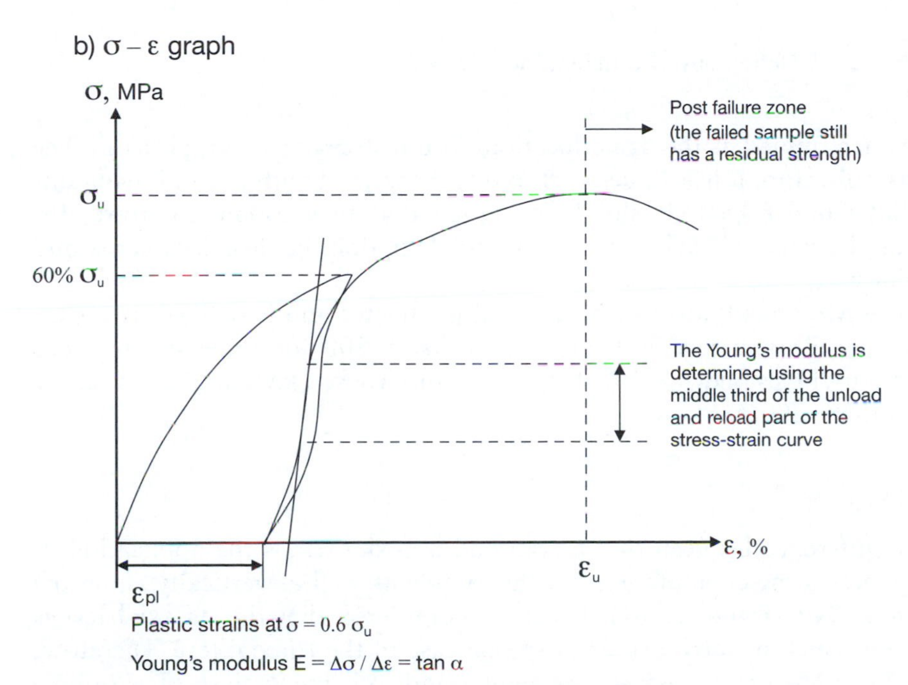

Figure 2.16 Uniaxial unconfined compression test

average, the value of the ground is 10 to 20 times less. The value of E for the ground is therefore an estimate and so it is always advisable to plan and do calculations based on a range of values.

The value of Poisson's ratio, μ, is important for structural analysis as this provides the ratio of horizontal strain to vertical strain (equation 2.8).

$$\mu = \frac{\Delta\varepsilon_{horiz}}{\Delta\varepsilon_{vert}} \qquad (2.8)$$

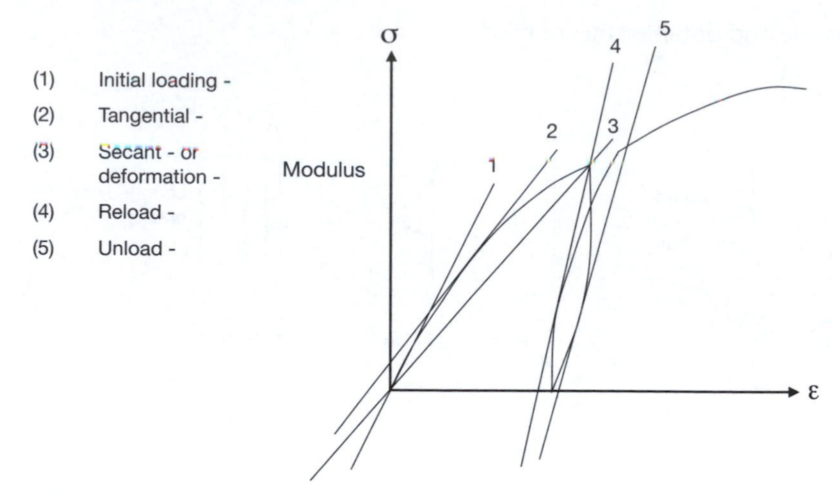

Figure 2.17 Definitions of different modulus values

It is determined at the same location on the stress/strain graph as E. The Poisson's ratio, μ has values of 0 to 0.5. A material with $\mu = 0.5$ maintains volume under load. It should be noted that in German literature, for example, Poisson's ratio has the inverse definition, i.e. it is between 2 and infinity.

The Modulus Ratio can be a useful parameter and is defined as E/UCS and is approximately 300 for most rocks; > 500 for some strong rocks and stiff limestones; < 100 for deformable rock, clays and some shales (Waltham 2002).

TRIAXIAL TEST

The difference between the triaxial and uniaxial test is the application of pressure to the circumference of the sample as well as vertically along the sample axis (Figure 2.18a). It can be considered that the uniaxial test is one in which $\sigma_2 = \sigma_3 = 0$, i.e. a special case of the triaxial test. Therefore, the experimental procedures are similar and so is the analysis. The sample size requirements, i.e. the h/d ratio, and the need for parallel end platens to be at right angles to the sample axis are exactly the same. In addition, these tests can be conducted under both stress and strain control. The only difference is that in a strain controlled triaxial test the axial and radial strains are increased equally until the axial strain has reached the desired value. In the stress controlled test the radial stress is kept constant while the axial stress is increased (it should be noted that in an indirect tension test, the radial stress is increased and the axial pressure is kept constant).

Triaxial tests are usually done when there is an interest in the shear strength parameters from which one can estimate the stand-up time for the ground, which is particularly important for weaker materials such as soils.

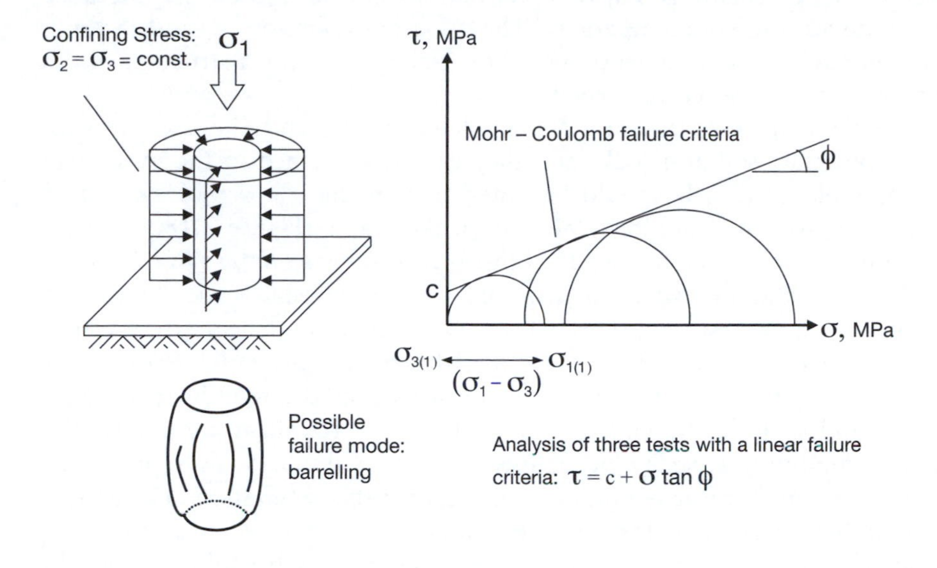

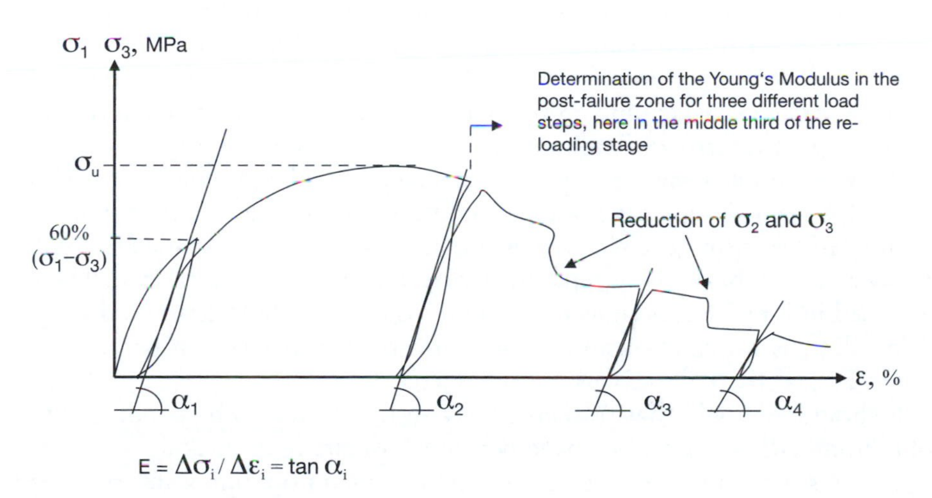

Figure 2.18 Triaxial test

The stand-up time allows an understanding to be obtained of the time that an open void can stand on its own without any support. Another reason to obtain the shear strength parameters of the material is to gain an idea of the deformation characteristics of the sample, i.e. deformations in the tunnel that are independent of E. Deformations that are dependent on E are elastic deformations, however, in addition to these deformations, there are plastic deformations, which are dependent on the apparent cohesion and internal friction angle of the material. These can get much larger than the elastic deformations. Plastic deformations in rock can develop because of crevasses and softening zones. The apparent cohesion, c, and internal friction angle, ϕ, give indications of how strong the matrix structure of the whole ground is when disturbed.

The triaxial compression test can determine E and the compression stability of the soil and rock, and they can be described in the same way as the uniaxial test. It should be noted that in the $\sigma - \varepsilon$ plot, the stress difference, $\sigma_1 - \sigma_3$, should be plotted on the y-axis (Figure 2.18c).

Figure 2.18b shows the results of three triaxial tests on the same material at different confining stress, σ_3 ($\sigma_1 = \sigma_3$ + the additional vertically applied stress during the test, and $\sigma_2 = \sigma_3$) on a $\tau - \sigma$ diagram (shear stress versus direct stress). For each pair of principal stresses (principal stresses are stresses acting on a plane where there are no associated shear stresses) Mohr circles can be drawn. In order to determine the shear parameters the failure condition needs to be defined, which in this case uses the linear Mohr-Coulomb failure condition as a basis for this estimation. The Mohr-Coulomb strength line forms an envelope for all the Mohr's circles by touching each circle at one point (i.e. a tangent to each Mohr's circle). In theory only two circles would be sufficient to construct this tangent, but, as we know, experimental results are subject to variability and so it is advisable to do at least three tests. The intercept with the shear stress axis, τ, is called apparent cohesion, c, and the gradient of the line is the internal friction angle, ϕ. Depending on the type of triaxial test conducted, different strength parameters can be obtained. For example, a quick undrained test will give undrained shear strength parameters (c_u and ϕ_u), which are useful for short-term design calculations in fine grained soils. A consolidated undrained test with pore water pressure measurement or a drained test will allow effective shear strength parameters (c' and ϕ') to be determined. These are used in long-term designs in fine grained soils or short-term (and long-term) designs for coarse grained soils. Effective shear strength parameters are indicated by the ′ next to the parameter.

It should be noted that triaxial extension tests can also be conducted to obtain information on the tensile behaviour of the rock or soil.

There is also more of an emphasis these days on obtaining information related to the small strain behaviour of soils. This is because the stiffness of soils is highly non-linear with increasing strain and is considerably larger at lower strains. Therefore, to get accurate indications of this relationship in triaxial tests, various techniques can be used. These include on-sample strain measurements, rather than using external measurements, and bender elements (where a vibrating element induces a shear wave into the sample that is picked up by another element on the opposite side of the sample; the travel time being indicative of the sample stiffness, in a similar way to seismic tests for the ground). This knowledge is important as input parameters for constitutive soil models used in numerical analyses (section 3.6).

The main questions that must be asked are: do these laboratory values determined from small samples of material relate to the ground *in situ*? What effect is there on these values from, for example, layering, fissures and water? It is ultimately the ground characteristics that are critical when designing a tunnel.

2.4 Ground characteristics/parameters

After determining the soil/rock material parameters from laboratory testing, for example, it is important to determine the ground parameters as these can be significantly different. It is only possible to do this in conjunction with the engineering geology report. The engineering behaviour of the ground is not solely controlled by the strength and stiffness of the rock/soil material. Discontinuities, such as joints, faults and bedding, act to reduce the *in situ* ground material properties compared to those of the rock/soil material. Only in the very rare case of a homogeneous and isotropic ground are the characteristics of the ground equal to those obtained for the rock/soil material. However, there is no 'recipe' as to how to determine ground parameters from field and laboratory experiments and geological descriptions.

It is necessary to take a holistic look at the ground, treating it as a matrix of several soils/rocks with, for example, dips and faults (defined below), and including any water. From the information obtained from a site investigation, the objective is to develop a geological model, and a hazard model, for the site, i.e. to highlight important information relevant for the design and construction (and potentially longer-term issues in the case of environmental aspects).

Material (soil or rock) tested in the laboratory is only a sample of the whole ground mass and these tests can only to some extent simulate the real conditions. A laboratory test of a rock sample that gives 40000 MPa for the Young's modulus, for example, may only yield 500 MPa on site (based on experience). This large reduction in Young's modulus can depend on large faults or lots of small faults, or faults filled with clayey material (the clay acts as a lubricant and the ground can move without exceeding the failure stress of the rock). Faults are fractures in the ground where relative displacements have occurred. Section 2.4.1 gives an example of how layering can affect the modulus of the ground. Reference should be made to Hoek and Brown (1997) for further information on how to practically estimate ground strength. Hoek and Diederichs (2006) also give information on empirically estimating the ground modulus. For a comprehensive list of geological terms, the reader is directed to a dictionary of geology, for example Whitten and Brooks (1972).

The descriptive engineering geology report resulting from the site investigation has a very important function as it contains information, for example, on the size and frequency of faults, about their characteristics, i.e. healed, closed, open, filled and the type of filling material and also the angle and direction of the faults relative to the tunnel axis in both the vertical and horizontal direction. The report also gives information on layering, folding of the strata (i.e. undulations), fabric, fragmentation, separation layers related to the ground, and also two angles of dip (azimuth and inclination relative to the tunnel location). The definitions for 'dip' and 'strike' are shown in Figure 2.19.

The positions of any discontinuities, any inhomogeneities and any anisotropy should be noted. Discontinuities are where the ground properties change abruptly. Inhomogeneities are where the properties of the ground change, either due to a change in material or to more subtle changes within the same material. Anisotropy is where the properties of the ground are different in various directions, for example different in vertical and horizontal directions. Unconformities are also important as these are planes or breaks between two sequences of rock with different dips.

Tables 2.7 and 2.8 provide information on some important geological descriptions about the ground used when designing tunnelling projects. These geological descriptions can affect the stress redistributions around the tunnel, influence the support requirements and also affect the loads

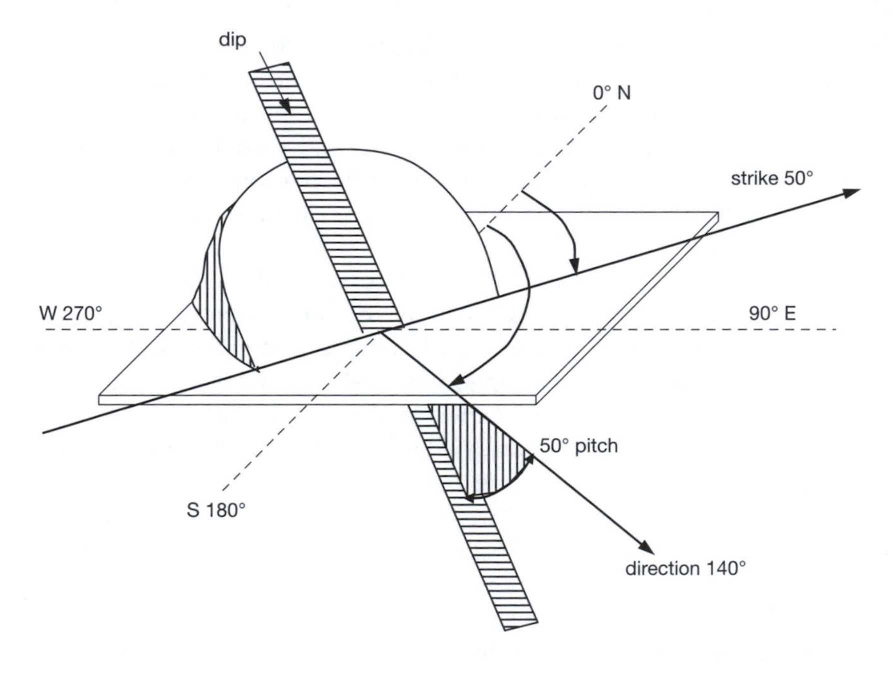

Definitions of dip and strike:

Strike is the direction in which a horizontal line can be drawn on a plane in relation to geographic north (Whitten and Brooks 1972).

Dip is the angle of maximum slope of the beds of rock measured from the horizontal at any point (Scott 1984).

Figure 2.19 Definitions of dip and strike

Table 2.7 a) Descriptions used for layer thickness and joint spacing (after BSI 1999, Prinz and Strauß 2006, and Anon 1977), b) Bedding inclination (after Forschungsgesellschaft für das Straßenwesen 1980)

a)	*Layer thickness*	*Spacing between joints*	*b) Ranges of inclination and description*
> 2 m	Very thickly bedded	Extremely wide	0° to 10° horizontal
0.6 to 2 m	Thickly bedded	Very wide	10° to 30° level
0.2 to 0.6 m	Medium bedded	Wide	30° to 60° inclined
0.06 to 0.2 m	Thinly bedded	Moderately wide	60° to 90° steep
0.02 to 0.06 m	Very thinly bedded	Moderately narrow	
0.006 to 0.02 m	Laminated	Narrow	
0.002 to 0.006 m	Thinly laminated	Very narrow	

Table 2.8 State of weathering (after BSI 1999, and Forschungsgesellschaft für das Straßenwesen 1980)

State	*Rock sample characteristics*	*Ground characteristics*
Unweathered	No effect of weathering visible	No loosening of the interfaces due to weathering
Partially weathered	Some noticeable weathering of individual mineral particles on freshly fractured surfaces	Partial loosening at discontinuities
Distinctly weathered (softened)	Softening due to weathering, but minerals still bonded together	Complete loss of strength at discontinuities

on the tunnel lining. Table 2.7 provides some descriptions related to the thickness of the layers, spacing and angle of dip for strata within the ground. Table 2.8 shows some weathering descriptions. Weathering weakens the ground through the action of water, wind and temperature.

Groundwater levels should be noted, together with the chemical composition of the natural groundwater with respect to concrete attack (sulphates and chlorides). Groundwater levels must be carefully assessed as there can be multiple levels depending on the relative permeability of the strata making up the ground. For example, in London, UK there is an upper and a lower aquifer and hence two groundwater levels.

Information on the geological history, age and mineralogy composition of the rock or soil (stratigraphy and petrography, i.e. the description and systematic classification of the rocks) is important. In clay soils, for example, it is the microstructure of the material that controls its properties such as strength and compressibility.

It is of primary importance for the tunnel builder to understand how all these factors influence the mechanical behaviour of the rock within the overall ground. Due to the variety of the influences there is no simple or definitive answer with respect to rock behaviour.

In addition, an extensive and well performed site investigation is only equivalent to a 'pin prick' as far as the problem is concerned and therefore there is always uncertainty and this can be found in the choice of the expert's words when writing the report. While reading the engineering geological report, be aware of the phraseology used: 'it can't be excluded that . . .', 'the layer boundaries could be uneven . . .', or 'expect fault thickness of, for example, 3 cm and more . . .'. Unfortunately, there is no definite mathematical formula, so when judging such non-committal language your own impressions of the cores and site can be very helpful. Due to the unavoidable uncertainties of site investigations, it is extremely important to check the predicted situation during the construction phase.

2.4.1 Influence of layering on Young's modulus

It often occurs that the ground is layered and there is a potentially large difference in material behaviour in these various layers. In the laboratory, layers can only be investigated separately, and in many cases soft layers are not present in the sampling core as they are washed out with the drilling fluid. In such cases it can be impossible to estimate the influence of layering on the overall behaviour of the ground. However, if it is assumed that the layering is uniform, it is possible to do a simple estimation. An example is shown in Figure 2.20 and the calculation is given below.

$$\varepsilon = \alpha \cdot \frac{\sigma}{E_S} + \beta \cdot \frac{\sigma}{E_C} \Rightarrow E = \frac{1}{\frac{\alpha}{E_S} + \frac{\beta}{E_C}} = \text{overall E} - \text{modulus}$$

where α and β are percentages of the soil material in the whole.

$$\alpha = \frac{d}{t+d} \quad \text{with} \quad d = \sum_i d_i \quad \text{and} \quad \beta = \frac{t}{t+d} \quad \text{with} \quad t = \sum_i t_i$$

$$\alpha = \frac{1.50}{1.53} \cong 0.98 \qquad \beta = \frac{0.03}{1.53} \cong 0.02$$

$$E = \frac{1}{\frac{0.98}{7000} + \frac{0.02}{70}} \cong 2350\text{MPa}$$

This illustrates that even though the clay is only 2% of the overall soil mass, it has a significant effect and reduces the overall E of the system by about two-thirds of the original sandstone value.

In this very simple example, no account has been taken of any crevasses, faults etc. which would reduce this value still further.

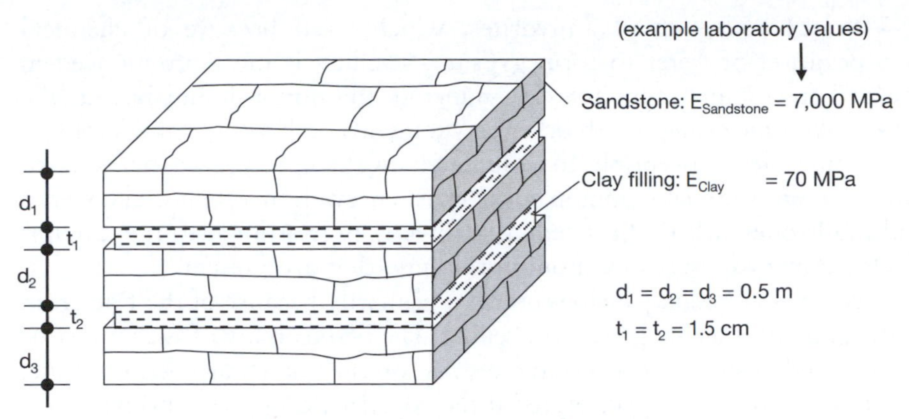

Figure 2.20 Example of how layering affects E-modulus

2.4.2 Squeezing and swelling ground

Squeezing and swelling of a ground can create high pressures on tunnel supports and such situations must be managed either within the construction process or by providing flexible or 'ductile' supports.

SQUEEZING

Squeezing rock is a plastic material that moves into an underground opening primarily because of pressures exerted by loads of overlying rocks. Although a clear distinction between swelling and squeezing ground is not always possible, squeezing ground differs from swelling in that it undergoes no appreciable volume increase owing to the penetration of water. However, ingress of water, even in small amounts, promotes plastic behaviour and contributes to the easy movement of squeezing ground (Wahlstrom 1973).

Unlike swelling ground, movements in squeezing ground involve materials outside the immediate area of the tunnel opening, and the volumes of material that have the potential of moving into the tunnel opening may be very large.

When the uniaxial compressive strength of the rock is less than 30% of the *in situ* stress, severe or extreme squeezing can occur (Thomas 2009a).

SWELLING

Swelling in mineral aggregates is caused by one or a combination of processes including (Wahlstrom 1973):

- adsorption of films of water attracted and held by surface forces of very small mineral particles;
- adsorption of free water by clay minerals such as montmorillonites;
- hydration;
- expansion of pore water as a consequence of the release of confining pressure.

With the exception of anhydrites, which swell because of chemical incorporation of water to form gypsum, swelling is most pronounced in rocks that contain abundant clay minerals (montmorillonite having the greatest volume change, with other clay minerals such as illite and kaolinite being much less susceptible to volume change), or clay-sized particles of other minerals. Some common materials that swell are shales, claystones and mudstones, where the swelling is directly proportional to the amount of clay minerals, especially montmorillonite, that are present.

Swelling is commonly a slow process, primarily because of the fine grain of the minerals that are prone to swelling. The permeability of such minerals is low, and penetration of water of any origin takes considerable time. Swelling is likely to be accelerated if the construction process brings water in contact with a soil, which is capable of swelling.

2.4.3 Typical ground parameters for tunnel design

Due to the many influences that determine the ground behaviour and the disparity that often exists between this and the properties of the individual soils/rocks it is not possible to determine binding parameters for a rock or soil type, i.e. it cannot be put into any 'Standard'. However, some typical values are provided in the following tables. These provide the reader with a 'feel' for the magnitudes of the values involved with certain parameters.

TYPICAL ROCK AND SOIL PARAMETERS

Tables 2.9 to 2.12 provide some typical values for the strength and deformation characteristics of rocks, generic hardness classes, and some strength and permeability values for soils respectively.

Table 2.9 Strength and deformation characteristics of some typical rocks (after Reuter 1992 and, Klengel and Wagenbreth 1987)

Ground	*Compressive strength (MPa)*	*E (MPa)*
Basalt	160–400	48000–105000
Dolomite	50–180	32000–100000
Gabbro	80–345	75000–120000
Gypsum	9–40	10000–29000
Granite	100–300	37000–72000
Chalk	20–240	16000–90000
Sandstone	10–290	6000–71000
Rocksalt	20–30	16000–24000
Slate	20–210	23000–85000
Concrete C20/25	25 (cube crushing strength)[a]	29000

Note: (a) For the influence of the shape of the specimens on the compressive strength see uniaxial compression test

Table 2.10 Generic hardness classes (after Reuter 1992 and Fecker and Reik 1987, simplified)

Ground	*Compressive strength (MPa)*	*Apparent cohesion (kN/m²)*	*Internal friction angle (degrees)*
Hard rock	high > 100	> 2000	> 40
	middle 20–100	200–2000	30–40
	low 5–20	20–200	20–30
Transitional rock	very low 1–5	< 20	< 20
Soft ground	very soft < 1	0– 10	0–10

Note: That the numbers in this table should only be treated as 'ball park' values.

Table 2.11 Example shear strength parameters for soils (after Waltham 2002, BSI 1986)

Soil type	*Apparent cohesion*[a] *(kN/m²)*	*Internal friction angle (degrees)*
Till and tertiary clays	Hard > 300	
	Very stiff 150–300	
	Stiff 75–150	28
Alluvial clay	Firm 40–75	
	Soft 20–40	19
Medium dense sand	–	32–36
Dense sand	–	36–40
Sandy gravel	–	35–50
Silty sand	–	27–34
Clayey silt	20–75	25
Consistency of clays	*(after BSI 1986)*	
Very soft	< 20	–
Soft	20–40	–
Firm	40–75	–
Stiff	75–150	–
Very stiff	150–300	–

Note: (a) Short-term or undrained shear strength (c_u), the long-term or drained (effective) shear strength (c') for clays is difficult to quantify, but is usually relatively small (< 10–15 kN/m²) *and decreases rapidly on disturbance and weathering.*

TYPICAL GROUND PARAMETERS

Table 2.13 provides some typical values for the shear strength of various rocks and Table 2.14 provides an example of a rock classification system. Further information on rock mass classification systems is given in section 2.4.4.

Table 2.12 Typical permeability values for soils

Type of soil	*Permeability, k (m/s) – limiting values*
Coarse gravel	10^{-1} to 5
Fine gravel	10^{-4} to 10^{-2}
Coarse sand	10^{-5} to 10^{-2}
Medium sand	10^{-6} to 10^{-3}
Fine sand	10^{-6} to 10^{-3}
Silt	10^{-9} to 10^{-5}
Clay	10^{-12} to 10^{-8}

Note: Below 10^{-8} is very low permeability, 10^{-6} to 10^{-4} is permeable and above 10^{-2} is very high permeability

Table 2.13 Shear parameters for several rocks (after Reuter 1992)

Rock	*Condition*	*Apparent cohesion (kN/m²)*	*Internal friction angle (degrees)*
Granite	–	200–3000	30–50
Sandstone	Parallel to joints	100	60
Limestone	Joints without fill	700	40
Limestone	Joints with loose material	100–300	22–27
Limestone	Joints with clayey fill	0–100	11–17

Table 2.14 Rock mass classification (after Bieniawski 1984 and Fecker and Reik 1987)

Description	*Very good rock*	*Good rock*	*Medium rock*	*Weaker rock*	*Very weak rock*
Average stand-up time	10 years with a 5 m span width[a]	6 months with a 4 m span width	1 week with a 3 m span width	5 hours with a 1.5 m span width	10 minutes with a 0.5 m span width
Apparent cohesion of the rock mass (kN/m²)	> 300	200–300	150–200	100–150	< 100
Internal friction angle of the rock mass (degrees)	> 45	40–45	35–40	30–35	< 30

Note: (a) The span width indicates the unsupported length in the direction of the tunnel advance.

2.4.4 Ground (rock mass) classification

In his book 'Engineering Rock Mass Classifications', Bieniawski (1989) states 'Rock mass classifications are not meant to be taken as a substitute for engineering design. They should be applied intelligently and used in conjunction with observational methods and analytical studies to formulate an overall design rationale with the design objectives and site geology.'

The objectives of rock mass classifications are therefore to (after Bieniawski, 1989):

- identify the most significant parameters influencing the behaviour of a rock mass;
- divide a particular rock mass formation into groups of behaviour, that is, rock mass classes of varying quality;
- provide a basis for understanding the characteristics of each rock mass class;
- relate the experience of rock conditions at one site to the conditions and experience encountered at others;
- derive quantitative data and guidelines for engineering design;
- provide a common basis for communication between engineers and geologists.

An early classification system for soft ground is the Tunnelman's ground classification as shown in Table 2.15 and provides information on the likely tunnel working conditions and some idea of the types of soils in which these conditions might occur.

For harder ground, a number of classification systems have been developed. Three of these classification systems are briefly described in this book: Rock Quality Designation (RQD), which is one of the simpler classification methods and is described in section 2.4.4.1, the Rock Mass Rating (RMR) system and the Rock Mass Quality Rating (Q-method) (sections 2.4.4.2 and 2.4.4.3 respectively). Further details are given in Appendix A. The reader is encouraged, however, to read the original, and subsequent, publications by the relevant authors of these systems in order to fully appreciate their usefulness and limitations.

2.4.4.1 Rock Quality Designation

The Rock Quality Designation index was developed by Deere in 1967 (Deere *et al.* 1967, Deere 1989) to provide a quantitative assessment of ground quality from drill cores. RQD was developed for assessing rock and can be misleading in soft ground. RQD is defined as the total length of 'solid' core pieces each greater than 100 mm between natural (not drill-induced) discontinuities expressed as a percentage of the total length of each core run, measured along the core axis. A solid core is defined as a core with at least one full diameter (but not necessarily a full circumference)

Table 2.15 Tunnelman's ground classification (after Thomson 1995, from Terzaghi 1950)

Classification	*Tunnel working conditions*	*Representative soil types*
Hard	Tunnel heading may be advanced without roof support	Very hard calcareous clay; cemented sand and gravel.
Firm	Tunnel heading can be advanced without roof support and the permanent support can be constructed before the ground will start to move.	Loess (e.g. wind blown soil deposits) above the water table; various calcareous clays with low plasticity such as the marls of South Carolina.
Slow ravelling	Chunks or flakes of material begin to drop out of the roof or the sides some time after the ground has been exposed.	Fast ravelling[a] occurs in residual soils or in sands with clay binder below the water table. Above the water table the same soils may be slow ravelling or even firm.
Fast ravelling	In fast ravelling ground, the process starts within a few minutes; otherwise it is referred to as slow ravelling	
Squeezing	Ground slowly advances into tunnel without fracturing and without perceptible increase of water content in the ground surrounding the tunnel. (May not be noticed in the tunnel, but will cause surface subsidence.)	Soft or medium-soft clay.
Swelling	Like squeezing ground, moves slowly into the tunnel, but the movement is associated with a very considerable volume increase in the ground surrounding the tunnel.	Heavily precompressed clays with a plasticity index in excess of about 30; sedimentary formations containing layers of anhydrite.

Table 2.15 (continued)

Classification	*Tunnel working conditions*	*Representative soil types*
Cohesive running	The removal of the lateral support on any surface rising at an angle of 34° to the horizontal is followed by a 'run', whereby the material flows like granulated sugar until the slope angle becomes equal to about 34°. If the 'run' is preceded by a brief period of ravelling, the ground is called cohesive running.	
Running		Running occurs in clean, coarse or medium sand above the water table.
Very soft squeezing	Ground advances rapidly into the tunnel in a plastic flow.	Clay and silts with high plasticity index.
Flowing	Flowing ground moves like a viscous liquid. It can invade the tunnel not only through the roof and sides, but also through the bottom. If the flow is not stopped, it continues until the tunnel is completely filled	Any ground below the water table that has an effective grain size in excess of about 0.005 mm.
Bouldery	Problems incurred in advancing shield or poling; blasting or hand-mining ahead of the machine may be necessary.	Boulder glacial till; rip-rap fill; some landslide deposits; some residual soils. The matrix between the boulders may be gravel, sand, silt, clay or combinations of these.

Note: (a) The term 'ravelling' is used to describe a situation when the ground collapses at the crown of the tunnel.

measured along the core axis between two natural discontinuities (Davis 2006). The procedure is shown on an example core in Figure 2.21.

The RQD provides a general assessment of rock quality and can be used as a basis for descriptive rock quality terms as shown in Table 2.16. However, it is limited to the mechanical structure of the rock and provides no information on discontinuity properties or strength (Davis 2006).

There are some potential issues with assessing RQD in the field or when reviewing borehole logs as it is frequently mis-logged. The key issues are (Davis 2006):

- natural discontinuities and drilling features are often not differentiated;
- drillers often only include cores of full circumference greater than 100 mm instead of full diameter cores.

Both of these issues lead to reduced RQD values.

Palmström (1982) suggested that if no core is available, but continuity traces are visible in surface exposures or exploration adits, the RQD may be estimated from the number of discontinuities per unit volume. The suggested relationship for clay-free ground is given in equation 2.9.

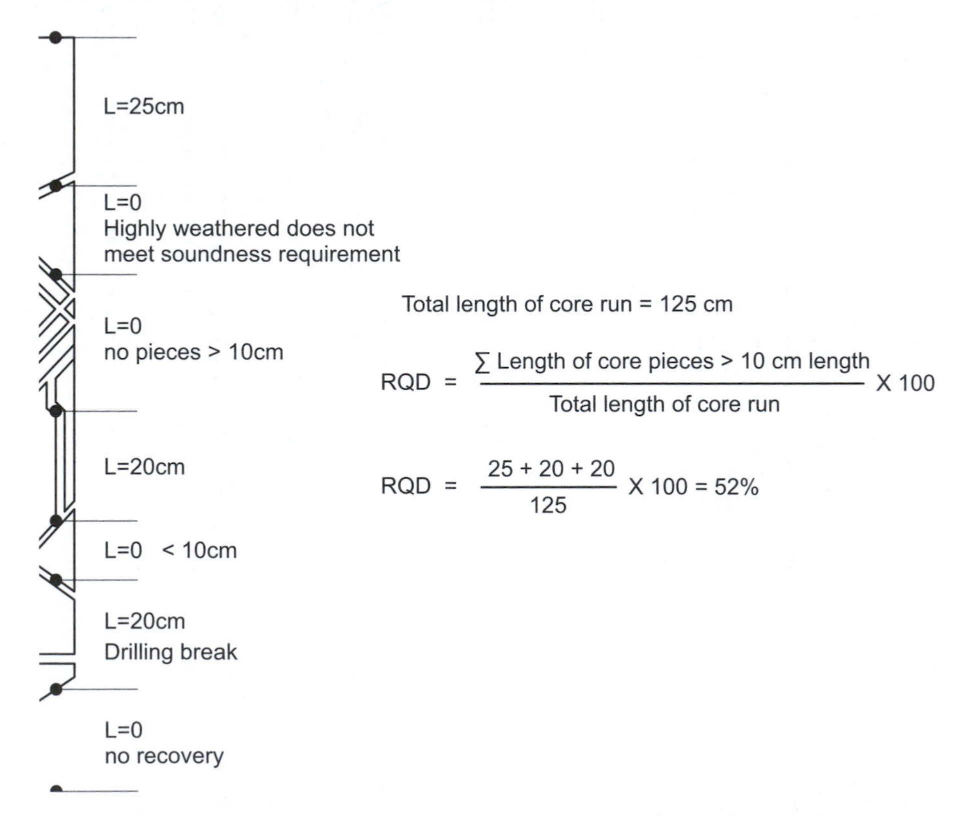

Figure 2.21 Procedure for measurement and calculation of RQD (after Deere and Deere 1989, used with permission from Don Deere)

Table 2.16 RQD values related to rock quality descriptions (after Deere and Deere 1989, from Deere *et al.* 1967)

RQD (%)	*Rock quality*
<25	Very poor
25–50	Poor
50–75	Fair
75–90	Good
90–100	Excellent

$$RQD = 115 - 3.3J_v \quad (2.9)$$

where J_v is the sum of the number of joints per unit length for all discontinuity sets.

2.4.4.2 Rock Mass Rating

The Rock Mass Rating system was developed by Bieniawski in 1972 and has been modified over the years as more data have become available (Bieniawski 1989). It should be noted that the RMR system was developed for hard rock conditions and it is of only limited use in soft ground.

The following six parameters are used to classify the ground using the RMR system:

- uniaxial compressive strength (UCS) of the rock material;
- RQD;
- spacing of discontinuities;
- condition of discontinuities;
- groundwater conditions;
- orientation of discontinuities.

The way to apply this system is to divide the rock into a number of structural regions in such a way that certain features are more or less uniform within each region. Appendix A (section A.1) provides details of the classification system used (Table A.1) and how to use this table to determine the RMR value for each region of the rock mass.

For tunnels, information can be obtained on stand-up time and maximum stable rock span for a given RMR (Figure 2.22).

In terms of application of the RMR system to tunnelling, Table 2.17 (after Bieniawski 1989) provides guidelines for the selection of rock reinforcement for tunnels in accordance with the RMR system (although it should be noted that this only applies to a 10 m span tunnel constructed using drill and blast, and no indication is provided as to how to extend this to other sizes of tunnel). These guidelines depend on such factors as depth to tunnel axis (*in situ* stress), tunnel size and shape, and the method of excavation.

Figure 2.23 shows how the rock mass classes in Figure 2.22 are modified for tunnel boring machines (TBM). It can be seen that the rock mass rating has to be higher to achieve similar stand-up times and roof span values. The letters indicate different TBM classes.

2.4.4.3 Rock Mass Quality Rating (Q-method)

The Rock Mass Quality Rating (Q-method) was proposed by Barton *et al.* (1974) for the determination of rock mass characteristics and tunnel support requirements. It is based on empirical data obtained from 200 tunnel construction projects in Scandinavia. It is probably the most widely used rock mass classification system today. It should be noted that the Q-method was developed for hard rock conditions and it is of only limited use in soft ground.

The numerical value of the index Q varies on a logarithmic scale from 0.001 to a maximum of 1000 and is defined by equation 2.10.

$$Q = \frac{RQD}{J_n} \times \frac{J_r}{J_a} \times \frac{J_w}{SRF} \qquad (2.10)$$

where

RQD = Rock Quality Designation ($0 \le RQD \le 100$)
J_n = joint set number ($0.5 \le J_n \le 20$)
J_r = joint roughness number ($1 \le J_r \le 4$)
J_a = joint alteration number ($0.75 \le J_a \le 20$)
J_w = joint water reduction factor ($0.05 \le J_w \le 1$)
SRF = stress reduction factor ($0.5 \le SRF \le 400$)

Appendix A (section A.2) provides Tables A.2–A.7 that give the classification of individual parameters used to obtain the Q-value for the rock mass. A detailed explanation of these parameters can be found in Barton *et al.* (1974)

The Q-value can be related to the stability of the excavation and support requirements. In order to do this, Barton *et al.* (1974) defined an additional parameter, which they called the equivalent dimension, D_e, of the excavation. This dimension is obtained from equation 2.11.

$$D_e = \frac{\text{excavation span, diameter or height in (m)}}{\text{excavation support ratio, ESR}} \qquad (2.11)$$

The value of ESR is related to the intended use of the excavation and to the degree of safety, which has an influence on the support system to be installed in order to maintain the stability of the excavation. Typical ESR values are given in Table 2.18.

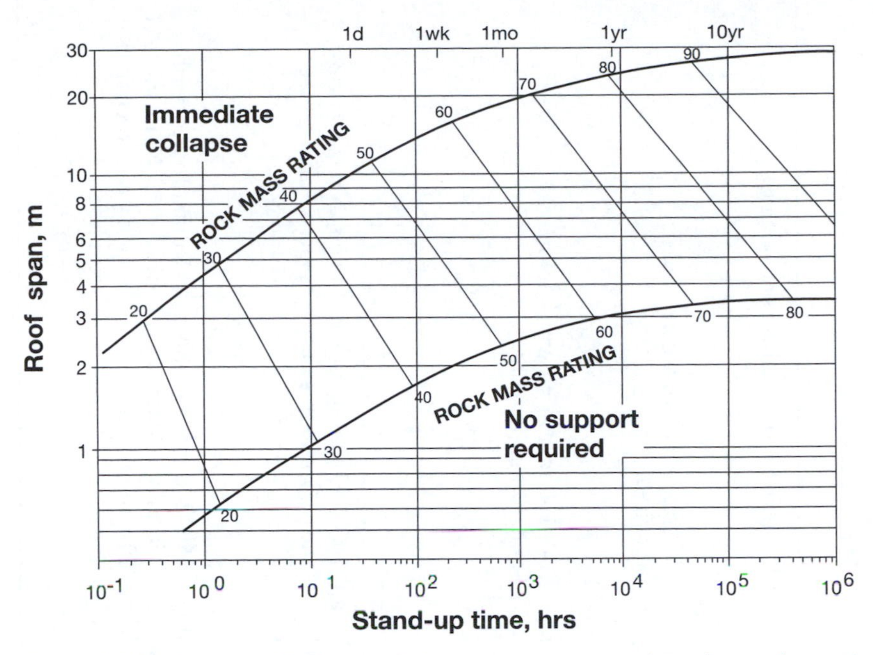

Figure 2.22 Relationship between the stand-up time and roof span for various rock mass classes (after Bieniawski 1989)

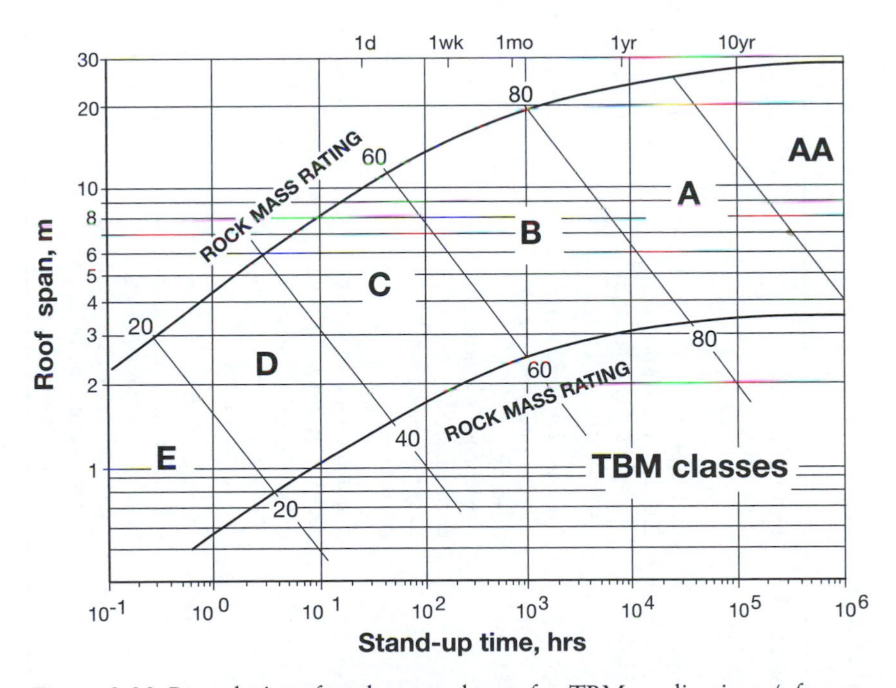

Figure 2.23 Boundaries of rock mass classes for TBM applications (after Bieniawski 1989, modified from a plot by Lauffer 1988, used with permission from VGT Verlag GmbH and taken from *Felsbau*)

Table 2.17 Guidelines for excavation and support of rock tunnels in accordance with the RMR system[a] (after Bieniawski 1989)

		Support		
Rock mass class	*Excavation*	*Rock bolts (20 mm dia. fully grouted)*	*Sprayed concrete*	*Steel sets*
Very good rock I RMR: 81–100	Full face. 3 m advance.	Generally, no support required except for occasional spot bolting		
Good rock II RMR: 61–80	Full face. 1.0–1.5 m advance. Complete support 20 m from face.	Locally, bolts in crown 3 m long, spaced 2.5 m, with occasional wire mesh.	50 mm in crown where required.	None
Fair rock III RMR: 41–60	Top heading and bench 1.5–3 m advance in top heading. Commence support after each blast. Complete support 10 m from face.	Systematic bolts 4 m long, spaced 1.5–2 m in crown and walls with wire mesh in crown.	50–100 mm in crown and 30 mm in sides.	None
Poor rock IV RMR: 21–40	Top heading and bench 1.0–1.5 m advance in top heading. Install support concurrently with excavation 10 m from face.	Systematic bolts 4–5 m long, spaced 1–1.5 m in crown and wall with wire mesh.	100–150 mm in crown and 100 mm in sides.	Light to medium ribs spaced 1.5 m where required.
Very poor rock V RMR: < 20	Multiple drifts 0.5–1.5 m advance in top heading. Install support concurrently with excavation. Sprayed concrete as soon as possible after blasting.	Systematic bolts 5–6 m long, spaced 1–1.5 m in crown and walls with wire mesh. Bolt invert.	150–200 mm in crown, 150 mm in sides, and 50 mm on face.	Medium to heavy ribs spaced 0.75 m with steel lagging and forepoling if required. Close invert.

Note: (a) Shape: horseshoe; width: 10 m; vertical stress: < 25 MPa; construction: drilling and blasting.

Table 2.18 Suggested excavation support ratios (ESR) (after Barton and Grimstad 1994, from Barton *et al.* 1974)

	Type of excavation	*ESR*
A	Temporary mine openings	2.0–5.0
B	Permanent mine openings, water tunnels for hydropower (excluding high pressure penstocks), pilot tunnels, drifts and headings for large openings, surge chambers	1.6–2.0
C	Storage caverns, water treatment plants, minor road and railway tunnels, access tunnels	1.2–1.3
D	Power stations, major road and railway tunnels, civil defence chambers, portals, intersections	0.9–1.1
E	Underground nuclear power stations, railway stations, sports and public facilities, factories, major gas pipeline tunnels	0.5–0.8

The equivalent dimension, D_e, plotted against the value of Q, is used to define a number of support categories in a chart published in the original paper by Barton *et al.* (1974). This chart has been updated a number of times to directly give the support requirements. Grimstad and Barton (1993), for example, modified it to reflect the increasing use of steel fibre reinforced sprayed concrete in underground excavation support. Figure 2.24 is reproduced from this updated chart.

In a further development, Barton (1999) proposed a method for predicting the penetration rate and advance rate for TBM tunnelling. This approach is based on an expanded Q-method of rock mass classification and average cutter force in relation to the appropriate rock mass strength. The parameter Q_{TBM} can be estimated during feasibility studies, and can also be back calculated from TBM performance during tunnelling. This method is briefly described in Appendix A (section A.2.1).

Barton (2002) provides some other useful correlations for the Q-value to assist site investigation and tunnel design. For example, a relationship between Q-value and seismic velocity (V_p) as used for some geophysical site investigation techniques (see section 2.3.2.1) is given in equation 2.12.

$$V_p \sim 3.5 + \log Q \qquad (2.12)$$

where V_p is in units of km/s. This relationship was developed from tests in hard rock, but this has been developed further for application to weaker and harder ground conditions. This has been achieved by normalizing the Q-value using 100 MPa as the hard rock norm. The relationship for the normalized Q-value, Q_c is shown in equation 2.13.

$$Q_c = Q \times q_u/100 \qquad (2.13)$$

where q_u is the unconfined compressive strength of the rock mass.

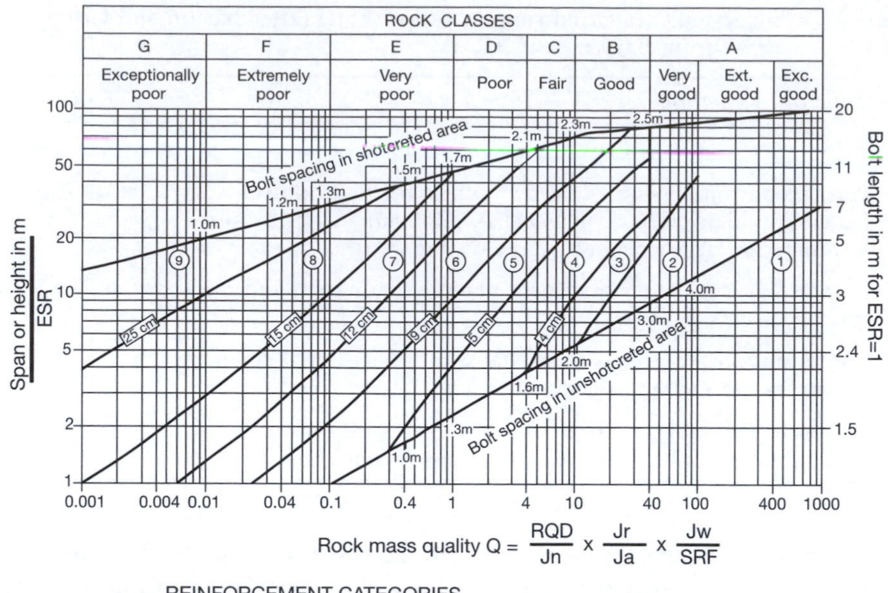

Figure 2.24 Estimated support categories based on the Q-value (after Grimstad and Barton 1993, reproduced from Palmström and Broch 2006)

Substituting equation 2.13 into equation 2.12, yields equation 2.14.

$$V_p \sim 3.5 + \log (Q_c{*}100/q_u) \tag{2.14}$$

Barton (2002) also proposed a relationship between the Q_c and the modulus, E, of the ground as shown in equation 2.15.

$$E = 10\ Q_c^{1/3} \tag{2.15}$$

Further details of these and other relationships can be found in Barton (2002).

2.4.4.4 *A few comments on the rock mass classification systems*

In this book there is only space to provide a brief overview of some of the rock mass classification systems currently in use. It is important to understand the basis of the systems and hence their applicability and limitations. It is therefore recommended that if the reader intends to use these systems, further, he/she conducts more detailed, background reading using the references provided.

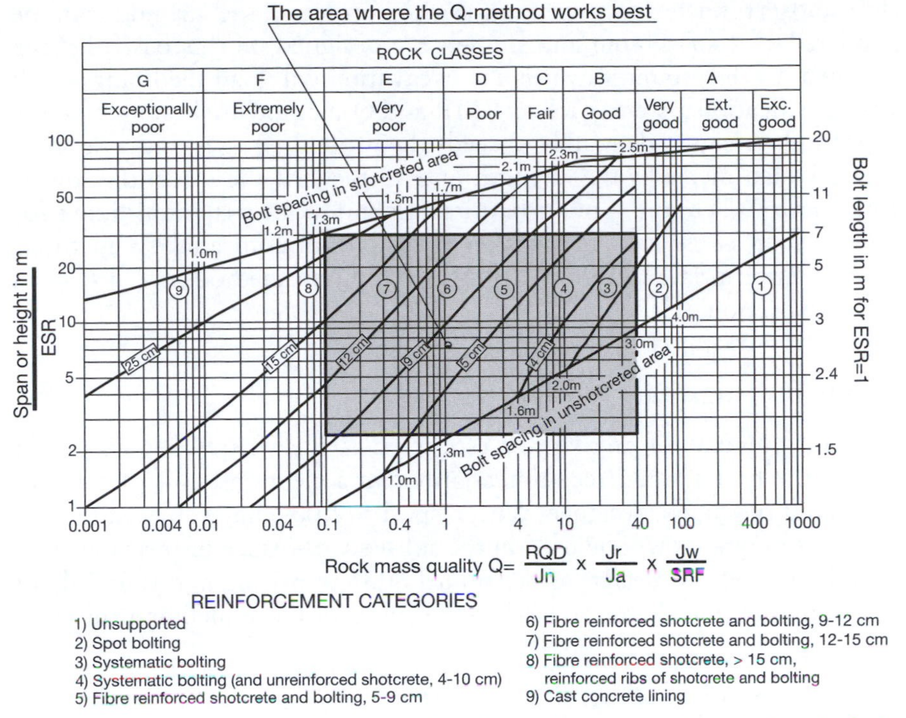

Figure 2.25 Limitations of the Q-method for rock support. Outside the shaded area supplementary methods/evaluations/calculations should be applied (reproduced from Palmström and Broch 2006)

One of the benefits of the RMR system is that it is relatively easy to use. The result produced by the RMR classification, however, is rather conservative. This can lead to an overestimation of the support measures (Maidl *et al.* 2008). As the RMR system and the Q-method are empirically devised they inevitably have their own deficiencies (as well as good points). As there are some reasonably consistent relationships between these systems, it is advantageous to apply both systems to the field data as a mutual check. There is an empirical relationship between the RMR and the Q-value as shown in equation 2.16 (Barton 2002).

$$\text{RMR} \sim 15 \log \text{Q} + 50 \tag{2.16}$$

Palmström and Broch (2006) investigated rock mass classification systems and particularly the Q-method, including Figure 2.24, and showed that actually the Q-method is most applicable within a certain range of parameters as shown by the shaded area in Figure 2.25. Outside this area, supplementary calculations and methods of evaluation are recommended.

For poorer quality ground these systems are less effective as shown in Figure 2.25. In these lower quality grounds the modulus values (or support criteria) are sensitive to small changes in the rating values.

For most tunnels for civil engineering projects, the ground can be considered as a continuum and tunnels are designed on this basis, i.e. the movement of the ground towards the excavation will load the lining. Rock mass classification systems such as RMR and Q-method are best used where the ground strength adequately exceeds the ground stresses and a support system, which increases the strength and stiffness of the discontinuities is appropriate. Where the ground requires a continuous structural lining for support, such is the case for weaker rocks, continuum analysis methods are more appropriate (BTS/ICE 2004). Continuum methods are discussed further in section 3.5.

2.5 Site investigation reports

The main outcome of any site investigation is the written report(s) that presents the findings and recommendations in a clear and concise manner so as to aid the tunnel designer. With respect to tunnelling projects, the site investigation reports will be used in the subsequent choice of the tunnelling method adopted, the design of the tunnel and the pricing and timescale of construction, and is therefore vital to the success of a tunnelling project.

2.5.1 Types of site investigation report

There are several types of report that can be produced from a site investigation and depending on the country will include a separate geology report, for example in Germany (after Hansmire 2007).

GEOTECHNICAL FACTUAL (OR DESCRIPTIVE) REPORT

This report should contain only factual information from the site investigation consisting of analysed data and objective consideration in accordance with existing standards, codes or specification. It does not have engineering interpretations. BS 5930 (BSI 1999) sets out, in general terms, the content of the Geotechnical Factual Report (GFR).

GEOTECHNICAL INTERPRETIVE REPORT

This is a report containing subjective considerations, interpretations, and comments from the engineer in charge; all in accordance with his knowledge and experience. The Geotechnical Interpretive Report (GIR) can be a project-specific report that presents the geological and engineering interpretation of the data. In its most simple form, it is a single report on a well-defined project. A geological profile is a geotechnical interpretation. In practice, many reports are written, revised, and in some cases superseded by later work. An interpretive report will address the project issues, and will often have design analysis, such as where rock mass classification is used to characterize the tunnel ground conditions upon which ground support and final lining requirements are established. The GIR is prepared primarily for use by the designers.

The GFR and GIR reports are often combined into one report in which the GIR makes up the main part of the report and the GFR the appendix. This report then forms the basis of the tendering process.

GEOTECHNICAL BASELINE REPORT (ESSEX 1997)

Within tunnelling contracts, a significant cause of cost overrun has historically been associated with contractors' claims for ground conditions significantly different from those expected at the time of tender. It has been difficult to assess these claims without well-defined benchmark conditions agreed at the outset between all the parties. The Geotechnical Baseline Report (GBR) has been designed as a tool to address this problem. The idea of a GBR is not new and has been the usual practice in the United States for many years. It is being increasingly more widely used in the UK and is a useful addition to tunnelling contracts irrespective of their type.

The GFR and GIR will form the basis for the GBR as appropriate. However, the GBR serves a different purpose and should be an entirely separate document. The GBR is intended to be contractual and to establish baseline conditions upon which a tender would be prepared. The GBR identifies the specific geotechnical data information from prior investigations or tests to be carried out in accordance with the contract that is in turn to be used to establish means and methods, and cost. It sorts out what data and past reports are relevant. It can also indicate specific previous work that is relevant, such as data obtained for different alignments or early interpretive engineering reports. An example of a 'baseline' is setting a maximum unconfined rock strength and rock hardness as the basis for the design of a TBM. During construction, the baselines in the GBR would be used to establish whether a change in geologic conditions has been encountered, resulting in financial consequences, for example the merit of additional payment to the tunnel contractor or benefit to the client.

2.5.2 Key information for tunnel design

Although certain information that is common to all tunnelling projects is required from a site investigation, there is some information that is particularly important depending on the type of tunnelling technique to be adopted. Some of these requirements are described below (after Kuesel and King 1996).

DRILL AND BLAST

Data are needed to predict the stand-up time for the size and orientation of the tunnel and the conditions for blasting during construction, i.e.

strength, stratigraphy, description and classification of the ground, water, gas, quartz content and abrasivity (this is obviously essential for all tunnelling techniques, except for possibly immersed tube tunnels).

HARD GROUND TUNNEL BORING MACHINES

Data are required to determine cutter costs and penetration rate. In addition, data to predict stand-up time are necessary to determine the type of machine which is to be used. Water inflow information is also important.

OPEN FACE SOFT GROUND TUNNEL BORING MACHINES

Face stability is important, i.e. stand-up time, and whether there is a need for mechanical devices to support the face built into the machine (face-breasting plates). Information is necessary to determine the requirements for filling the tail void. There is a need to characterize all potential mixed-face conditions.

CLOSED FACE SOFT GROUND TUNNEL BORING MACHINES

There is a need for data to make reliable estimates of the groundwater pressures, strength and permeability of the ground to be tunnelled. It is essential to predict the size, distribution and quantity of boulders. Mixed-face characteristics must be fully characterized.

PARTIAL FACE TUNNELLING MACHINES (FOR EXAMPLE ROADHEADERS)

Data are required on jointing to evaluate if the roadheader will be dislodging small joint blocks, or will grind away at the rock. Data on the hardness of the rock are essential to predict cutter/pick wear and hence costs. Quartz content and abrasivity are also important parameters.

IMMERSED TUBE TUNNELS

There is a need for ground data in order to reliably design the dredged slopes, to predict any rebound of the unloaded material and settlement of the completed immersed tube structure. Testing should emphasize rebound modulus (elastic and consolidation) and unloading strength parameters. There is also a need to ensure that all potential obstructions and/or rock ledges are identified, characterized and located. Any contaminated ground should also be fully characterized (*also important for all tunnelling techniques*).

CUT-AND-COVER TUNNELS

Exploration should be conducted over a sufficient plan area in order to define the conditions closely enough so as to reliably assess the best and most cost effective location to change from cut-and-cover to mine tunnels. The investigation should also evaluate the ground and groundwater conditions in order to aid design of the construction techniques and the excavation support systems to be adopted.

CONSTRUCTION OF PERMANENT SHAFTS

There should be at least one additional borehole for each shaft location. Data are required to design the construction method to be adopted and how to deal with groundwater conditions, both temporarily and permanently.

These tunnelling techniques will be described in detail in Chapter 5.

3 Preliminary analyses for the tunnel

3.1 Introduction

After obtaining ground characteristics from laboratory and field experiments, it is necessary to calculate the primary and secondary stresses in the ground, in order to assess the stability of the ground and likely loading on the tunnel lining. This will aid the selection of a suitable tunnelling method, assess whether ground improvement methods are necessary as well as provide the input parameters for preliminary analysis and modelling of the tunnel. This chapter focuses on obtaining the additional information, especially in soft ground, as well as the preliminary analysis techniques that may be employed.

The stability of a tunnel depends on certain key information:

- the tunnel depth and geometry;
- a detailed geological profile;
- the thickness and strength of the ground layers;
- the permeability of the ground and water pressures;
- the support provided during tunnelling.

3.2 Primary stress pattern in the ground

Primary stresses are the stresses in the ground prior to the construction of the void (tunnel). These stresses depend on the bulk unit weight and the depth at which they are determined as well as the coefficient of lateral earth pressure. The estimation of these initial or primary stresses is extremely important as it forms the basis of the loads that act on the combined tunnel support system, i.e. the ground and the tunnel lining. Commonly, the primary stresses are determined in the vertical, σ_v and horizontal, σ_h direction (Figure 3.1), which can be determined using equations 3.1 and 3.2.

Initial vertical stress: $$\sigma_v = \gamma z \qquad (3.1)$$

Initial horizontal stress: $$\sigma_h = K_0 \gamma z \qquad (3.2)$$

where γ is the bulk unit weight, z is the depth from the ground surface and K_0 is the coefficient of lateral earth pressure. An explanation of how to estimate K_0 is given in section 3.4.

In the structural analysis, either the full initial stress or a proportion of it is taken as the load on the tunnel.

If the tunnel is constructed in soft ground within the groundwater, two stress components have to be considered when determining the primary stress condition. First the effective ground stress, σ', and second the pore

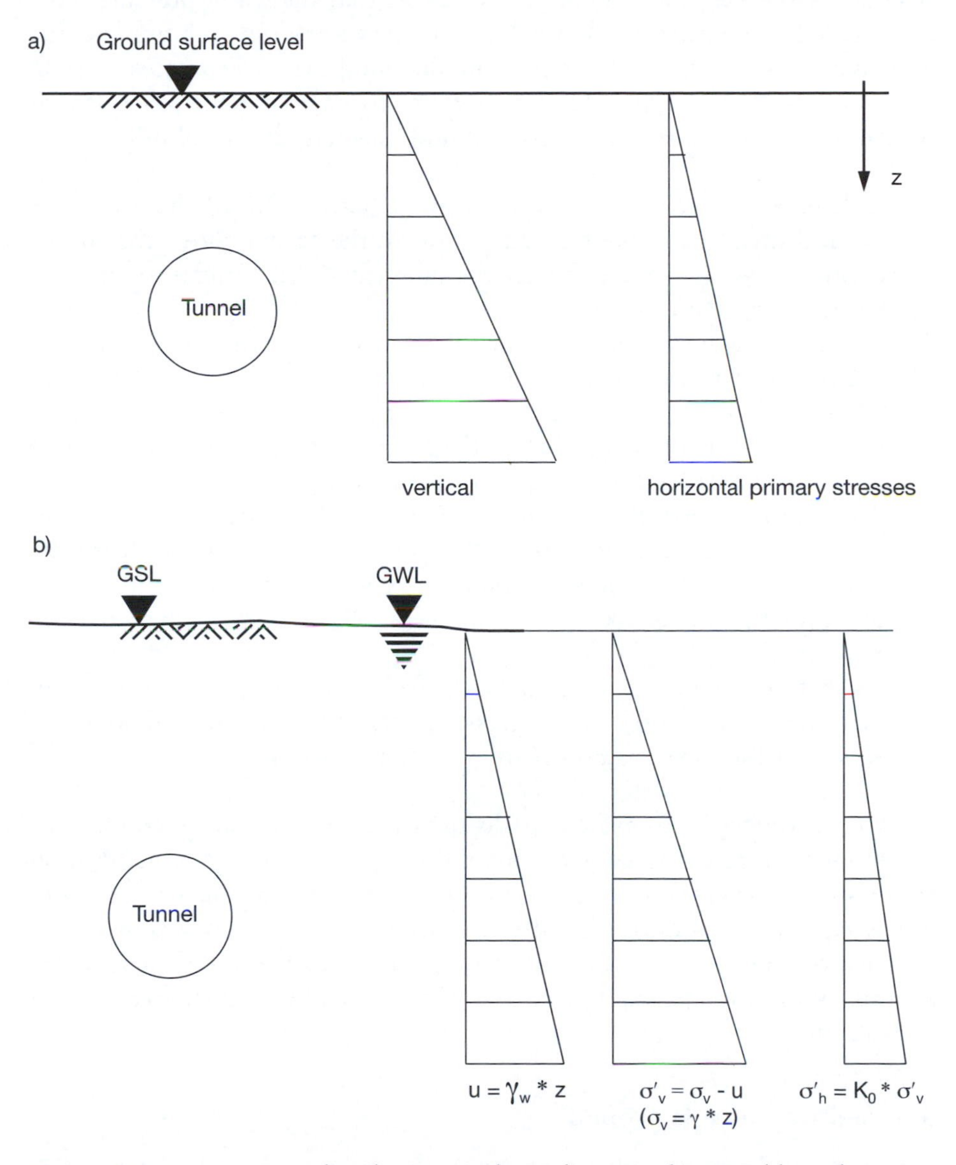

Figure 3.1 Primary stress distribution a) above the groundwater table and b) below the groundwater table

water pressure within the ground, u (below the groundwater level, GWL, this is equal to the water pressure in the ground). The total primary stress is then the summation of σ' and u (equation 3.3).

$$\sigma = \sigma' + u \tag{3.3}$$

It is often the effective stresses which dictate the behaviour of the ground in terms of shear strength as the pore water is assumed to have no shear strength. However, it should not be forgotten that the water pressure must be included when determining the loads acting on the tunnel lining, i.e. the effective stress generates bending and the total stress generates normal forces in the tunnel lining. The procedure to calculate the vertical and horizontal primary stresses (both total and effective) is as follows:

1. Calculate the vertical total stress using equation 3.1 (if the ground is layered then this is the summation of all the layers above the tunnel depth, i.e. $\gamma_1 z_1 + \gamma_2 z_2$ etc., where the subscripts 1 and 2 refer to different strata above the tunnel).
2. Calculate the pore water pressure at the tunnel depth. For example, if the tunnel is below the groundwater level and the water pressure can be assumed hydrostatic, then $u = \gamma_w z_w$, where γ_w is the unit weight of water and z_w is the depth below the level of the groundwater level (if the groundwater is flowing then u will be different).
3. The effective vertical stress, σ_v', is then calculated from $\sigma_v' = \sigma_v - u$, i.e. the effective vertical stress is the average stress acting between the particle to particle contacts within the ground material.
4. Multiplying the effective vertical stress by K_0 gives the effective horizontal stress, $\sigma_h' = K_0 \sigma_v'$.
5. To determine the total horizontal stress, σ_h, the pore pressure, u, determined previously (water pressure acts equally in all directions, i.e. K_0 is 1.0 for water) is added to σ_h', i.e. $\sigma_h = \sigma_h' + u$.

When a tunnel is excavated, it disturbs the primary stress conditions. Assuming that the tunnel construction is stable, this requires a redistribution of the stresses around the void. This is known as arching. The stresses form a new equilibrium and this is called the secondary stress condition. It can also happen temporarily for partial or separate construction phases, for example if there is a partial heading far in advance of the remaining heading construction.

3.3 Stability of soft ground

One of the key parameters that influences the choice of tunnelling technique is the stability of the ground as the tunnel is constructed. This is particularly critical around the tunnel heading. Depending on the stability of the ground itself, i.e. the stand-up time, a decision has to be made on the face support required, for example open face or close face (see Chapter 5). Furthermore, decisions on the ground improvement measures are made depending on the stability of the ground (see Chapter 4). This section provides guidance on how to estimate the stability of the face and the face support pressure required. There is a significant difference in how to estimate the stability between fine and coarse grained soils, and this is mainly due to the difference in the permeability of the soil (and with respect to the construction of the tunnel, the advance rate and geometry). In coarse grained soil (where the permeability is greater than approximately 10^{-7} to 10^{-6} m/s and construction advance rates 0.1 to 1 m/hour or less) any excess water pressures generated during construction will dissipate quickly and 'drained' conditions should be used in assessing stability. In fine grained soil with low permeability, 'undrained' conditions are more important, i.e. where the excess pore water pressures do not dissipate quickly, although if there is a stoppage in construction drained conditions may become more relevant (Mair and Taylor 1997).

3.3.1 Stability of fine grained soils

In saturated fine grained soils the short-term stability is dominated by the undrained shear strength of soil, c_u. Broms and Bennermark (1967), drawing on earlier work related to bearing capacities below foundations and field measurements, performed extrusion tests on a clay soil supported by a vertical retaining wall. They postulated the idea of a stability ratio N, which compared the overburden stress to the undrained shear strength of the soil in the form of a ratio (equation 3.4).

$$N = \gamma H / c_u \quad (3.4)$$

where H is the depth to tunnel axis (C+D/2), γ is the bulk unit weight of the soil and c_u is the undrained shear strength of the ground prior to excavation. The higher the value of N, the lower the stability. In the more general case where there is a surcharge at the ground surface and a support pressure is used at the face, for example as applied via an earth pressure balance machine (EPBM), the stability ratio, N, can be expressed as shown in equation 3.5.

$$N = (\sigma_s + \gamma H - \sigma_T)/c_u \quad (3.5)$$

where σ_s is the surcharge acting on the ground surface and σ_T is the support pressure applied at the face (note this is zero for a sprayed concrete lining). Figure 3.2 shows the parameters used, including P, which is the unsupported advance length of the tunnel (note this is zero for a shield TBM).

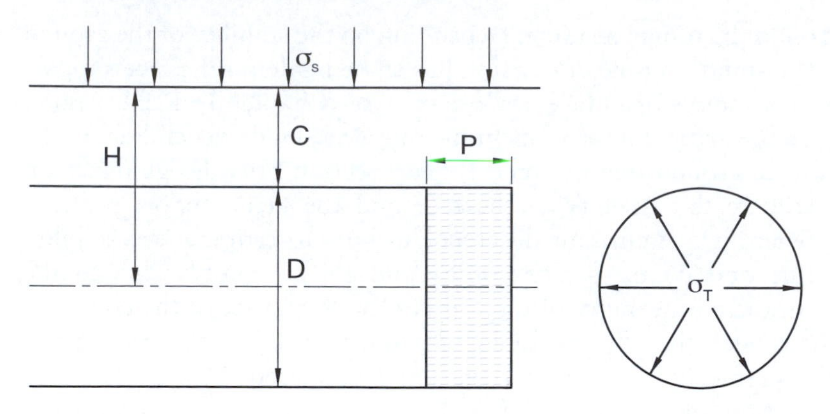

Figure 3.2 Stability parameters (after ITA/AITES 2007)

Various authors have published observations on the value of N, for example Peck (1969b) suggested N values ranging from 5 to 7. ITA/AITES (2007) suggests the following typical values:

- when $N \leq 3$, the overall stability of the tunnel face is usually ensured;
- when $3 < N \leq 6$, special consideration must be taken of the settlement risk, with large amount of ground losses being expected to occur at the face when $N \geq 5$;
- when $N > 6$, on average the face is unstable.

There are also other parameters that should be considered with respect to the stability of the face. These are:

- C/D, which controls the effect of depth on the stability condition, for a $C/D < 2$ a detailed face stability analysis is required;
- $\gamma D/c_{u,}$ which accounts for the possibility of localized failures occurring at the face, a value of $\gamma D/c_u > 4$ would indicate localized failure at the face is likely;

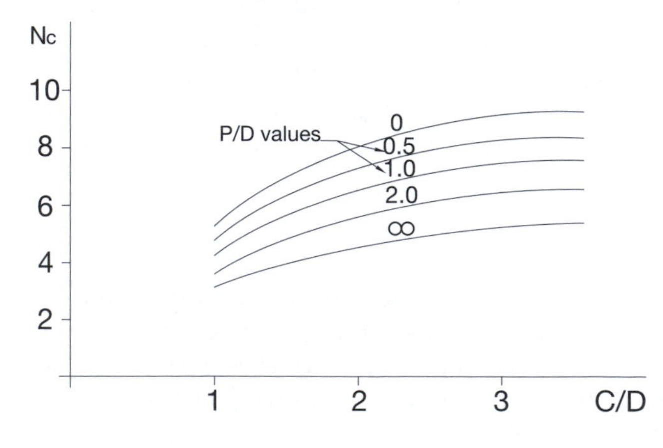

Figure 3.3 Critical stability ratio (N_c) (after Dimmock and Mair 2007b, used with permission from Thomas Telford Ltd and Professor R.J. Mair)

- P/D, which accounts for the distance behind the face until the lining is installed: an indication of the effect of this ratio on the critical stability ratio (N_c) is shown in Figure 3.3, using data from centrifuge tests.

Centrifuge modelling has been used to investigate many aspects of soft ground tunnelling including stability, ground movements and the effects of tunnelling on adjacent piles and structures. Centrifuge modelling involves constructing a small-scale physical model of the problem to be investigated. This is constructed in a strong box, which is then attached to the end of a beam and rotated at high speed. The rotation increases the gravitational forces on the model. This means that everything in the model weighs more and thus, for example a small depth of soil in the model simulates a much larger prototype depth in the field. Thus the dimensions, and many of the physical processes of the prototype, can be scaled correctly if an 'Nth' scale model is accelerated by N times the acceleration due to gravity. For example, lengths are scaled by 1/N and stresses are scaled 1:1 in the centrifuge, meaning that a dimension of 0.25 m in the model when spun at 100 g is equivalent to 25 m. An example of a centrifuge test apparatus would consist of a 1.7 m radius beam centrifuge, capable of spinning a 500 kg payload at 100 g, the equivalent to 230 rpm. Further information on centrifuge testing can be found in Taylor (1995a).

3.3.2 Stability of coarse grained soils

Atkinson and Mair (1981) describe a method for calculating the required tunnel face support pressure (σ_T) for coarse grained soil above the groundwater table and the general equation is shown in equation 3.6.

$$\sigma_T = \sigma_s T_s + \gamma D_s T_\gamma \tag{3.6}$$

where T_s is the tunnel stability number for the surface surcharge (σ_s), D_s is the diameter of the shield and T_γ is the tunnel stability number for the soil load. T_s and T_γ are equivalent parameters to the stability factor, N for fine grained soils. T_γ depends on the effective internal friction angle of the soil (ϕ') and can be determined from the graph shown in Figure 3.4a. T_s depends greatly on the depth of cover (C) as well as ϕ' and can be determined from the graph shown in Figure 3.4b.

If the tunnel is below the groundwater table, equation 3.6 should theoretically be modified to equation 3.7 (after Thomson 1995).

$$\sigma_T = \sigma_s T_s + [\gamma_d (C - h_w) + \gamma_{sat} h_w] T_\gamma + \gamma_w h_w \tag{3.7}$$

where γ_d is the bulk unit weight for the soil above the groundwater table, γ_{sat} is the bulk unit weight for the soil below the groundwater table, γ_w is the unit weight of water, h_w is the depth of the tunnel crown from the

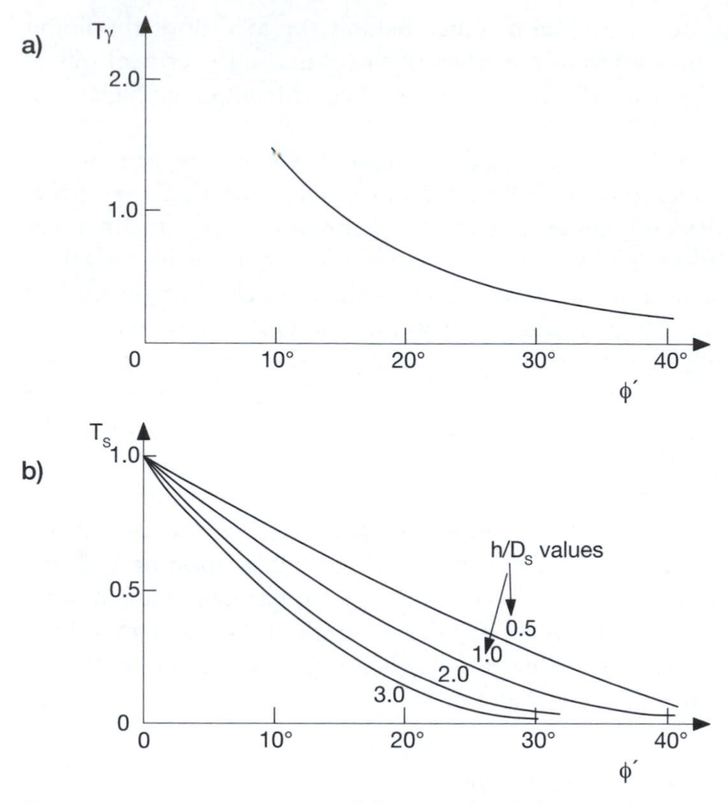

Figure 3.4 a) Determination of the tunnel stability number in coarse grained soils for the soil load, b) determination of the tunnel stability number in coarse grained soil for the surface surcharge (after Thomson 1995, from Atkinson and Mair 1981, reproduced with permission from Emap Ltd)

groundwater table and the other parameters have their usual meanings. It should be noted, however, that other factors influence the stability in this case, such as seepage forces towards the tunnel face and hence these must also be considered. This will result in considerably greater support pressures being required in order to prevent water inflows and provide drained stability. This can be achieved using compressed air support or slurry tunnelling machines. For coarse grained soils above the water table, there will probably be sufficient water content to enable some suction effects to develop (apparent cohesion), helping to stabilize the face. Stability solutions for face stability of slurry tunnel boring machines, based on limit equilibrium methods, have been developed by Anagnostou and Kovari (1996) and Jancsecz and Steiner (1994).

3.4 The coefficient of lateral earth pressure (K_0)

The coefficient of lateral earth pressure at rest can have a range of values ($0.1 < K_0 < 3$). In practice this parameter is difficult to obtain, but several aspects can be considered when estimating K_0. It should be emphasized again that the estimation is based on engineering judgement and any assumptions have to be checked by measurements as described in section 7.3.

LATERAL PRESSURE IN A SILO

Due to the difficulties of determining K_0 and because of the issues of determining the ground mechanics, there have always been experiments to estimate K_0. A common mistake, which even today leads to misunderstandings, is the determination of K_0 from the silo pressure. In a silo (Figure 3.5a) K_0 can be calculated from Poisson's ratio, μ, as shown in equation 3.8.

$$K_0 = \frac{\mu}{1-\mu} \tag{3.8}$$

However, because μ is generally in the range 0 to 0.5, using equation 3.8 would lead to K_0 values in the range of 0 to 1.0. The realistic range of μ for the ground is between 0.2 to 0.35, which leads to K_0 values of between 0.25 and 0.54. This example calculation shows that K_0 values of greater than 1.0 are not possible with this equation and values of K_0 of greater than 0.54 are only fully covered if one uses a μ value, which is not necessarily realistic for the ground. This equation represents a simplified case and is based on the assumption of elasticity in the ground and is only valid in rare circumstances in underground construction.

The value of K_0 is always going to be an estimation. To determine its value one needs to take into account the historical development of the earth, and hence the rock. The determination (or better estimation) of K_0 is part of the engineering geological survey. Five possible reasons are listed below for the variation in K_0, i.e. where K_0 has no relationship to μ (which is true for the majority of cases).

- **Ice age preloading.** It is possible that the lateral horizontal pressure of earlier times is impregnated into the ground and is still present today. The pressure of huge glaciers from past ice ages is an example of this. In this case K_0 can be higher than 1.0 (Figure 3.5b).
- **Layering, synclines, anticlines (saddles and troughs)**. Figure 3.5c shows the influence of layering and layering with saddle structures depending on the position of the tunnel relative to the geological formation and hence the need to use different K_0 values. In this case a potential rotation of the principal stress conditions would be expected, i.e. the assumption that the vertical and horizontal stresses are principal stresses is no longer valid, or limited. This also applies if the layers are dipping.
- **Crevasses.** In open crevasses K_0 is very small (Figure 3.5d). This is the same as for the cases where the crevass contains soft soil or water. If K_0 were high in this situation, the crevass would be most likely closed.

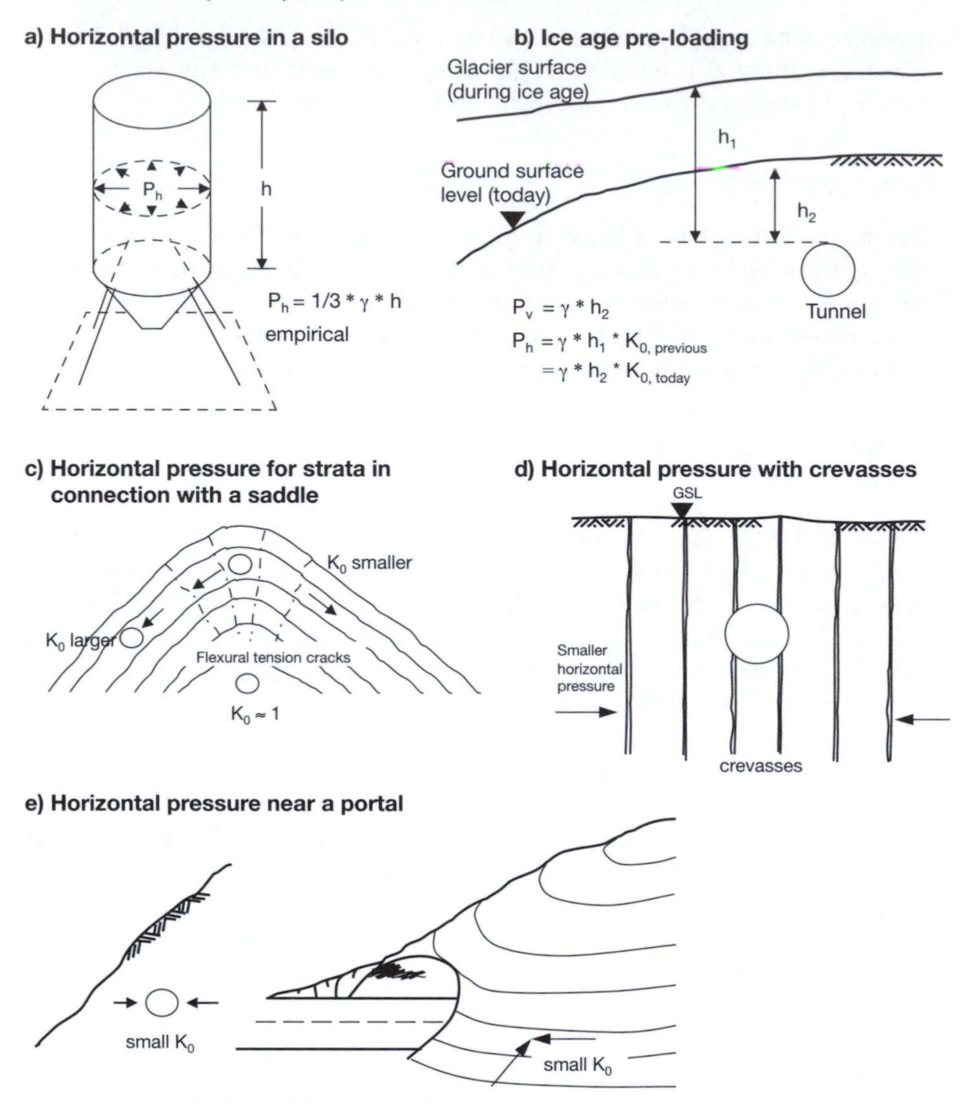

Figure 3.5 Coefficient of lateral earth pressure

- **Depth.** Close to the ground surface K_0 would be expected to be small due to weathering. In addition, for a high K_0 value the tension of the ground is missing, for example at a slope (Figure 3.5e).
- **Tunnel in groundwater.** If the tunnel is constructed within the groundwater, at least two components need to be considered when estimating the primary stresses. These components are the effective ground stress and the water pressure ($K_0 = 1$) as described in section 3.2.

For the reasons mentioned above, there is no definitive value for K_0. However, one can statistically define the range of K_0 as 0.1 to 3.0. For

Table 3.1 Typical values for K_0

Ground material	K_0
Sand	0.4–0.5
Clayey soil (between rock layers)	0.6–0.8
Slurry	1.0
Soft rock	0.4–0.6
Hard soil/rock	(0.2) 0.5–0.8 (1.2)
London Clay	0.6–1.5

normally consolidated soils, i.e. a soil that has not experienced greater stresses acting on it in the past than are acting on it now, K_0 can be estimated based on the internal friction angle, ϕ', of the material, for example $K_0 = (1 - \sin \phi')$. For overconsolidated clays, i.e. where the soil has experienced larger stresses in the past than it is experiencing now, K_0 is likely to be greater than 1.0. Some examples of typical values of K_0 are shown in Table 3.1.

3.5 Preliminary analytical methods

3.5.1 Introduction

It is impossible to take all the influences, parameters and boundary conditions that are dependent on the geology and construction phases into account in a calculation. Therefore, analytical models have been developed which simplify reality to such an extent that the remaining parameters can be dealt with in a calculation and at the same time lead to sensible results.

In the following discussion, three common analytical methods are briefly described; the bedded-beam spring method, the continuum method and the tunnel support resistance method. The assessment of which method to use, depends on the tunnel depth. In soft soil, two conditions can be defined as:

- shallow, $C<2D$, i.e. where the ground above the tunnel crown in assumed to have no bearing capacity;
- deep, $C>3D$, i.e. where the ground above the tunnel crown is acting as a support;

where C is the tunnel crown depth and D is the tunnel diameter.

$C<2D$: The excavation process creates a softening zone in the crown area, which for shallow tunnels in soft ground reaches the ground surface. As a result, no arching can develop over the crown. The ground in this area has no bearing capacity and acts only as a load on the tunnel lining. For the unsupported area, on average, an angle of 90 degrees is assumed at the tunnel crown (Figure 3.6a). This is a very conservative approach.

C>3D: For a supporting crown, the ground is capable of creating a supporting ring, i.e. the ground can form an arch and transfer loads around the tunnel void.

In the range 2D< C <3D, the ground above the tunnel crown can be acting either as a support or not depending upon the geological conditions, i.e. bedding arrangements.

Further details on the design of shield tunnel linings, segmental linings for example, can be found in ITA (2000). Further details on structural design models for tunnels in soft ground can be found in Duddeck and Erdman (1985).

3.5.2 Bedded-beam spring method

The tunnel support is idealized as an elastically supported circular ring. The elastic bedding is achieved through radial and potentially tangentially arranged springs. The spring stiffness simulates the support behaviour of the ground. The important parameters of the ground are the stiffness modulus E_s (which is included in the spring stiffness) and the coefficient of lateral earth pressure K_0 (which is included in the loading). The calculation is carried out elastically. As the ground is only represented by springs, the analysis cannot provide any information with regard to the settlement at the ground surface and to the possible stress and deformation behaviour of the ground (secondary stress situation). Figure 3.6a shows the model used within the bedded-beam spring method with an unsupported crown, the so called ‘partially bedded method’ for shallow tunnels (ITA 1988). For deep tunnels the bedded-beam spring method is generally not used because even with a supporting crown area, the supporting nature of the ground is not sufficiently taken into account.

The bedded-beam spring method is the fastest and simplest calculation method. Therefore it is often applied even though it has limited potential for interpretation with respect to the real situation due to the many simplifications made. It is often used to determine thickness and required reinforcement of the supporting circular ring following the results of a more sophisticated calculation method. The usage is mainly for shallow tunnels in soft ground or weak rock.

3.5.3 Continuum method

The ground, in which the tunnel is constructed, is idealized as a continuum, i.e. there are no discontinuities in the material. The method assumes that the ground is an infinitely large thin section with a hole at the centre (Figure 3.6b). This calculation method allows the interpretation of the deformation and strains in the ground. In addition, this method allows the construction phases to be simulated. The elastic modulus, E, is required as a parameter for the ground. The structural system can be established for both an

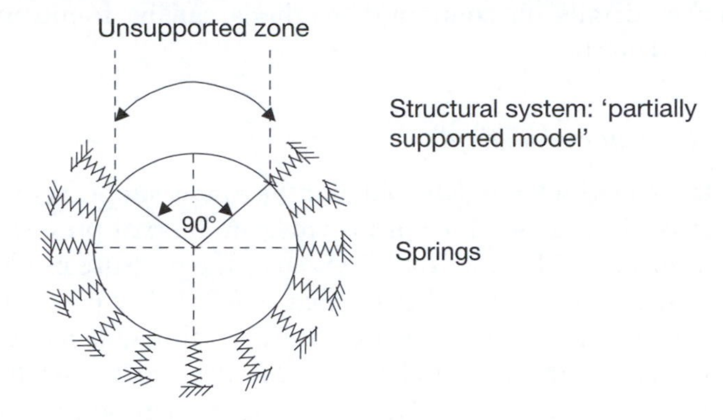

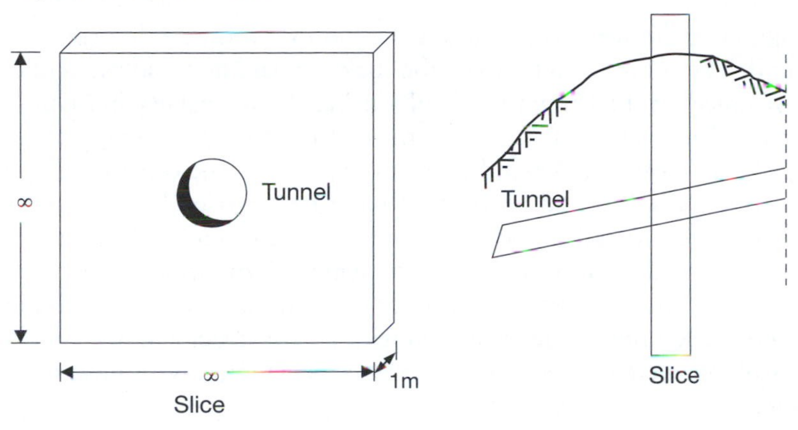

Figure 3.6 Calculation models for tunnels in soft ground

unbedded as well as a bedded crown area. However, the assumptions for this calculation method for deep tunnels are often unfavourable; with increasing depth, the load on the tunnel lining grows linearly and with this its thickness. In combination with an elastic calculation, this leads to a potentially unrealistically large lining thickness. In the calculation, the load acting on the tunnel lining is limited, i.e. the total overburden pressure is not considered. Instead, only the weight of the disturbed zone, which develops over the tunnel crown as a result of the tunnel construction, is used in the calculation. The biggest difficulty lies in the estimation of the height of this disturbed zone and thus the load for which the thickness of the tunnel lining has to be calculated. Estimating the size of the disturbed zone, which acts as the overburden on the tunnel, is based on experience, engineering judgement and the ground characteristics. The method is used in shallow tunnels in weak rock as a partial continuum method and for deep tunnels in weak rock as a continuum method with a bedded crown area.

An example calculation using the continuum method is provided in Appendix B. Further details on continuum methods can be found in Duddeck and Erdman (1985).

3.5.4 Tunnel support resistance method

For the support resistance method, it is assumed that the tunnel support constrains the deformations of the ground, i.e. it provides an internal pressure (resistance) against the ground. The resistance is taken as the pressure inside the tunnel in the calculation and is defined as P_T (Figure 3.7a). The pressure inside the tunnel is dependent on the deformations. This 'thought' model (tunnel support as an internal pressure) can be applied to deep rock tunnels. The tunnel support resistance method is also a continuum method and in addition to the elasticity modulus, the cohesion and the friction angle of the ground are required.

The design criterion for this method is the limitation of the deformation of the ground. The connection between the rock deformation and the tunnel support resistance can be shown pictorially using the Fenner-Pacher curve, as shown in Figure 3.7b (w is the settlement of the tunnel crown).

$w < w_{crit}$: The more the ground deforms (distresses) before the tunnel support is placed, the lower the load that has to be carried by the tunnel lining and the higher the self supporting element of the ground. The required tunnel support resistance reduces with increasing deformation.

$w > w_{crit}$: When the deformation reaches a certain amount, it results in softening and weakening of the ground fabric. To construct a stable tunnel beyond this point, increasing support resistance is essential with increasing deformations.

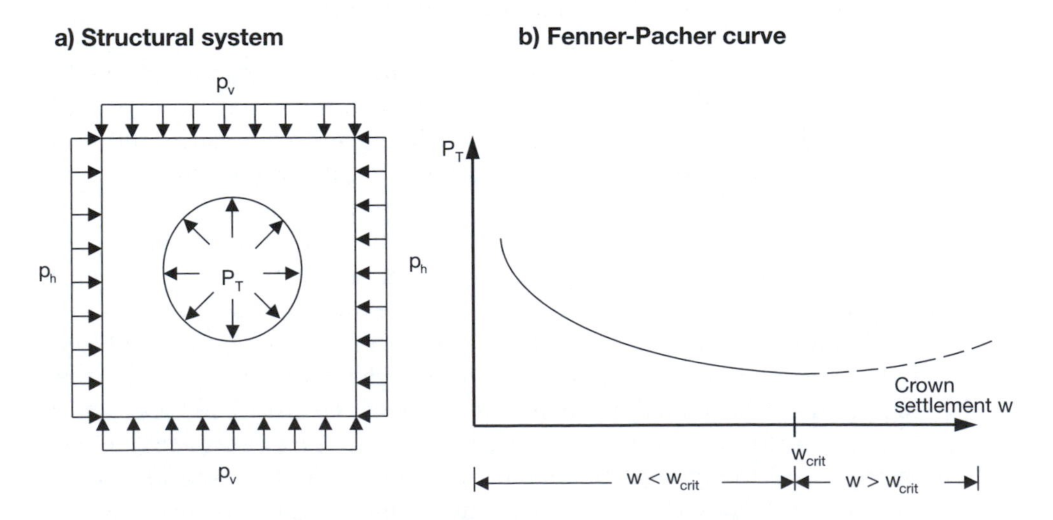

Figure 3.7 Tunnel support resistance method

Thus, there is a deformation value for the ground at which the required tunnel support resistance is minimal. This deformation should be reached when all the stress redistribution has finished. By keeping the deformation to w_{crit}, it would be possible to have the optimal support system both from an economical and rock behaviour point of view. The relationship between the support system resistance and the deformation is dependent on the geology. This means that for every ground there is a different Fenner-Pacher curve and a different critical deformation. This leads to a number of problems.

- First: how big is w_{crit}? If a lot of experience exists in comparable geological conditions with similar underground construction methods, it could be possible to put a quantitative boundary on the critical deformation. However, in an unknown ground this is nearly impossible.
- Second: even if the critical deformation is known, the difficulty remains to ensure that the construction phases result in a final value of w_{crit}. Many of the factors that influence the development of the deformations are not linear and are time dependent, for example the curing of sprayed concrete when using sprayed concrete lining (see section 4.3.2). Furthermore, there is the problem of checking the rock deformations with measurements. This is particularly relevant for the rock deformation ahead of the tunnel drive. (The topic of deformation measurement is looked at in section 7.3.)
- Third: the tunnel support is not calculated but assumed as an inner pressure. Hence there are no internal forces. Therefore the problem exists to translate the w_{crit} and the associated required support resistance into a sprayed concrete lining thickness and related reinforcement.

The support resistance method is consequently not suitable for an analysis in the traditional sense (structural system with load → determination of internal forces → proof of stresses). The ground is the main support element and is in the forefront of the analysis. The tunnel support system, in this case a sprayed concrete lining, supports the disturbed boundary areas of the void and is, by comparison with the methods for soft ground, of lower importance.

The advantages of the theory of support resistance are as follows:

1 The full overburden can be assumed. The stress redistributions and overstressing in the ground can be determined, which is not possible with the other analytical methods. In the support system resistance method, the creep of sprayed concrete is taken into account. This means that the calculated deformations are greater and closer to reality compared with the methods using elastic analysis.
2 The choice of the calculation method therefore also depends on the type of ground: soft ground or rock. Principally it has to be decided

how much self-support the ground possesses, i.e. whether one builds uneconomically (dimensioning the tunnel support too large) or unsafe (assuming the self-support of the ground to be too large). This depends mainly on how valid the estimation is. *It should also be noted that without estimation no structural analysis functions in underground construction!*

3.6 Preliminary numerical modelling

3.6.1 Introduction

The previous section described simple analytical methods, which can be used to estimate the stresses in the tunnel lining or the required lining thickness for a given deformation of the ground. In recent years, as a result in the increases in computing power and the fact that there are many commercial packages available, the use of numerical models has increased significantly. This section describes the use of some of these numerical models. However, it is not the intention of this section to fully describe how to carry out tunnel analyses, but to briefly describe some aspects of the problem. The reader is directed to other books such as Potts and Zdavkovic (1999 and 2001) for further details. The benefits of numerical methods over analytical or closed form solutions (as described in section 3.5) are highlighted by Potts and Zdavkovic (2001) as being able to:

- simulate the construction sequence;
- deal with complex ground conditions;
- model realistic soil behaviour;
- handle complex hydraulic conditions;
- deal with ground treatment, for example compensation grouting;
- account for adjacent services and structures;
- simulate intermediate and long-term conditions;
- deal with multiple tunnels.

It must be remembered, however, that numerical analyses are only as good as the user's experience, the input values and the numerical modelling package. They are not a panacea and must be treated as any other engineering tool. Many assumptions are still required in order to produce workable analyses and these require good engineering judgements and a clear understanding of their implications.

Numerical methods can be divided into different types depending on the computation methods adopted in the software package. For modelling continua, such as soils, the most common numerical methods adopted for analysing tunnelling projects are the finite element method or finite difference methods. For modelling discontinua, such as rocks, the most common numerical models adopted are the discrete element method and the boundary

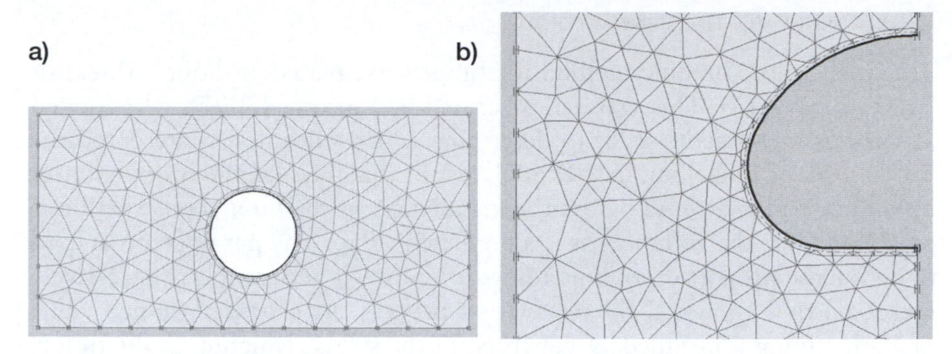

Figure 3.8 Example 2-D meshes using finite elements, a) 2-D plain strain analysis of a bored tunnel construction, and b) 2-D plain strain analysis (with symmetry associated with the tunnel centreline) of a tunnel being constructed using NATM, as produced using the PLAXIS® software (courtesy of Wilde FEA Ltd)

element method. It should be noted that there is overlap between these latter methods and the modelling of continua.

Tunnelling is a three-dimensional problem. Even with more powerful computers, however, these models can be computationally demanding. In addition, three-dimensional models for tunnelling problems are not easy to set up, even though modern commercial software packages are making this easier for routine problems. This means that two-dimensional models are still very common. Adopting a two-dimensional model for a tunnelling problem immediately implies that a number of assumptions are needed with respect to the construction process. In particular, the fact that the three-dimensional arching effect, which is so important for the behaviour of the ground to allow economic tunnels to be constructed, cannot be modelled directly.

There are a number of ways to represent the tunnel in 2-D models. When modelling shallow tunnels or if the ground surface response is key to the analysis, then a plane strain analysis is required. Typical finite element meshes for the 2-D plane strain analyses are shown in Figure 3.8a and b.

3.6.2 Modelling the tunnel construction in 2-D

In 2-D there are a number of ways of modelling the construction method. These include the following:

'GAP' METHOD

In this method a predefined void is introduced into the finite element mesh that represents the total 'volume loss' expected. The 'gap' is greatest at the crown of the tunnel and zero at the invert (Rowe *et al.* 1983).

CONVERGENCE–CONFINEMENT METHOD

This is the most suitable method for tunnels excavated without a shield or TBM, e.g. NATM. This was demonstrated by Karakus (2007), who looked at various methods in order to determine which ones best represented the 3-D effects of tunnelling in 2-D analyses. In this method the proportion of unloading of the ground before the installation of the lining construction is prescribed, i.e. the volume loss is a predicted value. The parameter λ is used to define the proportion of unloading. Initially λ is zero and is progressively increased to 1 to model the excavation process. At a predetermined value of λ_d the lining is installed, at which point the stress reduction at the tunnel boundary is λ_d times the initial soil stress. The remainder of the stress is applied to create the lining stress, i.e. the stress imposed on the lining is $(1-\lambda_d)$ times the initial soil stress (Potts and Zdavkovic 2001).

PROGRESSIVE SOFTENING METHOD

This was developed for NATM (or sprayed concrete lining) tunnelling by Swoboda (1979). The method involves reducing the ground stiffness in the heading by a certain amount. The lining is installed before the modelled excavation is complete. The method can cope with crown and invert construction or side drifts.

VOLUME LOSS CONTROL METHOD

This is similar to the convergence–confinement method, in this case, however, the expected volume loss at the end of construction is prescribed. This is useful if the volume loss can be estimated with a reasonable degree of certainty, and is also useful for back analysis of tunnelling operations. In this method the support pressure at the tunnel boundary is reduced in increments, and the volume loss generated can be monitored. Once the prescribed value is achieved, the lining is installed. Depending on the stiffness of the lining, further deformations and hence volume loss may occur, so it may be that the lining is installed before the prescribed volume loss is reached to allow for this additional value.

It is also important when setting up the model to use appropriate boundary conditions, both for far field conditions, for example the restraints applied at the edges of the mesh area, including hydraulic conditions, and the near field conditions associated with, for example the lining. Normally, in a simple 2-D plane strain analysis, the restraints to movement at the far field conditions are that the ground surface is not restrained from moving, the base of the mesh is restrained vertically and horizontally and the edges are restrained horizontally, but not vertically.

MODELLING THE LINING

As mentioned previously, in 2-D the analysis does not recognize the 3-D support from the lining already installed behind the face, into which the stresses arch. So called wished-in-place lining occurring in a single increment in the analysis is common, for example when using the volume loss or convergence–confinement approaches. There are two ways of modelling the lining using solid element or shell elements. Solid elements are standard elements used for representing most materials within finite element meshes and hence there are a wide range of constitutive models available for these elements. However, solid elements have the problem that the element shape can be an issue (defined by the aspect ratio of length to width). Linings are relatively thin in relation to the tunnel diameter and therefore a large number of elements are required to maintain acceptable aspect ratios. Shell elements in contrast have zero thickness and curved shell elements can be used to model tunnel linings. This removes the problem of aspect ratio and allows more flexibility with respect to the mesh definition. There are many issues to consider when modelling tunnel linings, particularly segmental linings, and the reader is encouraged to read more detailed literature on this subject, for example Potts and Zdavkovic (2001).

3.6.3 Modelling the tunnel construction in 3-D

3-D numerical analyses allow the possibility of modelling the tunnel operation more realistically, particularly the behaviour of the ground ahead of the tunnel face and the 3-D arching effects that occur around the tunnel face. Although these analyses are more costly in terms of computation time, it is still not possible to model accurately every aspect of the tunnel construction in detail and assumptions are still required. However, modern software packages do offer the possibility of doing this type of analysis with relative ease. It must be remembered, however, that it is important to understand what you are doing and to consider the limitations and assumptions that are made in these analyses. Figure 3.9 shows an example of a finite element 3-D mesh.

An example of three-dimensional numerical modelling was reported by Ng *et al.* (2004) who carried out a series of three-dimensional finite element analyses to investigate multiple tunnel interactions for sprayed concrete lined tunnels in stiff clay using ABAQUS®. In parallel, Lee and Ng (2005) studied the effects of tunnels on an existing loaded pile using three-dimensional finite element modelling, again using ABAQUS®. Bloodworth (2002), following on from previous work by Burd *et al.* (2000), conducted a detailed study to investigate the effects of new tunnelling on existing structures in 2-D and 3-D. This work highlights many of the issues associated with simulating both the tunnelling operation and the realistic modelling of, in this case, the buildings at the ground surface. The numerical

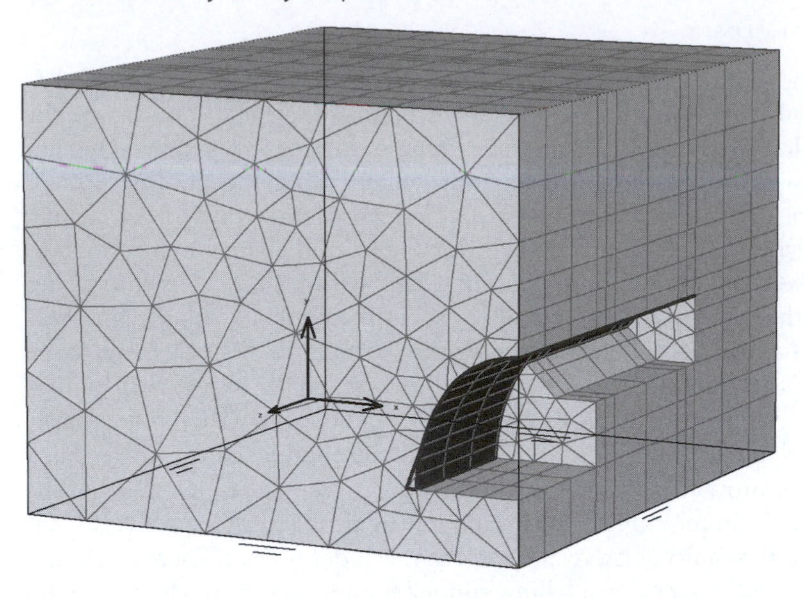

Figure 3.9 Example 3-D analysis of a tunnel being constructed using NATM, as produced using the PLAXIS 3-D Tunnel® software (courtesy of Wilde FEA Ltd)

modelling generally over-predicted the damage to the buildings at the ground surface. It was suggested that this was due to the level of detail that could sensibly be modelled within the numerical models whilst working with the computer power available at that time.

3.6.4 Choice of ground and lining constitutive models

One of the most critical aspects of any numerical modelling is the choice of constitutive models for the ground and the lining, i.e. how the material behaviour is simulated. Many people still use linear elastic or elasto-plastic (e.g. Mohr-Coulomb) constitutive models to analyse the soil behaviour. However, it has been shown by many researchers that the soil model has a large impact when modelling tunnelling operations. The construction operations involve unloading as well as loading stress conditions and so any model must be able to cope with this. In addition, the strains around tunnelling operations are often small and for soft ground this means that the stiffness of the material is extremely nonlinear. Therefore, if pre-yield behaviour dominates the ground response, it is essential to model the nonlinear elasticity at small strains. The reason for people choosing simpler constitutive models for the ground is generally related to the choice of input parameters. The more sophisticated the soil model the more parameters that are required. Obtaining these parameters from available site investigation information can be difficult and often requires assumptions to be made.

Ideally, to model the ground behaviour successfully, it is important to consider its nonlinear stress-strain behaviour, variable K_0 values, anisotropy and consolidation characteristics.

For modelling sprayed concrete lining there is a need to model the age- and time-dependent behaviour of the material, as well as its nonlinearity. Typical models include (Thomas 2009a):

- *Hypothetical Modulus of Elasticity (HME),* which uses reduced values of elastic stiffness for the lining to account for 3-D effects, ageing of the elastic stiffness, creep and shrinkage. It is largely an empirical based model.
- *Age-dependent elastic models*, which, as an alternative to the HME model, models the ageing stiffness explicitly.
- *Age-dependent nonlinear mod*els, which takes into account the non-linear stress-strain behaviour of concrete when loaded to more than 30% of it compressive strength. As sprayed concrete can be loaded heavily early on, i.e. at low strength, this nonlinearity could be relevant.

Further details of modelling sprayed concrete can found in Thomas (2009a).

4 Ground improvement techniques and lining systems

4.1 Introduction

This chapter is divided into two sections. The first describes techniques of improving and stabilizing the ground, with respect to both strength and also permeability. The second describes the various lining techniques commonly employed in tunnel construction. It should be noted that many of the stabilization techniques and lining methods are intimately linked with the tunnel construction methods described in Chapter 5, so the reader is advised not to treat these chapters in isolation, but to treat both as part of the tunnel construction process.

4.2 Ground improvement and stabilization techniques

This section describes a number of techniques that can be used to improve the stability of the ground to aid construction of the tunnel, and in soft ground to reduce/control ground displacements and hence mitigate the effects of the tunnelling operation on adjacent structures.

With respect to settlement control, it is obviously better if the choice of tunnel alignment avoids the necessity of using settlement control measures (ITA/AITES 2007). Increasing the depth of the tunnel to provide a larger cover depth will reduce the magnitudes of the displacements reaching the ground surface and shallow subsurface structures including existing tunnels and services. It is important to choose an alignment for the tunnel so that the tunnel passes through the strata which have the most favourable mechanical properties. Choosing the smallest cross section for the tunnel can help as this provides a more stable face. This may mean, in the case of transportation tunnels, choosing between a single larger diameter tunnel and a twin-tube tunnel. Twin-tube tunnels are often recommended for safety reasons as the second tube can act as an emergency exit in case of an accident, such as fire. If a TBM is used, choosing an alignment that is as straight as possible is beneficial. However it may be necessary to use artificial ground improvement measures, and some of the more common techniques are described below.

Many of the techniques described in this section can generally be applied either from the ground surface or from within the tunnel during construction. The latter will obviously slow the rate of advance of the tunnel.

4.2.1 Ground freezing

Although perceived as a relatively expensive last resort, in cases where something goes wrong and no other solution is available, this can be a powerful technique as it can be used across the whole range of ground types, depending on the groundwater flow rate. In fact, in shallow tunnelling where access can be gained from the ground surface, it is used relatively frequently (Pelizza and Piela 2005).

The freezing method is only applicable when the ground contains water, ideally still, fresh water. A ground with a moisture content greater than 5% will freeze. Water can be added via a fire hose, a sprinkler system, a borehole or injection device to raise the moisture content in the ground. The principle of ground freezing is to use a refrigerant to convert *in situ* pore water into a frostwall, with the ice bonding the soil particles together.

As a rule, if used from within the tunnel, freezing lances are installed from the tunnel in the direction of tunnel excavation as the frozen ground should create an arching mechanism (Figure 4.1). The lances are situated in the crown and, if necessary, at the springline. In order to achieve a closed frozen body, the distances between the lances are limited, e.g. 1 m, in combination with a length of 20 m or more. It is important that the frozen areas overlap to provide an impermeable barrier. Cooling fluid is pumped through the freezing lances. Examples of cooling materials are brine (salt solution) with a temperature of –50 °C to –20 °C, or liquid nitrogen which evaporates at –196 °C.

For excavations from the ground surface, a cylindrical freeze wall is formed around the periphery of the planned excavation or a layer of ground above the tunnel roof is frozen. The refrigerant pipes are equally spaced at approximately 1 m apart and, in order to ensure a continuous freeze wall, they need to be accurately drilled with minimal deviation.

Advantages of ground freezing:

- The strength of ground can be increased.
- An impermeable barrier is created. (Although it should be noted that if the freezing process is conducted from within the tunnel as opposed to from the ground surface, it is normal only to extend the frozen ground from the crown to the tunnel springline (or above) and hence this just extends the flow path for the water and does not make a completely impermeable barrier: See Figure 4.1. It is normal in this case to use ground freezing in combination with pressurized tunnelling.)
- It is non-toxic and noiseless.

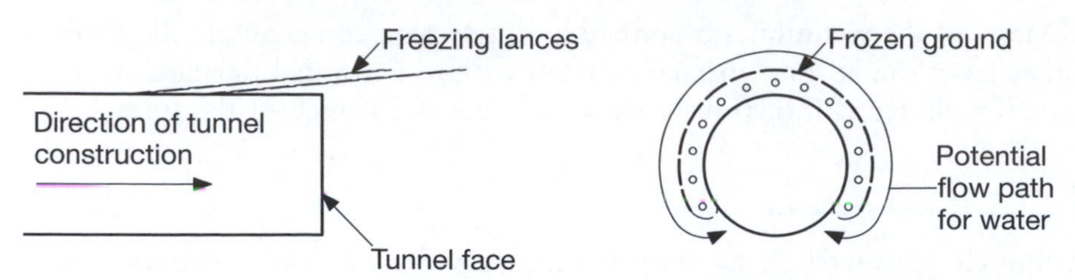

Figure 4.1 Potential flow paths when in-tunnel ground freezing is used at the tunnel crown

- It is totally removable (unlike grouting) – although there can be an adverse reaction in some soils.

Limitations:

- The time required to achieve ground freezing can be many weeks depending on the ground and groundwater conditions.
- Flowing water causes heat drain and can prevent the ground freezing. The limiting flow rate depends on the type of freezing being used (see below). For example, if a two phase brine freezing process is used, a maximum flow rate of 2 m/day can be tolerated, whereas for a direct process using liquid nitrogen, the maximum flow rate is 20 m/day.
- The boreholes must be accurately positioned to create a continuous frozen zone.

Care must be taken as there is the potential for the ground to heave during the freezing process and subsequent settlement at the end of the freezing process (ITA/AITES 2007). Ground heave is related to the frost susceptibility of the ground. In coarse grained soils, the frost susceptibility is low as the permeability is high. This means there is less heave because the water can drain as the freezing progresses. Conversely, in fine grained soils the frost susceptibility is high as these materials have a low permeability and therefore there is more heave as the water does not drain during the freezing process. Ground heave can be limited by controlling the speed of the freezing process and the sequence of freezing.

There are two methods of ground freezing:

- Two phase method (closed) – *Figure 4.2a*. In this method, a primary refrigerant (ammonia or freon) is used to cool a secondary fluid (usually brine).
- Direct process (open) – *Figure 4.2b*. In the direct process liquid nitrogen is used to freeze the ground. The nitrogen is passed down the freeze pipes and then allowed to evaporate into the atmosphere. This direct process is good for short-term or emergency projects. Liquid nitrogen is likely to be the only effective method for freezing pore water in fine grained soils.

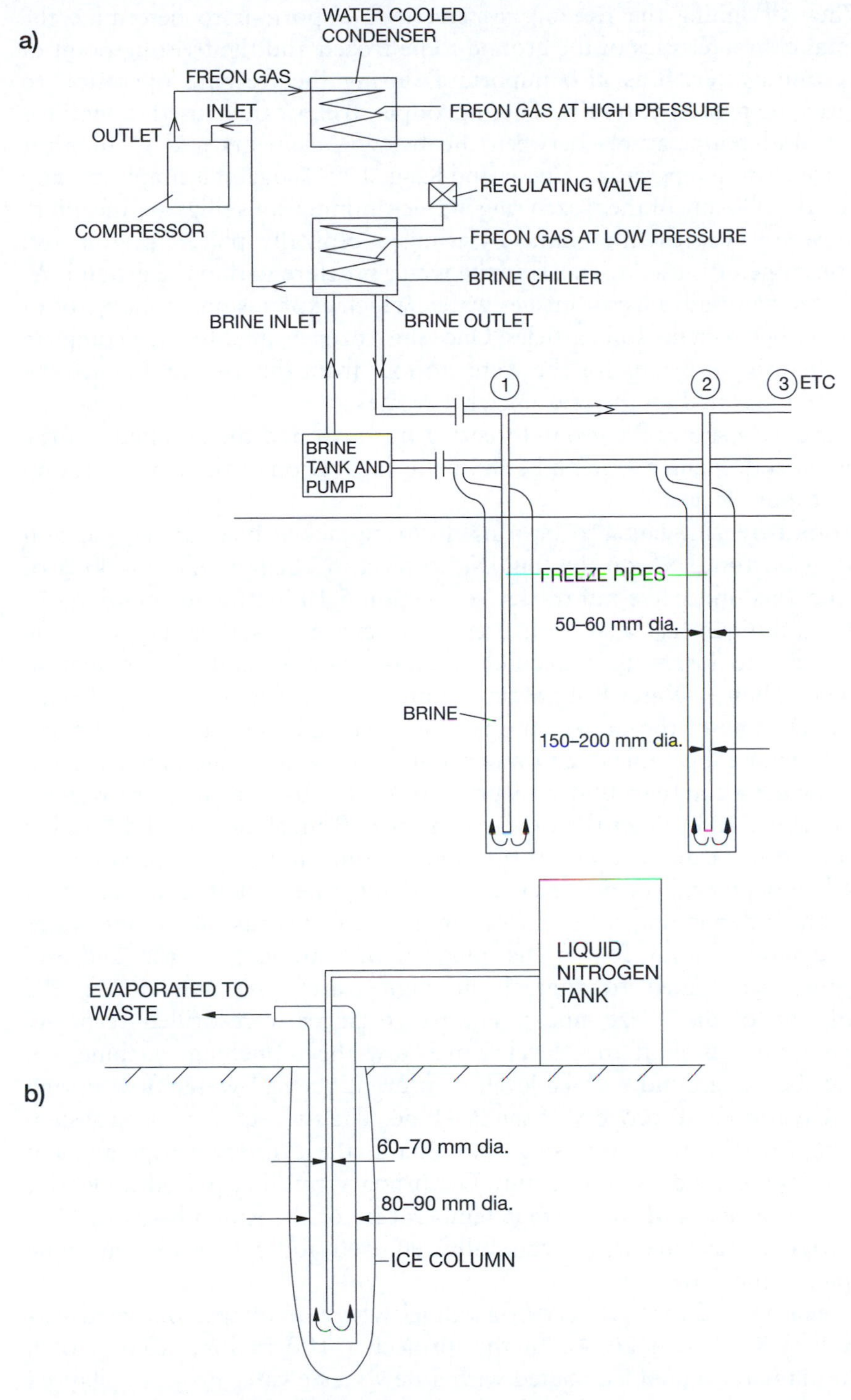

Figure 4.2 Ground freezing methods, a) two phase method, b) direct process (after Harris 1983)

When designing the freezing system, it is important to determine the thermal characteristics of the ground to be frozen and the freezing point of the groundwater. It is also important during the freezing operation to monitor the process carefully. Thermocouple strings can be used to monitor the ground temperatures between the freezing elements and to monitor the refrigerant temperature. Kuesel and King (1996) suggest a simple method of ensuring closure of the frozen ring in free-draining soils (high permeability soils) when constructing shafts by using a centrally placed piezometer. The piezometer is used to measure the water pressure within the ground. As the freeze front advances it pushes water, which expands on cooling, out of the pores between the soil particles. Once the frozen ground forms a complete ring, there is no means for the water to exit from the area and hence the pressure measured by the piezometer increases.

Figure 4.3a shows the ground freezing tubes around the perimeter of the tunnel portal. Figures 4.3b and c show the excavation of the frozen ground at the tunnel face.

Ground freezing has also been used during jacked box tunnels (section 5.10) in Boston, USA on the 'Big Dig' project in 2001 to allow jacking of box sections under live rail tracks (see section 5.10.3.2 for further details).

The technique has also been used to rescue TBMs that have become flooded due to adverse ground conditions, for example on the 2.6 m internal diameter Thames Water Ring Main in London, UK (Clarke and Mackenzie 1994). On one of the drives an open face machine with a backhoe excavator became inoperable due to water inundation when it hit an unexpected water-bearing sand stratum at a pressure of 4.5 bar. In order to remove this machine and restart the drive using an EPB machine, ground freezing was found to be an economical solution. The machine was approximately 55 m below ground surface level and a 7.6 m diameter shaft was excavated using an underpinning method (see Figure 4.26), i.e. just above the water bearing sand stratum. It was then plugged with a concrete base, and precautions were taken to control the high water pressures during the installation of the freeze lances. The freeze lances were drilled vertically from within this shaft to a level 5 m below the tunnelling machine, i.e. 61.3 m below ground surface level, to prevent vertical water flow during the excavation and recovery of the machine. The two stage freezing system was employed in this case using ammonia as the primary refrigerant and brine as the secondary refrigerant. The primary freezing period took four and a half weeks and the average temperature of the ground was –12 °C. The original machine was successfully removed and a new EPB machine completed the drive.

An example of the recovery of a tunnel where a collapse occurred was in Hull, UK (Brown 2004). In this project a 100 m long section of a 3.6 m diameter tunnel associated with a new wastewater project collapsed and ground freezing using liquid nitrogen was used to stabilize the ground and provide an impermeable barrier to allow reconstruction to take place.

Figure 4.3 Examples of ground freezing used in tunnelling, a) horizontal freezing to rescue a broken down TBM in Cairo, b) freezing at the portal of a 13 m diameter road tunnel in Du Toitskloof, South Africa, through decomposed granite, c) excavation of the tunnel crown through frozen sand and gravel in Dusseldorf (courtesy of British Drilling & Freezing Co. Ltd)

The tunnelling took place through predominantly alluvial granular deposits at a depth of 15.5 m. On this project the maximum consumption of liquid nitrogen over any 24 hour period was 165,000 litres.

Further details on artificial ground freezing can be found in Harris (1995), Holden (1997) and Woodward (2005).

4.2.2 Lowering of the groundwater table

If groundwater lowering can be achieved successfully, a marked improvement is possible in the ground properties. However, groundwater table lowering is, even as a time limited measure, not always possible. Under running streams, in settlement-critical inner city areas, in areas where there may be an influence on existing water supply aquifers, or in areas where there could be a potential adverse effect on the flora, this measure should not be used. Furthermore it requires intensive installations for holding the extracted water, which may have to be treated before it can be disposed of.

In permeable strata where the permeability, k, exceeds about 10^{-3} cm/s, or where an aquifer can be dewatered below less permeable strata, the level of the water table over a wide area can be drawn down by pumping from boreholes and deep wells. These processes are widely used in open excavations and are suited also to cut-and-cover tunnels and shallow bored tunnels, for example Lainzer Tunnel LT31, Vienna, see section 8.3 (Megaw and Bartlett 1982).

There are two principle methods of groundwater lowering: wellpoints and deep filter wells. Wellpoints, although one of the most versatile methods of dewatering, are limited to dewatering to a depth of about 6 m (limited by the effective vacuum lift of a pump), although staged wellpoints can be used to go deeper, but a greater excavated plan area is required. Wellpoints are installed at between 1 to 3 m intervals by wash boring, i.e. using high pressure water jetting to form the borehole, but the spacing depends on the permeability of the ground. Figure 4.4a shows a typical arrangement for a wellpoint system. Wellpoints can also be used from inside the tunnel. In this case they should be directed upwards.

Deep wells can be used to dewater to greater depths. These consist of 300 mm or greater wells sunk at an average spacing of 3 m or more to below the level required for the dewatering. A filter is used at the base of the well around perforated suction pipes, above which a submersible pump is located (Figure 4.4b). It is important to establish a detailed conceptual model from the site investigation and pumping test data, preferably with distance/drawdown/time results. Further details on the design of wells can be found in Woodward (2005).

Drawdown of the groundwater level can cause consolidation settlements in the surrounding ground and hence affect adjacent structures, and therefore it should be closely monitored. The extent of the drawdown zone depends on the depth of the well and the type of ground.

Further information on groundwater lowering and dewatering can be found in Preene *et al.* (2000), Cashman and Preene (2001) and Powers *et al.* (2007).

4.2.3 Grouting

Grouting involves the process of injecting a material into the ground with the following two principal objectives:

- to reduce the permeability of the ground;
- to strengthen and stabilize the ground. In soft ground this leads to an increase in its 'strength' and in jointed rock in its 'stiffness'.

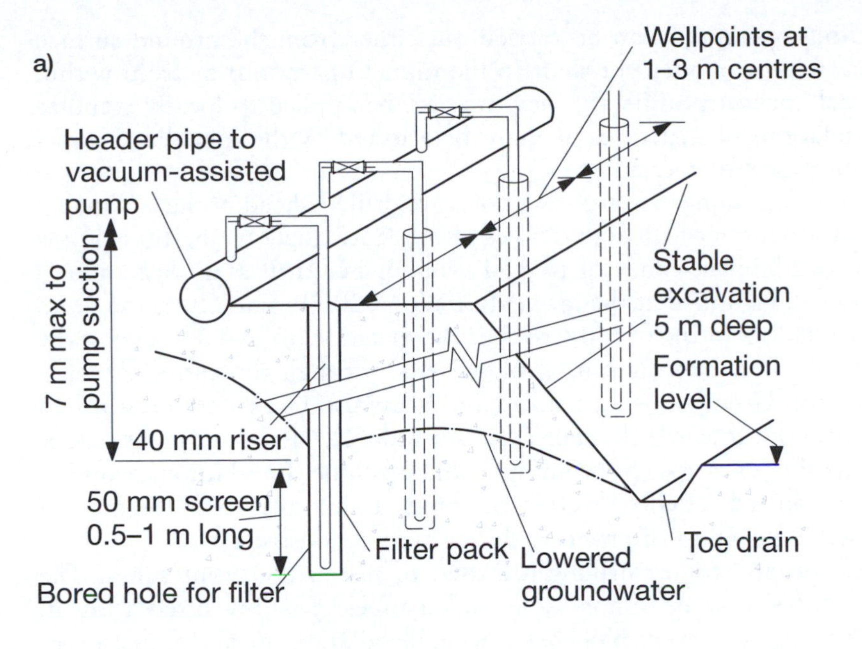

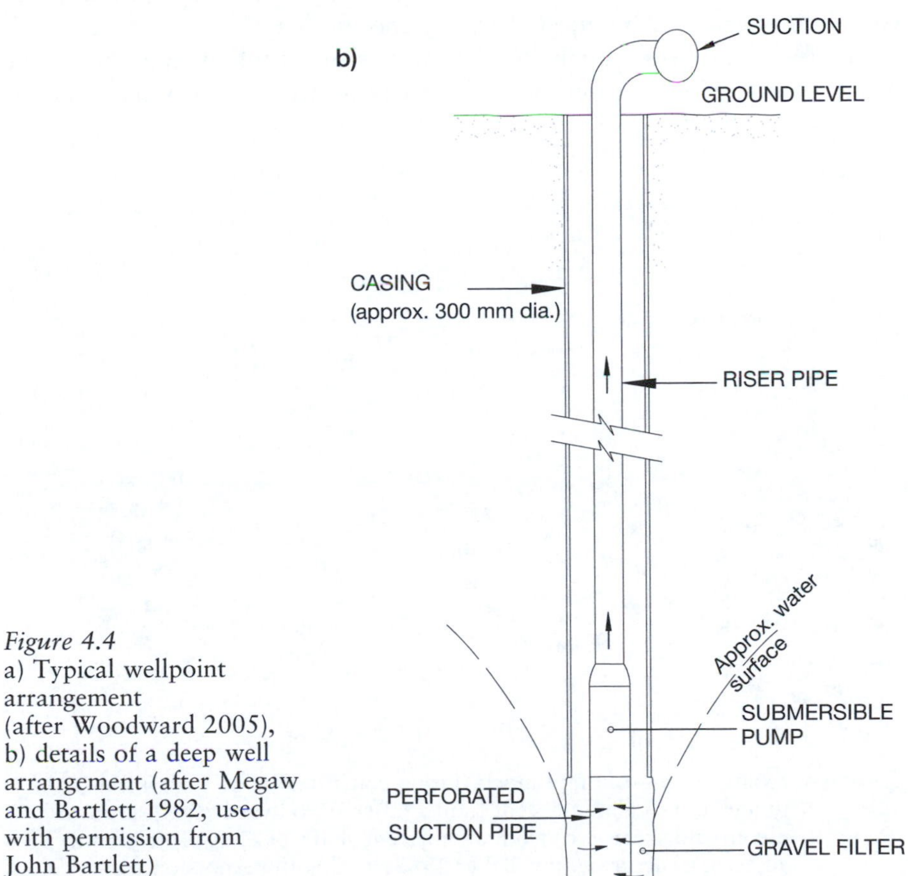

Figure 4.4
a) Typical wellpoint arrangement (after Woodward 2005), b) details of a deep well arrangement (after Megaw and Bartlett 1982, used with permission from John Bartlett)

Grouting operations can be carried out either from the ground surface (or from within an adjacent shaft to the tunnel operation) or from within the tunnel construction itself. They can also be applied to locally stabilize the foundations of structures likely to be affected by the tunnelling works in the form of settlements.

For tunnel grouting, the grouting holes are drilled ahead of the advancing tunnel in a pattern of diverging holes at an acute angle to the tunnel axis to form overlapping cones of treated ground. For drill and blast tunnels the holes can be drilled at the face (Muir Wood 2000). For TBMs the holes can be drilled forward from the rear of the machine, to avoid affecting the cutter wheel, but direct grouting of the face through the cutter wheel is also possible. Grouting using a shield TBM can also be carried out through the shield, both towards the face and also radially. However, great care is needed as there is a risk of grouting-in the machine. In addition, grouting can be conducted radially through the lining to fill any voids. Figure 4.5 shows some examples of grouting during tunnel construction.

Percussion and rotary drilling are used to install the grout tubes. The grouting tubes may be simple open-ended tubes, possibly fitted with an expendable tip to prevent blockage during installation, or perforated tubes which allow grout to be injected over a specific length.

The use of a tube-a-manchette (TAM) or sleeved tube makes successive injections at specific locations possible. Perforations at appropriate intervals

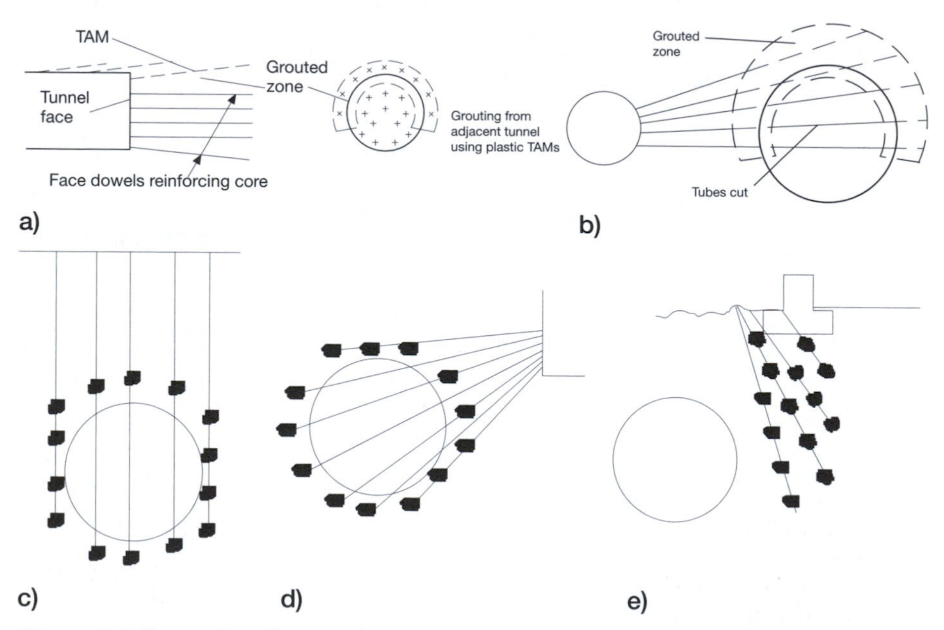

Figure 4.5 Examples of grouting tunnels during construction, a) from within a tunnel, b) using an adjacent tunnel (after Woodward 2005), c) from the ground surface, d) from an adjacent shaft, or e) as protection to adjacent structures (after Baker 1982, used with permission from ASCE)

along the tube are closed by an external elastic sleeve which can be opened by the internal pressure of the grout. The grout is passed to the injection point by a movable separate internal tube. The grout is contained within the location of the perforation using seals either side of the end of this internal tube. Figure 4.6 shows details of a tube-a-manchette (called a sleeve port pipe in the US).

There are several types of grouting technique and these can be described as permeation grouting, jet grouting and compaction grouting.

PERMEATION GROUTING (CHEMICAL GROUTING)

This technique fills the voids in the soil with either chemical or cement binders with the intention of not disturbing the fabric of the ground. The range of particle sizes over which it can be applied is from sands (0.06 mm) to coarse gravels (60 mm). Further information on permeation grouting can be found in Karol (1990).

JET GROUTING

This technique uses high pressure jets to break up the soil and replace it with a mixture of excavated soil and cement. The range is wider than for permeation grouting, extending from clays (< 0.002 mm) to fine gravels (10 mm). Jet grouting may be used in pre-bored holes or the 'jets' can be self-drilled. Once the jet has reached the required depth, it is rotated and the jetting fluids are pumped at high pressure to the jetting tip as the system

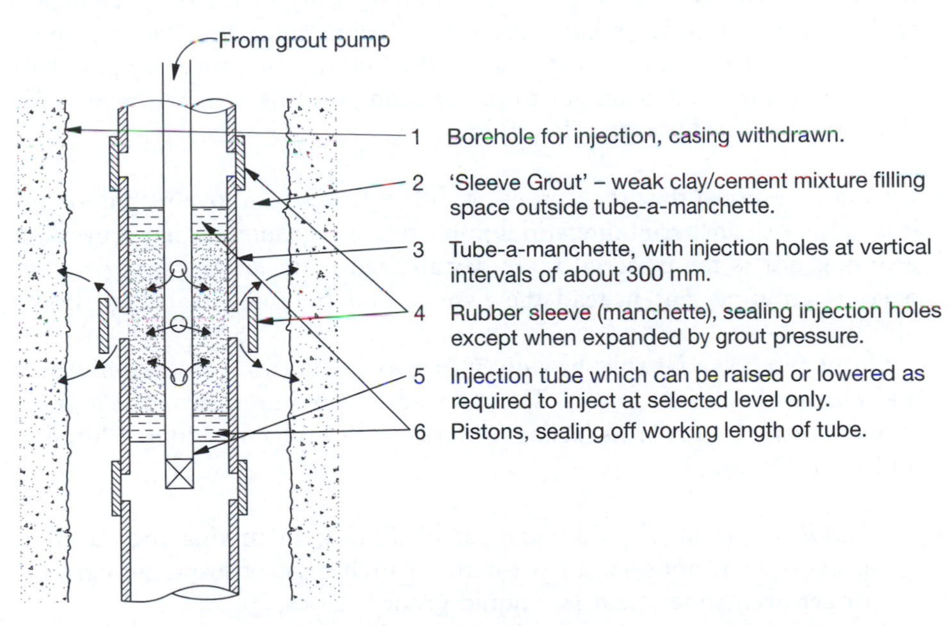

Figure 4.6 General arrangement of a tube-a-manchette (after Megaw and Barlett 1982, used with permission from John Bartlett)

is withdrawn from the hole at a controlled rate to form an *in situ* column (Woodward 2005). There are three basic jetting systems, a single jet which uses just grout, a double jet system involving grout with an air shroud and a triple jet system where the grout is discharged through one hole and just above this is a second jetting point where an air-water mixture is injected. It should be noted that if the hole blocks with debris, a sudden pressure can build up with bursting pressures developing, which can damage adjacent services and even flood cellars with grout. This is a real risk and the operator has to carefully monitor the return flows and pressures. The system used depends on the ground type, with the single system being suitable for sands with $N_{SPT} < 15$ (where N_{SPT} is the standard penetration test blow count, as described in section 2.3.2.2), and the other systems used for finer grained soils (Woodward 2005). If the jet is not rotated then more of a 'panel' shape is produced rather than a column. Further details on jet grouting can be found in BSI (2001a).

COMPACTION GROUTING

This technique differs from both permeation and jet grouting in that it is a ground improvement technique rather than a ground treatment technique. Compaction grouting is essentially the injection of a low slump (typically 25–100 mm) grout, i.e. stiff grout, such that an expanding bulb forms. This expansion causes deformation and densification around it and ultimately improves the ground. The method is carried out by either drilling or driving small diameter casings (89–114 mm typically) to the required depth, withdrawing the rods or knocking off the drive point and then pumping the grout to the bottom of the hole (Essler 2009). The range of applicable soils for this method is similar to permeation grouting ranging from sands (0.06 mm) to medium gravels (30 mm).

It should be noted that rock grouting differs from the above techniques as it is neither the material interstitial pores that are grouted as in permeation grouting nor is the material body destabilized as in jet grouting or compaction grouting, but instead the fissures and fractures are filled (Essler 2009).

Grout types can be split broadly into two categories, suspension grouts and chemical solution grouts. There are several requirements that a grout should meet in terms of its basic properties as listed below (after Whittaker and Frith 1990).

- *Stability* – grouts should remain stable during the mixing and injection processes and not separate prematurely in the case of suspension grouts, or set prematurely if it is a liquid grout.
- *Particle size* – for a suspension grout this sets the lower limit of the grain size of the soil that it can penetrate.

- *Viscosity* – this is basically a measure of its ability to penetrate soils. Other flow properties and the gelling time determine the maximum injection radius.
- *Strength when set or gel strength* – this depends on whether the grout is being used to strengthen the ground or reduce its permeability.
- *Permanence/durability* – the grout, when set, should resist chemical attack and erosion by groundwater.

Suspension grouts basically consist of cement slurry with a cement/water ratio of approximately 0.1 to 0.4, and an optional clay component. The purpose of the clay is to reduce the cement consumption and to improve the stability and viscosity of the suspension. Sand can be added to grout suspensions when large fissures are to be injected. Additives such as plasticizers (comprising metal salts, such as lithium, sodium and potassium salts) can be used in suspension grouts to prevent the clay particles flocculating (i.e. clumping together) and this will give different properties to the grout. Suspension grouts are best suited to injection into fissured rocks and granular media with large voids and porosity (down to a particle size of approximately 0.2 mm). A suspension grout containing fine to coarse sand, cement and a plasticizer is technically known as a mortar, which can be used to plug large fissures and cavities (Whittaker and Frith 1990).

Chemical grouts usually consist of solutions and resins which form gels. They reduce the permeability by void filling and strengthen the ground. These grouts have a major advantage over suspensions in that they can be injected into very fine grained soils, since some liquid grouts, such as resin types, have viscosities approaching that of water (down to a particle size of approximately 0.02 mm). The strength of chemical grouts is generally low compared to cement grouts.

The most common types of grouts are either cement bentonite (suspension grout) or silicate based (chemical grout). The type of grout depends on both the ground type and the grouting technique adopted. For filling large voids, materials such a pulverized fuel ash (PFA, a waste product from coal-fired power stations) can be used.

Further details on grouting techniques and grout materials can be found in Xanthakos *et al.* (1994) and Moseley and Kirsch (2004).

4.2.4 Ground reinforcement

There are three distinct types of ground reinforcement methods (Whittaker and Frith 1990, Woodward 2005):

ROCK DOWELS

These are reinforcing elements with no installed tension. They consist of a rod, faceplate and nut (a conical spacer is sometimes used if the angle between the dowel and the face plate differs significantly), and can be made from deformed steel bars, glass fibre or plastic, depending on whether a permanent or a temporary installation is required. The rod is usually embedded in a mortar or grout filled tube, although resin capsules are also used extensively. Dowels can be used as a systematic reinforcement of the ground or in hard rock can be placed at discrete locations to prevent unstable parts of the ground falling into the excavation. (Note: in Austria and Germany cemented rock dowels are commonly know as 'SN-anchors', named after 'Store-Norfors', the Norwegian city where they were first used. In the US dowels are known as nails.)

Another development is inflatable rock dowels (Swellex®, Atlas Copco). These consist of folded steel, closed at the end, and inflated by water. The steel expands and is pressed against the wall of the borehole providing close contact between the dowel and the ground, resulting in no need for grout or resin.

ROCK BOLTS

These are reinforcing elements which are tensioned during installation. They consist of a rod and mechanical or grouted anchorage (resin capsules or cement) coupled with some means of applying and retaining the rod tension. Mechanical fixings are suitable for hard rock, whereas grouted, fixed length bolts can be used in most rock types. The length varies between 2 to 8 m for resin capsule grouted bars, and 3 to 20 m for an expanding shell fixing on a bar. Figure 4.7 shows some diagrams of typical rock bolts and dowels. (Note: in some countries the term 'bolt' is also used for untensioned systems.)

ROCK ANCHORS

These are reinforcing elements which are tensioned following installation and are of higher capacity and generally of greater length than rock bolts. They consist of high strength steel tendons usually in the form of cables to which is fitted a stressing anchorage at one end and means of transferring a tensile load to the cable at the other end. These can be used in most rock types. Double corrosion protection is required for permanent anchors and conducting proof loading tests of each anchor is normal during tensioning. As mechanical anchors slacken with time, and hence could allow movement of the ground, fully bonded anchors should be used.

There are four generally accepted mechanisms by which rock reinforcement can improve the stability of the ground (Whittaker and Frith 1990).

1 By stabilizing individual blocks of material that may detach due to gravity in relatively competent and well-jointed rocks, by using rock bolts with an anchorage force capacity greater than the weight of the block.

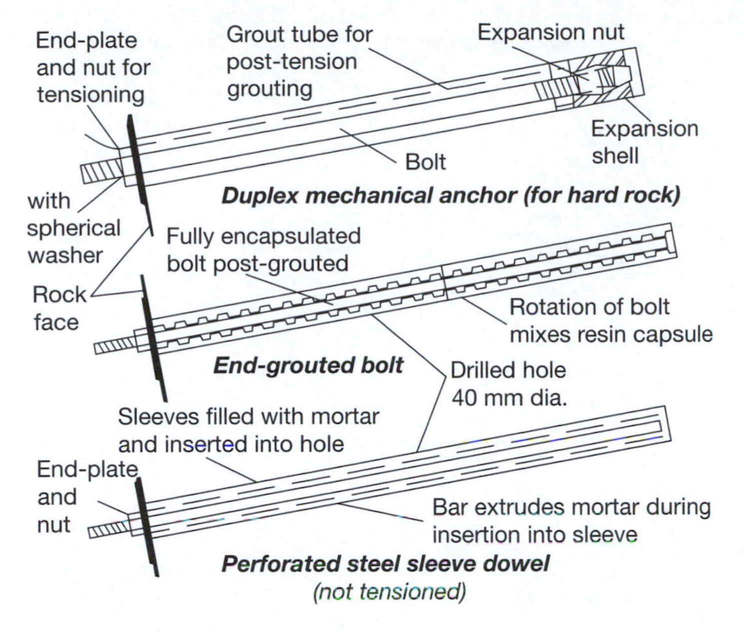

Figure 4.7 Examples of typical rock bolts and dowels (after Woodward 2005)

2 By using tensioned or untensioned bolts to maintain the shear strength of the ground along discontinuities in weaker fractured ground conditions.
3 By using fully grouted untensioned rock bolts in laminated or stratified rocks to preserve the inter-strata shear strength.
4 By using tensioned rock bolts installed relatively quickly after excavation to improve the degree of confinement or the minor principal stress (this is normally perpendicular to the tunnel wall) in overstressed rocks.

Rock reinforcement alone is unlikely to be appropriate if (Woodward 2005):

- the support pressure required is greater than 600 kN/m^2;
- the spacing of dominant discontinuities is greater than 600 mm;
- the rock strength is inadequate for anchorages;
- the RQD is low or there are infilled joints or high water flow.

Figure 4.8 shows some typical examples of the arrangement of rock bolts/dowels within tunnels.

An example of a typical specification for supporting blocks for short (5 m) spans within the ground is given below (Woodward 2005):

- minimum bolt length, 0.5 × span or 3 × width of an unstable block;
- maximum spacing, 0.5 × bolt length or 1.5 × width of a critical block, and 2 m when using mesh restraints;

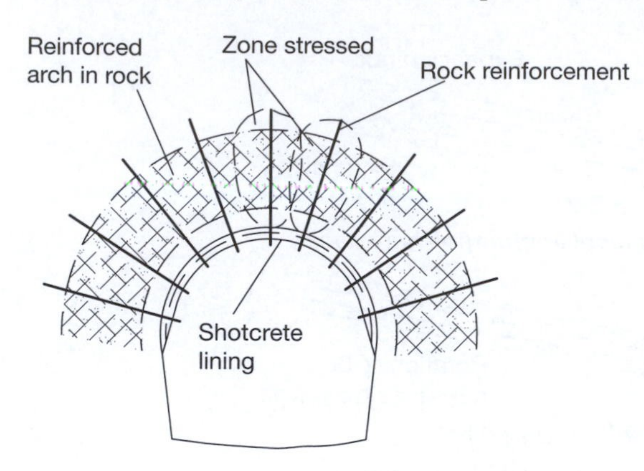

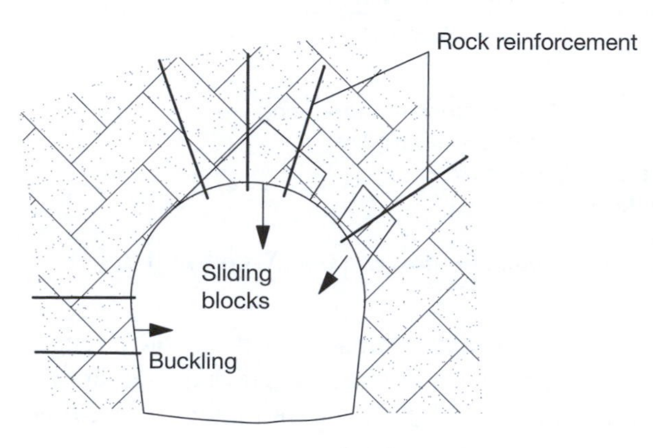

Figure 4.8 Typical examples of the arrangement of rock bolts/dowels within tunnels (after Woodward 2005)

- larger spans in fractured rock will require primary, secondary and even tertiary reinforcement.

(Note: dowels and bolts are also applicable for soft ground.)

Figure 4.9 shows an anchor installation associated with sprayed concrete lining in the Heidkopf Tunnel (HKT), Göttingen, Germany. This tunnel was constructed through sandstone and limestone and consisted of a twin tube, 2-lane road tunnel, 1720 m long (each tube) and a cross section of 88–129 m^2 (approx. width 12 m). Figure 4.10 shows the load testing of an anchor as part of the construction of the Lainzer Tunnel LT31, Vienna (see section 8.3 for further details of this tunnel).

4.2.5 Forepoling

This technique is aimed at limiting the decompression in the crown immediately ahead of the face (ITA/AITES 2007). Longitudinal bars (dowels) or steel plates (forepoling plates) are installed ahead of the tunnel from the periphery of the face, typically over the upper third or quarter of the

Figure 4.9 Anchor installation associated with sprayed concrete lining, HKT Tunnel, Germany (courtesy of ALPINE BeMo Tunnelling GmbH Innsbruck)

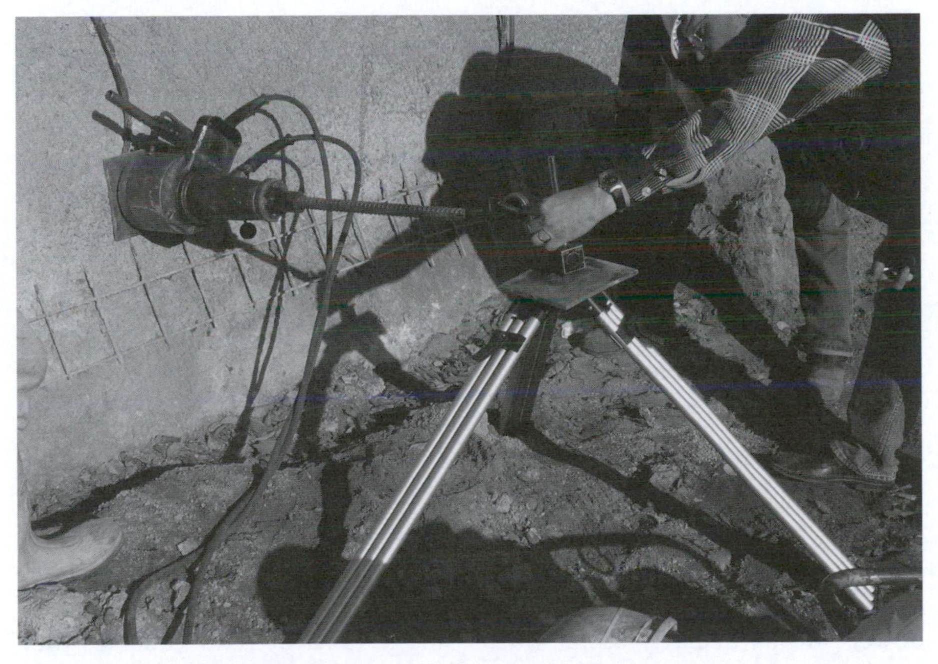

Figure 4.10 Load testing of an anchor, Lainzer Tunnel LT31, Vienna, Austria

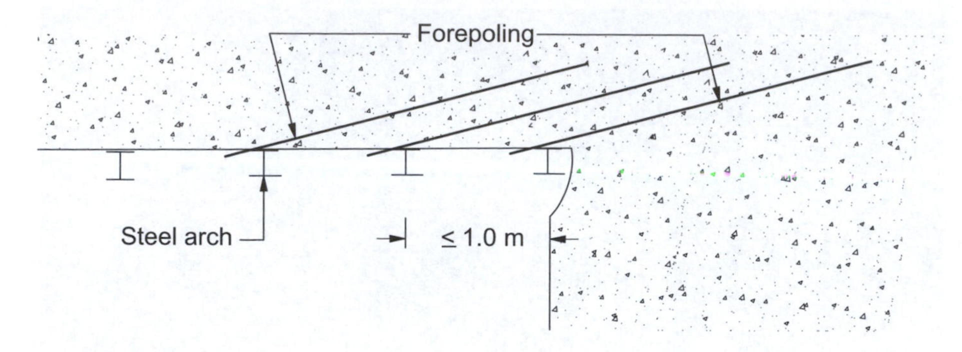

Figure 4.11 Basic arrangement of forepoling using dowels (after ITA/AITES 2007)

excavated profile. In rock, the plates or dowels driven ahead of the excavation are also known as spiles.

DOWELS

The material used for dowels is the same as that used for rock dowels (section 4.2.4). Dowels are also installed from within the tunnel, and positioned on lattice girders at an angle to the direction of tunnelling (Figure 4.11). If the ground is too dense or too hard, dowels are placed in pre-drilled holes. In this case the hole can be filled with grout before the dowels are pushed further into the ground. They are designed to protect the crown area against afterfall (i.e. falling blocks of material from the tunnel roof). The separation of the dowels is dependent, amongst other things, on the size of the blocks in the ground, and as a rule is greater than 20 cm. The length of the dowels is approximately 3 to 4 m (about three to four times the heading advance).

PLATES (SHEETS)

Forepoling plates, mainly made from steel, are pushed forwards individually, but close together in the same way as described for the dowels. Plates are used with coarse grained soils such as sands or gravels, which would fall through the spacing between dowels.

4.2.6 Face dowels

Face dowels can be used to improve the stability of an excavated tunnel face. The technique involves installing an array of dowels over the cross section of the tunnel face (Figure 4.12). The dowels should be made of material that can be easily excavated, for example fibreglass. Fibreglass dowels are particularly useful when using a TBM or roadheader as they can be cut through easily. However, fibreglass dowels can produce sharp,

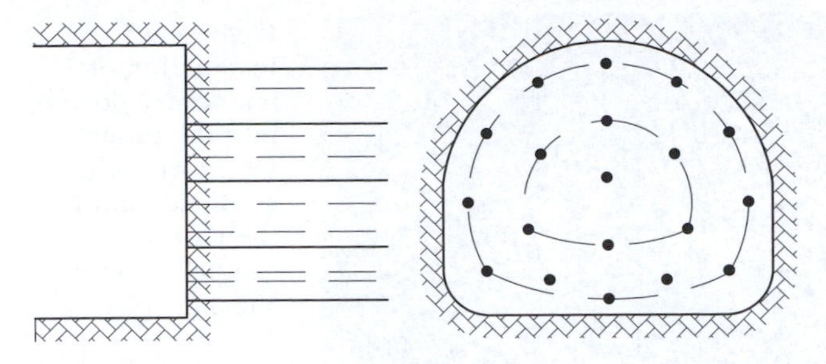

Figure 4.12 Schematic showing a typical arrangement of dowels used in the tunnel face (after ITA/AITES 2007)

and potentially dangerous, ends when excavated due to the brittle manner in which they break. Therefore, steel dowels can prove more useful even though the excavation process needs more care. Steel dowels can be easily cut using handheld cutters. Ideally, the dowels should provide a continuous stability as the excavation advances. This is achieved by the dowels being shortened as the face advances until a minimum length is reached. At this point a new set of dowels is installed in the face. These new dowels overlap with the previous dowels by a few metres.

Figure 4.13 shows an example of temporary face dowels being used during the construction of the LT31 Tunnel in Vienna, Austria. In this case the dowels were 12 m long and overlapped longitudinally by 5 or 6 m. There was also a plate (12 cm × 50 cm) attached to the end of the dowels that distributed the stress and avoided overstressing the sprayed concrete, which was relatively thin (approx. 10 cm) compared to the side walls (approximately 30 to 35 cm thick).

4.2.7 Roof pipe umbrella

In the roof pipe umbrella method steel pipes are drilled from within the tunnel in the direction of tunnelling around the perimeter of the tunnel roof, as described in the artificial ground freezing section 4.2.1. The steel pipes have a diameter of approximately 70 to 150 mm. After drilling, the holes are filled with grout. The spacing of the pipes ranges from approximately 20 to 50 cm. The length of the pipes is often 15 m or more. The previous umbrella overlaps with the subsequent umbrella by at least 3 to 5 m. Roof pipe umbrellas act like forepoling, i.e. they are supposed to protect the crown area against afterfall. However, due to their larger diameter and length, roof pipe umbrellas are a lot more robust than forepoling.

Figure 4.14 shows an example of forepoling (dowels) in association with sprayed concrete lining being used as a roof pipe umbrella on the LT31 Tunnel, Vienna (see the case study in section 8.3 for further details of this tunnel).

Figure 4.13 Installation of temporary dowels into the tunnel face during the construction of the Lainzer Tunnel LT31 in Vienna, Austria

4.2.8 Compensation grouting

Compensation grouting is a technique developed to control the settlement of structures in the vicinity of tunnels constructed in soft ground and is one of the most specialized forms of ground treatment. Settlements can occur around tunnels as a result of stress changes during construction and are discussed further in section 7.1. It is therefore not a stabilization technique to aid tunnel construction as such, but to avoid the tunnel construction adversely affecting adjacent structures. It can also be used as a method of maintaining or re-levelling structures or ground subject to on-going settlements, for example due to the consolidation of clay soils. Generally, compensation grouting is only considered after it has been determined that the ground displacements cannot be reduced to an acceptable level by increasing the support to the ground from within the tunnel during construction.

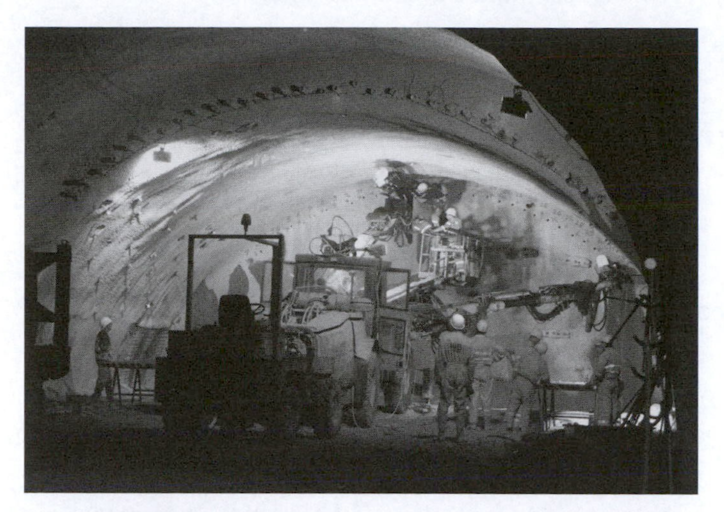

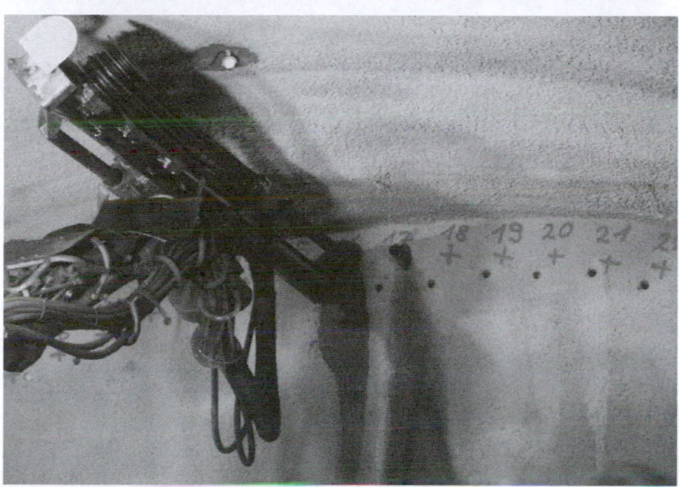

Figure 4.14
Forepoling using dowels as a roof pipe umbrella on the Lainzer Tunnel LT31 in Vienna, Austria

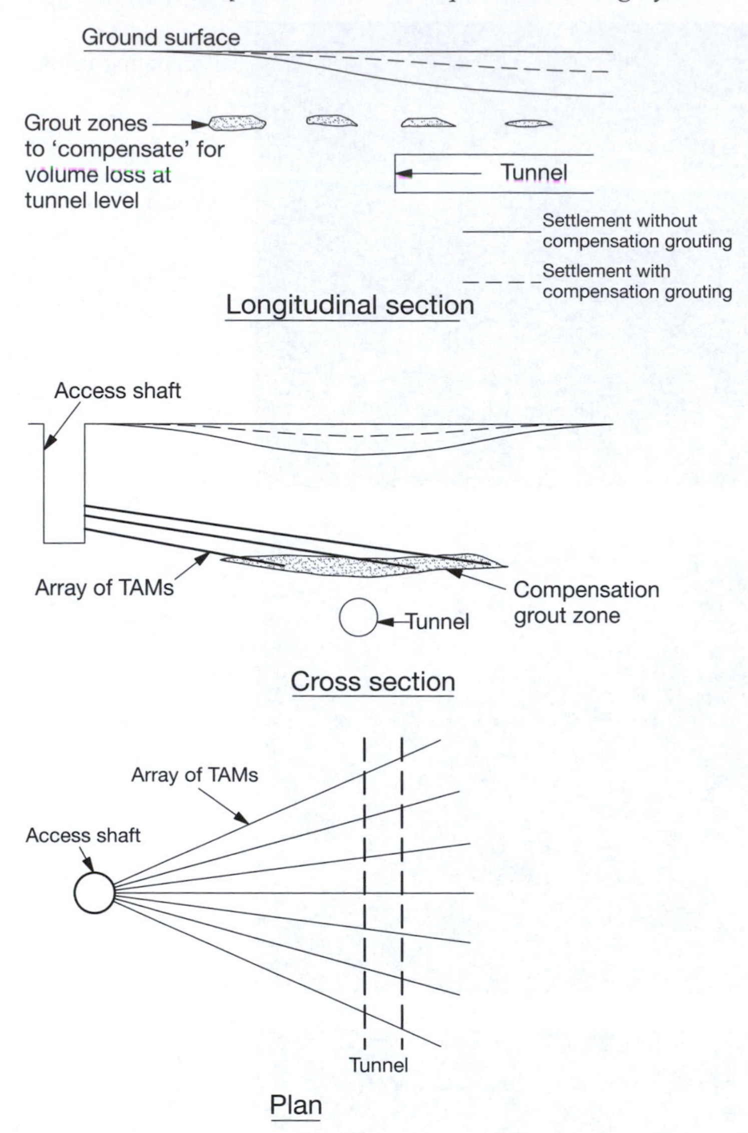

Figure 4.15 Compensation grouting for tunnel construction (after Woodward 2005 and Burland *et al.* 2001a)

This method involves injecting grout into the ground at a level between the tunnel crown and the structure to be protected (Figure 4.15). This is commonly done from grout holes drilled radially from a shaft. Tube-a-manchette (TAM) grouting (section 4.2.3) or similar techniques are used to inject the grout in controlled amounts after an initial preconditioning of the ground locally. Preconditioning means that grout is used to locally compress the soil, possibly fracturing the ground, so that subsequent injections have an immediate effect on the ground. Controlled volumes of grout are used to 'compensate' for the occurring ground displacements. The volumes involved are usually low, for example 20–100 litres, to ensure that grout does not travel out of the compensation grouting zone. The required grout depends on the ground efficiency, i.e. the ratio of grout volume to be injected to ground volume change. Typically, the ground efficiency in London Clay might be better than 50%, in a soft clay 10–15% and in sands 20–30% (Essler 2009).

It is important that there is real time feedback during the grouting operations from instrumentation on the structures being 'protected' so that the compensation grouting can be accurately used. This is a good example of the need for a clear observational approach (section 7.3.3). Injecting the grout into the ground stops the ground moving downwards, which is essentially a jacking operation. Therefore a reaction force is necessary to generate this support, which is directed downwards. Hence, it is important to position the drill holes at a level so that this downward force will not adversely affect the tunnel heading (and the subsequently installed lining). Careful control is also needed when using this technique near existing tunnels as there is a danger of the outward forces generated by the grout injection causing deformation or damage to the existing tunnel lining.

4.2.9 Pressurized tunnelling (compressed air)

If a tunnel is constructed within the groundwater, special precautions have to be taken or tunnel construction methods chosen which prevent water from getting into the tunnel as this would make the works impossible. One such method is by using air under high pressure within the tunnel during construction. Air pressure can be used to control water flow, and hence stability, below the groundwater table and is one of the oldest pressurized face support methods used in tunnelling.

The disadvantage of constructing a tunnel under air pressure is that, in order to maintain the air pressure at the face, all the materials and spoil, as well as the workforce, have to be passed through an airlock system. The maximum working pressure and the time that workforce can spend working in compressed air have to be strictly controlled. Originally the whole tunnel length was put under air pressure, i.e. from the face to the pressure chamber, which was most often near the starting shaft. However, developing from the idea of no longer wishing to put the whole tunnel under pressure, pressure bulkheads (airlocks) started to be placed closer to the rear of the tunnel face (Figure 4.16).

Pressurized air is compressed into the tunnel by using compressors installed on the surface until the required overpressure is established (atmospheric overpressure acts in addition to air pressure). In order to achieve this, the necessary amount of air has to be determined prior to, and during, tunnel construction. The danger of 'blowouts' has to be considered and, as far as possible, minimized.

Figure 4.16 Airlock arrangement on a pipe jacking project for a new sewer in Germany (see section 5.11 for a description of pipe jacking)

BLOWOUTS

A blowout is where the pressurized air finds a pathway to the ground surface, blows out suddenly and the pressure at the face drops and can no longer be maintained. This can result from the seepage of air into the ground and the consequent loosening of the ground matrix (for example lifting of sand). This can increase the porosity enormously resulting in a sudden loss of pressure within the tunnel. Obviously, this situation is particularly dangerous for tunnellers.

WORKING UNDER PRESSURE

The normal air pressure at the surface of the earth, or more precisely at sea level, is 1013 millibars, which is around 1 bar or 100 kN/m^2.

The critical level for working under high pressure begins with an atmospheric overpressure of approximately 1.0 bar. If someone is put in a situation where the pressure is higher than atmospheric, the amount of soluble nitrogen (N_2) increases in the body. If the outside pressure is suddenly reduced the nitrogen comes out of solution instead of being exhaled and this can result in the formation of bubbles in the blood. The bubbles can result in blockage of the arteries. It is possible that the excess nitrogen can settle in the joints and result in pain. Complaints, which can be traced back to both phenomena, often only occur after many years. The illness is known under many different names, for example caisson or decompression illness (this is the same condition that can be experienced by divers).

In order to avoid endangering tunnellers medically, the decompression chamber times on a tunnel construction site are strictly regulated. Different regulations and guidelines exist in various countries, for example in the UK 'The Work in Compressed Air Regulations 1996' (UK Government 1996) and in Germany the pressurized air regulation of 1972 in the Bundesgesetzblatt Teil 1, Nr. 110, last updated in 2008 (air pressure regulation).

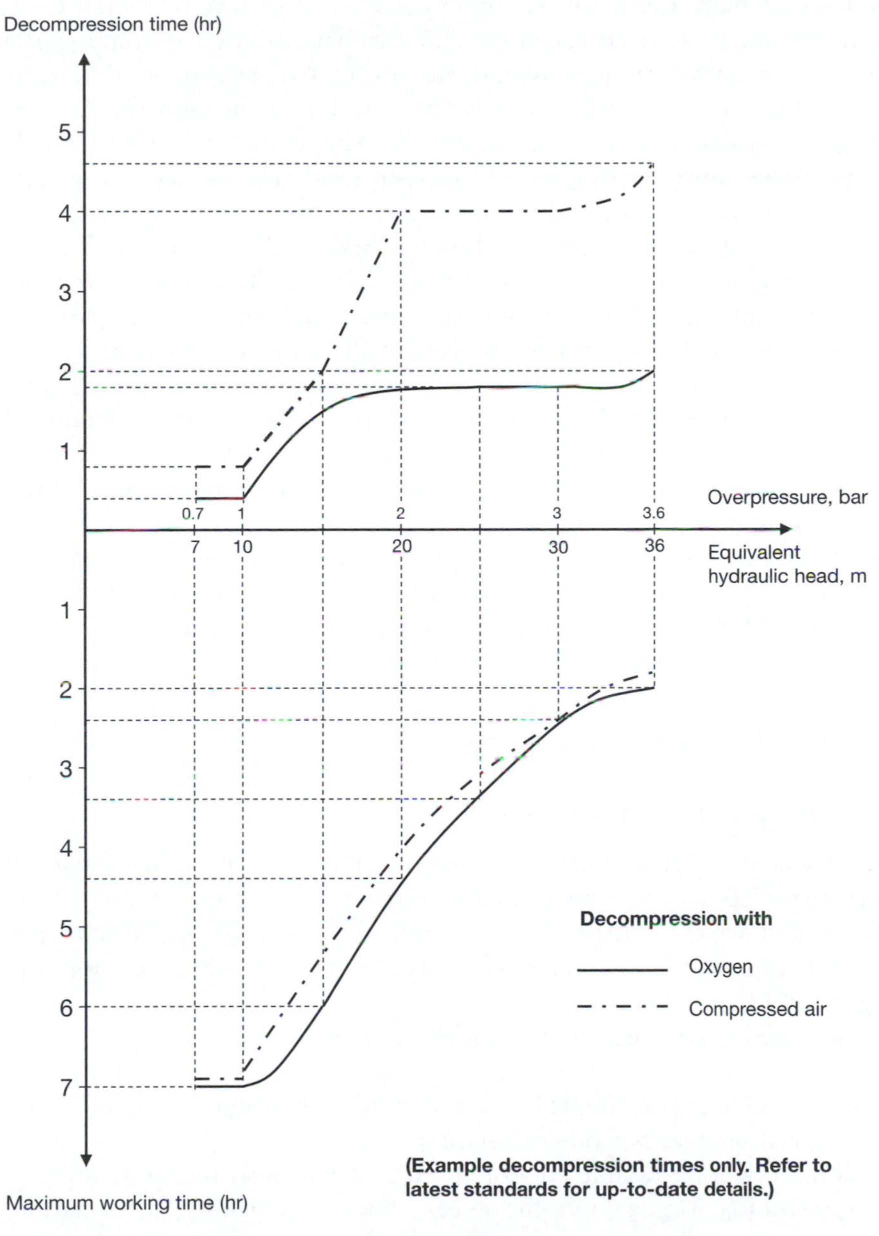

Figure 4.17 Simplified decompression times after Bundesgesetzblatt (2008), Germany

Generally the situation can be summarized as:

- the higher the air pressure, the longer the decompression time;
- the longer the working exposure, the longer the decompression time.

Figure 4.17 shows, in a simplified form, the recommendation of the Bundesgesetzblatt, Germany. Maximum working and decompression times are shown up to an overpressure of 3.6 bar. The use of a decompression chamber is required from an overpressure of 0.7 bar. Decompression using oxygen through face masks has now become the norm with the exhaled oxygen being discarded outside the airlock. This makes it easier to exhale any excessive nitrogen. The level of oxygen inside the airlock is carefully monitored to ensure no build up of oxygen as this poses a fire risk.

Material and personnel are put through the airlock separately. The personnel airlocks must consist of two chambers. In the UK, when the working pressure is above 0.7 bar a medical lock must be provided for the recompression and treatment of any person showing symptoms of decompression illness (UK Government 1996). In some countries this is only necessary if no suitable hospital facilities are available in the vicinity of the site. After decompression the personnel are required to stay on the construction site for 30 minutes as most of the symptoms appear within this timeframe. With the aid of precautionary measures it is possible to practically rule out any delayed symptoms. Although there is an opinion these days that working under high pressure is no riskier than other tunnelling works, there are health and safety implications and consequently these risks have to be managed (see section 6.1.15).

4.3 Tunnel lining systems

4.3.1 Lining design requirements

The design of a permanent tunnel lining solution is influenced by the full range of project-specific operational requirements, i.e. not only the ground loading and water control. These include the electrical and mechanical operation and maintenance and safety aspects such as resistance under fire (Legge 2006).

Design issues that need to be considered are:

- functionality, for example lining and material type and lining thickness, material properties, constructability;
- durability, for example corrosion, adverse chemical reactions, fire;
- appearance, for example the effect of water, cracking, deformation, surface texture.

CHARACTERISTICS OF LINING BEHAVIOUR

A list of common characteristics that pervade all lining systems are (after Kuesel and King 1996):

- The processes of ground pre-treatment (e.g. grouting), excavation and ground stabilization (e.g. rock bolting) alter the pre-existing state of stress in the ground, before the lining comes into contact with the ground.
- A tunnel lining is not an independent structure acted upon by well-defined loads. The loads acting on a tunnel are not well defined, and its behaviour is governed by the properties of the surrounding ground. Design of a tunnel lining is not a structural problem, but a ground-structure interaction problem, with the emphasis on the ground. Defining the loads on the tunnel lining is one of the most challenging aspects for a civil engineer on a tunnel project.
- Tunnel lining is a four-dimensional problem. During construction, the ground conditions at the tunnel heading involve both transverse and longitudinal arching, or cantilevering from the excavated face. All the ground properties are time-dependent, particularly in the short-term, which leads to the commonly observed phenomenon of stand-up time, without which most practical tunnel construction methods would be impossible. The timing of the lining installation is an important variable. In addition, some tunnel linings such as sprayed concrete lining (shotcrete) can itself have time-dependent characteristics. In the case of sprayed concrete lining the stiffness is extremely time-dependent and the effects of early creep have to be taken into account before the lining reaches its full strength.
- The most serious structural problems encountered with actual lining behaviour are related to the absence of support rather than to the intensity and distribution of the load, for example inadvertent voids left behind the lining. However, exceptions exist and poor ground conditions could result in additional, unexpected loading on the lining.
- In most cases in hard rock, the bending strength and stiffness of structural linings are small compared with those of the surrounding ground. The properties of the ground therefore control the deformation of the lining, and changing the properties of the lining will not significantly change this deformation. It is important that the lining has adequate ductility to conform to the imposed deformations, and adequate strength to resist bending stresses is therefore secondary. The lining therefore forms a flexible ring confined by the ground.

4.3.2 Sprayed concrete (shotcrete)

Sprayed concrete is concrete which is conveyed under high pressure through a pneumatic hose and projected into place at high velocity, with simultaneous compaction (DIN 2005). Sprayed concrete can also be called 'shotcrete' and both terms are used in this book.

Sprayed concrete is an effective material for tunnel linings as (after Thomas 2006):

- it is a structural material that can be used as a permanent lining;
- it can be applied as and when required in a wide range of profiles and it can be adjusted to suit a wide range of ground conditions. Sprayed concrete is particularly suited to lining shafts, junctions, non-circular tunnels and tunnels of variable shape;
- it is soft when sprayed, but rapidly increases in stiffness and strength, thereby providing an increasing amount of support to the ground with time. This helps to limit movements in the ground, but also allows a degree of stress re-distribution to occur;
- it is possible to mechanize the shotcreting process, thus providing potential health and safety benefits.

Sprayed concrete consists of water, cement and aggregate, with various additives. The mix, compared to conventional cast concrete, has more sand, a higher cement content, smaller sized aggregate and more additives. This leads to a faster increase in strength and other properties with age, a lower ultimate strength and more pronounced creep and shrinkage behaviour. The creep behaviour may be important, particularly when loaded at an early age, and could become 'overstressed', i.e. loaded to a high percentage of its strength (Thomas 2006). This behaviour has to be considered very carefully when modelling the tunnel. Furthermore, it has to be taken into account when analysing observed displacements as the sprayed concrete will deform initially without any stresses being induced. It is critical to judge when the displacements are exceptionally large and the tunnel is in danger of collapse. There is no threshold value for the collapse and it depends on each situation. Measurements from previous cross sections can be consulted (if available), but often it is down to the experience of the civil engineer. This is discussed further in section 5.7 on NATM tunnels and section 7.3.4 on in-tunnel monitoring.

Additives such as microsilica have been found to improve durability of sprayed concrete linings and this forms a more dense concrete. Steel fibres can also be used as reinforcement for sprayed concrete and further details on this can be found in Thomas (2009a).

Waterproofing of sprayed concrete linings can consist of sheet membranes where complete watertightness is required. Where criteria for watertightness are less onerous then spray-on membranes or simply the inherent impermeability of concrete itself can be used to prevent water ingress. However, it should be noted that sprayed concrete is not as watertight as cast concrete as the joint between the sprayed concrete layers and construction advances are not completely watertight.

The specification of sprayed concrete works is straightforward since there are several published guidelines, for example ÖBV (1999) and

EFNARC (1996). Target strengths should always be specified for the early age period, i.e. less than 24 hours. It is important that the sprayed concrete gains sufficient strength to carry the anticipated loads at all ages (Thomas 2006).

There are two ways of producing sprayed concrete: the *dry mix* process and the *wet mix* process (Thomas 2009a).

The *dry mix* process uses a mixture of naturally moist or oven dried aggregates, cement and additives, which is conveyed by compressed air to the nozzle where it is mixed with water and liquid accelerator. The water/cement ratio is controlled by the 'nozzleman' during spraying.

A few reasons for using a dry mix process are (Thomas 2009a):

- higher early-age strength;
- lower plant costs;
- small space requirement on site;
- more flexibility during operation, i.e. it can be available as required as there is less equipment cleaning needed.

This means it is suited to projects requiring small to intermediate volumes of sprayed concrete and where there are space constraints on site (Thomas 2009a). The main disadvantages of the *dry mix* process are the higher levels of dust (health and safety) and the potential variability of the product due to the influence of the nozzleman.

The *wet mix* process involves conveying ready-mix (wet) concrete to the nozzle by either compressed air or pumping. Liquid accelerant is added at the nozzle. This is either controlled by the nozzleman (old system) or at a separate accelerator pump. The water/cement ratio is fixed when the concrete is batched outside the tunnel (Thomas 2009a).

There is a trend to use the *wet mix* process for a number of reasons (Thomas 2009a):

- there is greater quality control, i.e. less human variability;
- higher outputs can be achieved compared to the *dry mix* process as in the wet mix process robotic spraying techniques are required due to the weight of the nozzle;
- less rebound of the sprayed concrete off the excavated surface;
- less dust;
- it is easy to keep records of the exact mix and quantities sprayed due to the use of ready-mix batches and robotic spraying.

This means that the wet mix process is suited to projects requiring large volumes of sprayed concrete at regular intervals. In terms of cost, there is very little difference between the two processes (Thomas 2009a).

Figures 4.18 and 4.19 show some examples of sprayed concrete application during the construction of the LT31 Tunnel in Vienna, Austria (see section 8.3 for further details).

Figure 4.18 Sprayed concrete application of the invert of a side wall drift during the construction of the Lainzer Tunnel LT31 in Vienna, Austria

Figure 4.19 a) Manual spraying and b) Spraying equipment used during the construction of the Lainzer Tunnel LT31 in Vienna, Austria

Table 4.1 Comparison between single and double shell linings (after Legge 2006, from Sala 2001, used with permission from Alex Sala)

	Single shell	*Double shell*
Advantages	• Reduced lining thickness • Smaller excavated profile • Reduced total costs	• Inner lining installed well behind the face • Less effect on excavation process • Greater ability to control quality
Disadvantages	• Groundwater control required • Reduced watertightness compared with double shell linings • Groundwater is in direct contact with the permanent lining	• Blocking of waterproofing and build-up of water pressure behind lining • Location and repair of leaks when watertight membranes are used can be difficult (leaks within watertight concrete are easy to identify, i.e. they are visible, and hence easy to repair) • Additional cost of approx. 5–25%

Modern lining designs for sprayed concrete may not be finalized before construction, i.e. not 'fully engineered' at the detailed design stage, and they are refined during construction following the assessment of monitoring results (see the observational method in section 7.3.3).

Traditionally, sprayed concrete linings have been constructed using a 'double shell' or two pass lining approach. This involves a sacrificial primary lining being installed followed by a permanent secondary lining (Legge 2006). The function of the primary lining is purely to stabilize the tunnel following excavation and avoid loose material falling on the workforce; it generally has no long-term load carrying design function. The primary lining can have a long-term function if the groundwater is not aggressive to concrete and the inner lining has purely the function to keep the tunnel watertight. Double shell linings have advantages in poorer and variable ground conditions because the primary lining can be installed quickly and more time spent on creating the secondary lining. The alterative to the 'double shell' approach is the 'single shell' or one pass lining, which forms the final lining. A single shell lining is installed with the advancing face and the initial support required to stabilize the ground following excavation is an integral part of this final lining. It is therefore important that the quality of the installed lining material is higher than for the primary lining of a double shell approach. However, the single shell system can potentially reduce the overall cost and time. Table 4.1 shows some advantages and disadvantages of single and double shell systems.

It should be noted that sprayed concrete can also be used to construct shafts. Further details on sprayed concrete can be found in Thomas (2009a) and Franzen *et al.* (2001).

4.3.3 Ribbed systems

Support systems based on steel ribs have been used for many decades. This technique involves rolled steel sections being placed around the circumference of the excavated tunnel profile at specified intervals. It is inevitable that there will be gaps between the steel ribs and the ground and it is important that these gaps are suitably wedged to prevent excessive deformations. The importance of ensuring that the loads carried by the steel supports are evenly distributed around the tunnel profile is well recognized. Point loading of the steel supports significantly reduces their ultimate load bearing capacity.

It is common these days to combine steel ribs with sprayed concrete. If a layer of sprayed concrete is applied prior to erecting the steel ribs, this helps to overcome some of the problems with wedging. A subsequent layer of sprayed concrete is then applied to integrate the steel ribs into the lining and provide additional stability.

The legs of the steel arches are often set into concrete blocks to help distribute the loads into the ground and prevent settlement.

Lattice girders, rather than rolled steel sections, combined with sprayed concrete are also commonly used these days.

Figure 4.20 shows the mesh being installed ready for shotcreting, in combination with lattice girders. This two-lane road tunnel is the 2nd tube of the Katschberg Tunnel, Austria, and was constructed through hard rock (gneiss). It had a width of approximately 12 m, a length of 4300 m and a cross section of 88–111 m^2.

Figure 4.20 Mesh installation and a lattice girder arrangement

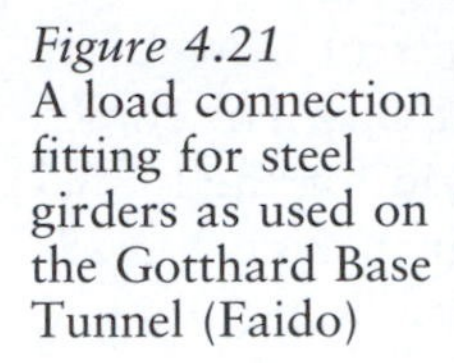

Figure 4.21
A load connection fitting for steel girders as used on the Gotthard Base Tunnel (Faido)

Figure 4.21 shows an example of a load connection fitting for steel girders.

4.3.4 Segmental linings

Segmental linings systems support the ground with a structure made up of a number of preformed interlocking structural elements. Together these elements form a continuous self supporting structure in the ground, which is most commonly circular in shape.

Although segmental linings are commonly used for soft ground conditions, the design principles are equally applicable to hard ground conditions. The permanent loading and the load developments with time are the main differences.

The design loads on segmental linings can be classed as either temporary or permanent. Temporary loads include demoulding, storage/stacking, transportation, handling, erection and grouting pressures. Permanent loads include external ground loads, external water pressure, imposed loads (traffic, adjacent foundation/pile loads), internal pressures (water pressure), external construction (adjacent tunnel construction) and flotation forces (King 2006).

Tunnels in soft ground are often designed to resist the full overburden of the ground and associated external water pressures. However, this is particularly conservative in stiff ground, such as overconsolidated clays. In overconsolidated soils it is often the ratio of horizontal to vertical stress (K_0) that is important (section 3.4), but this is difficult to assess as it is both time and construction related. It should be noted that the largest loads acting on the segments are often from the jacks moving the tunnelling shield or TBM, and hence this must be taken into account in the design.

Segmental linings may be connected together using bolts or dowels, or may have no physical connection, for example expanded linings. The choice of connections is closely related to the construction method for the tunnel (King 2006). The principle of expanded linings is shown in Figure 4.22. Figure 4.23 shows an example of an expanded segmental lining for one complete ring.

When using segmental linings with a tunnel boring machine (see section 5.5 on TBMs), the lining is erected in the tail of the TBM. An example of a segmental lining erector in the tail of an earth pressure balance machine (EPBM) (section 5.5.3.3) is shown in Figure 4.24. For a picture of a completed segmental lining, see Figure 5.31d.

Segmental tunnel linings are commonly made from unreinforced concrete, steel or fibre reinforced concrete, spheroidal graphite (cast) iron (SGI) (Figure 4.25), and steel. Table 4.2 indicates some of the main advantages and disadvantages of these different segmental lining types.

The durability of the segments is also a major design consideration. Clients these days require a 100 year design life or greater and this is in excess of commonly adopted building design codes (King 2006). It should be noted that this is the case for all tunnels regardless of their construction method. Maintaining tunnels is also costly and difficult due to limited access. It is therefore necessary to conduct a durability risk assessment to internal and external environments (see section 6.2 on risk assessment). The durability of segmental linings can be improved and the associated risks reduced in various ways. These include: avoiding cast-in metallic components in concrete linings, increasing the cover to reinforcement, minimizing the

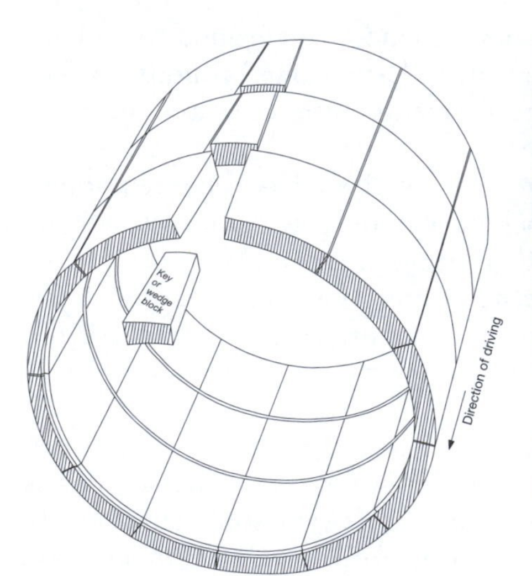

Figure 4.22
Principle of expanded wedge block segmental lining (after Whittaker and Frith 1990)

Figure 4.23
Example of an expanded segmental lining set for one complete ring

Figure 4.24 Segmental lining erector in the tail of an EPB tunnel boring machine (the photograph also shows the jacks for moving the tunnelling shield forward off the erected tunnel lining)

permeability and increasing the density (high cement contents are common in concrete lining segments), using coatings, removing reinforcement if at all possible and using unreinforced or fibre reinforced segments, and using cathodic protection (King 2006). It should be noted that steel fibres, when used as reinforcement, still have the potential to corrode.

Figure 4.25 Example of bolted SGI lining as used in London, UK (courtesy of ALPINE BeMo Tunnelling GmbH Innsbruck)

Fibre reinforced concrete segments were used on the Thames Tunnels as part of the Channel Tunnel Rail Link in the UK. Burgess and Davies (2007) state that the fibre-reinforced pre cast segments in this case made the manufacturing process easier as the space normally required for erecting the steel reinforcement cages was not required. The manufacturing process also employed steam curing which meant that the segments could be de-moulded after only 4 hours. Due to the 4.5 bar water pressure expected within the highly fissure chalk on this project, each segment had two rows of sealing

Table 4.2 Comparison of different materials used for segmental linings (after King 2006, used with permission)

Lining material	*Advantages*	*Disadvantages*
Unreinforced concrete	• Inexpensive compared to reinforced sections • Readily available • No corrosion concerns	• Low bending strengths • Low tensile bursting resistance • Low shear strength • Lead time for mould manufacture
Steel reinforced concrete (RC)	• High bending and tensile bursting resistance • High shear resistance • Readily available	• Expense of supply and fabrication of reinforcement cages • Corrosion considerations • Lead time for mould manufacture
Fibre reinforced concrete	• No major corrosion concerns • Moderate strengths • More ductile than unreinforced concrete	• Relatively new in tunnels and there have been difficulties getting the necessary approvals • Lower tensile and flexural capacity than RC • Lead time for mould manufacture
Spheroidal graphite cast iron (SGI)	• High tensile and compressive strength • Very high tolerance control • Lighter than equivalent concrete sections	• Expensive • Lead times for pattern manufacture • Repainting may be a possible maintenance requirement
Steel	• High tensile and compressive strength • Very high tolerance control • Lighter than equivalent concrete sections • Fabrication time shorter than SGI	• Expensive • Corrosion (repainting may be a possible maintenance requirement) • Mass production slower than SGI and tolerance control more labour intensive (except pressed steel – used for temporary works)

gaskets, one synthetic and one hydrophilic (see below). The tunnel boring machines used for the tunnelling on this project were slurry machines designed to be used in 'mix-shield' mode, whereby the face would be supported by a combination of pressurized slurry and a balancing air bubble. This had two advantages for the tunnel lining segments. The first was that the bentonite in the area behind the cutterhead (plenum) of the machine helped to reduce the jacking pressures on the segments and resulted in very little jacking damage. The other advantage was that the cutterhead torque was greatly reduced enabling a better control of the roll of the rings, thus avoiding any shearing failure of the bolts in the previously constructed ring.

WATERPROOFING

Waterproofing is important in tunnel lining construction to prevent excessive water flow into the tunnel. This is a particular problem if the tunnel is constructed below the groundwater table where it can act like a drain. Waterproofing of segmental linings has traditionally been by the use of caulking (applying a sealing material to the inside of the lining at the joints), but these days is generally achieved by the use of preformed gaskets. There are two basic forms (BTS/ICE 2004, King 2006):

- *compression seals* – these are manufactured from man-made rubbers (ethylene-propylene-diene monomer (EPDM) or neoprene) and are fitted around individual precast concrete or SGI segments;
- *hydrophilic seals* – these are made from specially impregnated rubbers or specially formulated bentonite compounds that swell on contact with water.

These waterproofing systems are not used for waterproofing the segments themselves, but to prevent water from penetrating between adjacent segments. The gaskets require a compression force to be applied to the lining as it is erected (compression seals more so than hydrophilic seals), which creates a line load on the segment that needs to be considered in the design.

TOLERANCES OF SEGMENTS

This needs careful consideration as they have practical implications for the constructability of the ring and performance of the gaskets. Herrenknecht and Bäppler (2003) recommend the following dimensions/tolerances:

- segment width ±0.6 mm;
- segment thickness ±3.0 mm;
- segment length ±0.8 mm;
- longitudinal joint evenness ±0.5 mm;
- ring joint evenness ±0.5 mm;
- cross-setting angle in longitudinal joints ±0.04°;
- angles of the longitudinal joint taper ±0.01°.

SEGMENTAL LINING RINGS USED FOR SHAFT CONSTRUCTION

Shafts can also be constructed using segmental linings. One method of construction commonly employed, if the ground can remain unsupported for a suitable amount of time, utilizes an 'underpinning' technique (Figure 4.26a). This involves excavation starting at the ground surface by an excavator (depending on the diameter of the shaft) lowered into the construction area (Figure 4.26b). Trimming of the sides can be carried out

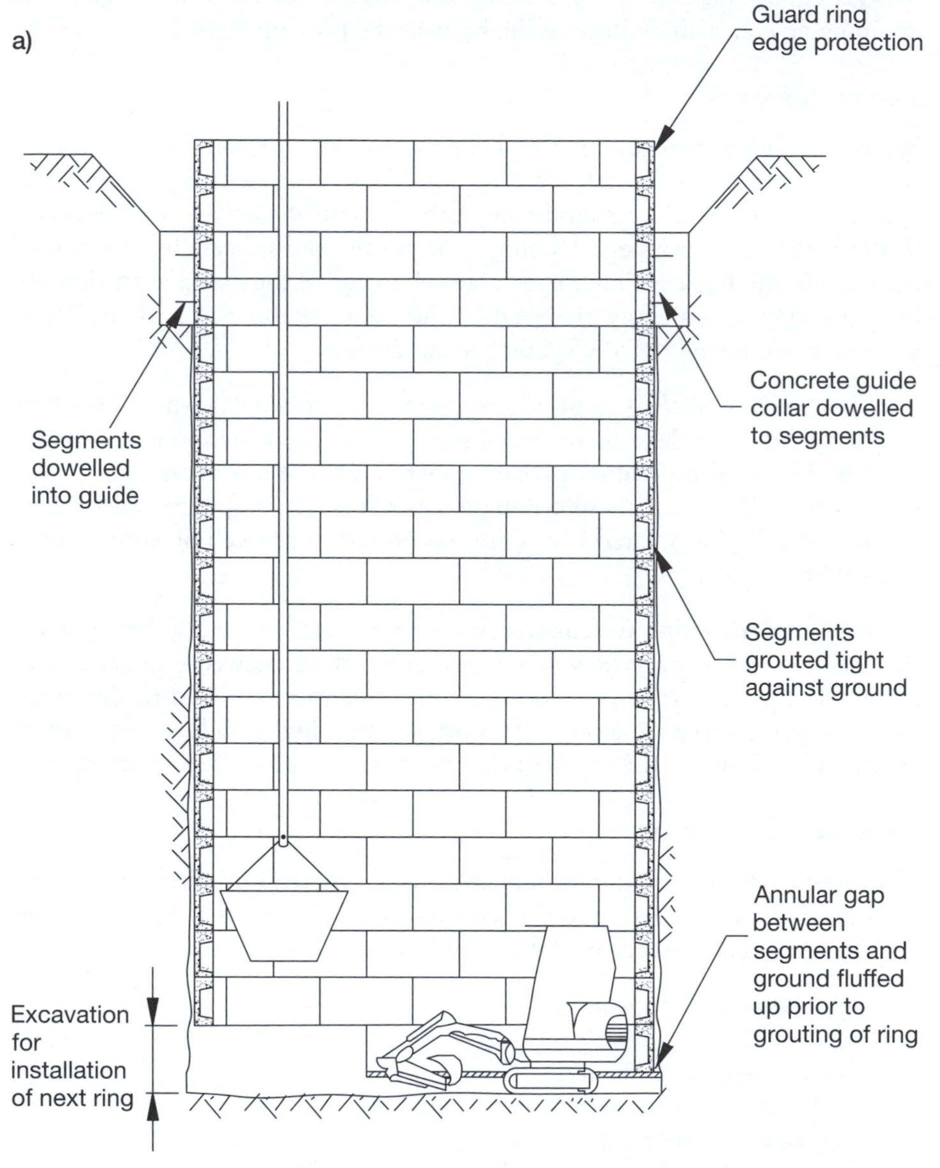

Figure 4.26 a) 'Underpinning' technique for constructing shafts (after BTS/ICE 2004, used with permission from Thomas Telford Ltd)

manually by using, for example, handheld clayspades (noting the health and safety issues associated with these devices, i.e. hand-arm vibration syndrome, see section 6.1 on health and safety). The excavation continues until a level is reached whereby a complete ring of segments can be installed, and the gap behind the lining is grouted immediately (Figure 4.26c). It should be noted that these segments may need to be installed by hand and so need to be of a manageable size and weight. The excavation is then

Figure 4.26 (continued) An example of the 'underpinning' technique used to construct a shaft, b) shows the excavation process, and c) shows the installation of the concrete segments

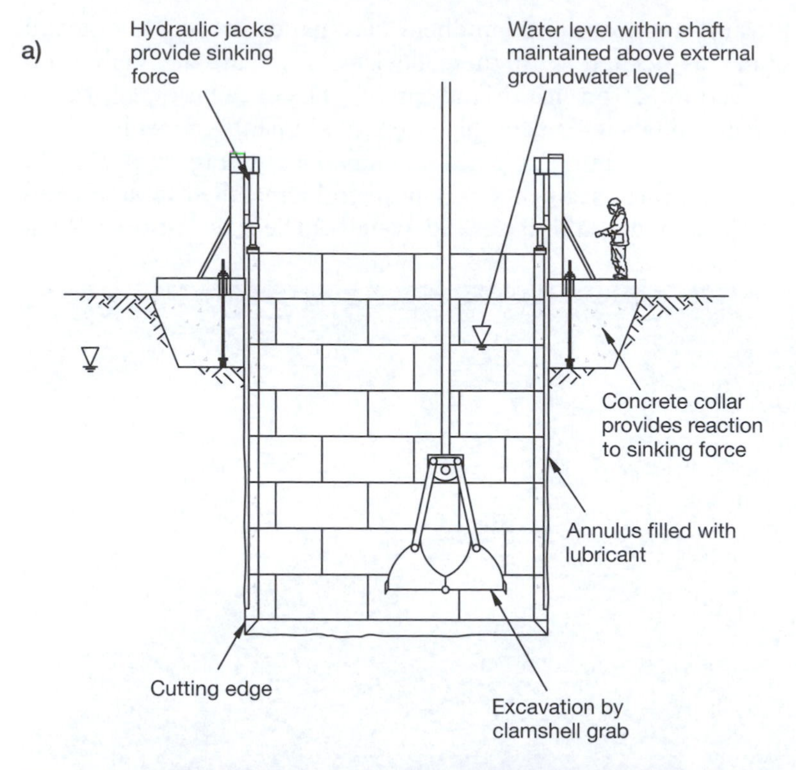

Figure 4.27 a) Caisson-sinking method to construct shafts (after BTS/ICE 2004, used with permission from Thomas Telford Ltd), b) shows the construction of a reception shaft for a pipe jacking operation as part of the Terminal 5 project at Heathrow, UK

continued until the next complete ring can be installed below the previous one. Further details of a deep shaft construction in London where this technique was used and the associated issues involved are discussed in Morrison *et al.* (2004).

An alternative technique which uses segmental linings is called the caisson-sinking method (Figure 4.27). This is generally used in ground where the stand-up time is poor or where base stability is of concern due to water pressure (BTS/ICE 2004). This technique uses a jacking process whereby a segmental lining is erected at the ground surface and then sunk using hydraulic jacks or weights (at the ground surface) to assist the self-weight of the shaft in order to overcome the ground friction. The excavation of the shaft may be conducted using an excavator within the shaft if the excavation is dry or from the ground surface using a grab on a crane in wet conditions. The friction on the outside of the shaft is reduced by using lubrication. Maintaining verticality of the shaft is critical during the sinking process as a small deviation from the vertical can cause an onerous loading condition.

4.3.5 **In situ** *concrete linings*

In situ concrete linings can either be used in self supporting ground (for example rock) as the main lining, or as a second-stage lining where a temporary support system has already been placed during the excavation process (for example steel arches, sprayed concrete or rock bolting). In both cases the lining is cast *in situ* using a formwork system, which provides a gap between the ground, or initial support system, and the formwork into which wet concrete is placed (Figure 4.28a–c). The concrete lining can either be plain or reinforced. Once the concrete has reached a suitable early stage strength the formwork is 'struck', i.e. removed. In 'wet' ground, either a waterproof membrane is used between the initial support system and the cast *in situ* lining (Figure 4.29), or a watertight cast *in situ* lining can be used.

These linings are often used with 'system formwork' where travelling steel or wood forms are advanced, often as separate 'invert' and 'arch' forms, in tunnels with a suitable length of regular cross section and where the operation can be developed around a 24 hour cycle. Although expensive, 'system formwork' can become economically viable with extended use (Winter 2006). One benefit of using formwork is that it can be built as required to any shape and it is therefore highly adaptable. Thus it can be used in tunnels where there are junctions or tunnels of varying cross section.

Figures 4.28 and 4.29 show examples of formwork and falsework for the inner lining construction as used on the Heidkopf Tunnel, Germany (see Figure 4.9 for details of this tunnel).

These techniques can also be used for the construction of shafts. For example, a variation of the system formwork is 'slip form' lining system. In this case, the shaft is constructed and supported with the lining

Figure 4.28 Construction of the inner lining, Heidkopf Tunnel, Germany, a) and b) setting up the shuttering at the tunnel portal, c) shuttering in the main tunnel

Figure 4.29 Inner lining construction, including waterproofing as used for the emergency cross-passage, Heidkopf Tunnel, Germany (courtesy of ALPINE BeMo Tunnelling GmbH Innsbruck)

construction starting at the bottom of the shaft and the formwork moving up continuously as the lining is cast *in situ*. The slip form technique can also be used in the tail of a tunnel boring machine. As the machine moves forward the concrete tunnel lining is formed continuously *in situ* behind the machine.

4.3.6 Fire resistance of concrete linings

Fire resistance of concrete linings is an important design criterion and the relevant standard for the design of structural linings should be used. In the UK the standard is Eurocode 2, BS EN 1992–01–02:2004 (BSI 2004b). In this standard there are at least three methods to determine the fire resistance of reinforced concrete members, including the '500 degree isotherm' method and the 'zone' method.

In all these methods the size and shape of the element together with the minimum thickness and cover to reinforcement influence the fire resistance. Allowance is also made for the moisture content of the concrete, the type of concrete and the aggregate used and whether any protection is provided (BTS/ICE 2004).

There are two basic options for fire protection of linings, either external or internal protection. External protection can be provided for relatively low temperature fires by the application of boarding or sprayed-applied coatings. Internal protection can be provided by adding polypropylene fibres to the concrete mix. In this case the polypropylene fibres melt and the resulting capillaries provide an escape path for moisture in the concrete, which can help to reduce spalling (Thomas 2009a). Further details on the fire resistance of concrete linings can be found in BTS/ICE (2004).

The importance of fire resistance was highlighted by the major lorry fire in the Mont Blanc road tunnel through the Alps in 1999. Although the fire caused significant damage to the tunnel, the immediate stability of the tunnel during the fire, which reached temperatures of up to 1000 °C, was not affected. In addition, three fires have so far occurred in the Channel Tunnel linking France and the UK in 1996, 2006 and 2008. All of these fires were caused by lorries catching fire on the heavy goods vehicle trains. All three caused damage to the tunnel lining meaning that repair and replacement of the damaged sections were required. The 2008 fire damaged 650 m of tunnel lining and cost approximately €60M to repair.

5 Tunnel construction techniques

5.1 Introduction

Tunnels can be constructed by using a number of different techniques. This chapter discusses these techniques and highlights the applications and limitations of the different construction methods. It should not be read in isolation, but is integrally linked with the support and ground improvement methods described in Chapter 4. Further, in order to choose the most appropriate construction technique, aspects such as the ground characteristics, the impact of the tunnel construction on the surroundings, as well as the economic and health and safety issues need to be considered, i.e. the construction technique(s) depends on many factors all of which are project specific. A useful analysis was carried out by Thuro and Plinniger (2003) to estimate the performance of different tunnel construction techniques (TBM, roadheader and drill and blast).

Although there is a difference between soft ground and hard ground (rock), tunnelling techniques are being used in a wider range of ground conditions and this boundary is becoming increasingly blurred. However, there are many examples in engineering literature where a distinction is made between soft and hard ground and hence in this book, for convenience, this distinction is continued (see section 1.4). It should be noted that from a practical point of view, if the ground is stable, the construction of the tunnel can focus on economics and be driven by the limits of the tunnelling equipment. For soft ground, which needs immediate support, construction is driven by the need to support the ground immediately after creation of the void.

With most tunnelling methods the ground is mined in 'advances' (even when tunnelling with shields and tunnel boring machines, TBMs), i.e. a tunnel section is created as part of the total tunnel void in a single cycle. The section length, which is approximately 0.8 to 4.0 m, depends, amongst other things, on the stand-up time of the ground. The stand-up time being the time the void remains stable without any support. Soft ground only has a very short, or no, stand-up time. It is therefore necessary to support the void immediately. During the time that is needed for the installation of any support method, or until it takes effect, the construction method has to ensure the stability of the created void. This can be achieved using the techniques described in section 4.2, or by using a tunnel shield as described later in this chapter. It should be noted that the total cross section of the tunnel is not necessarily excavated in a single advance, but staged advances are common, especially when using sprayed concrete lining, as described in section 4.3.2.

5.2 Open face construction without a shield

5.2.1 Timber heading

One of the oldest methods of tunnel construction involves the use of timbering to provide the temporary support for the ground during tunnelling and is still used extensively in all underground works (Figure 5.1). In particular, this technique is very useful in soft ground (ground that has some stand-up time, such as a stiff clay) for constructing short sections of relatively small diameter access tunnels as part of metro station upgrading projects, for example passenger access tunnels between platforms, and storage or plant rooms. In the case of these one-off excavations it is often not worthwhile employing such techniques as sprayed concrete or tunnelling machines.

5.2.2 Open face tunnelling with alternative linings

In stiff clays, such as London Clay in the UK, where the ground has some stand-up time, open face tunnelling can be adopted in association with segmental linings, such as SGI segmental linings. Figure 5.2 shows a pilot bore and enlargement tunnel construction using SGI lining segments in

Figure 5.1 Timber heading as used in London Clay for constructing a connecting passage between two other tunnels (approximately 30–80 m^2), King's Cross Station upgrade, London (construction period 2002–2008) (courtesy of ALPINE BeMo Tunnelling GmbH Innsbruck)

Figure 5.2 Construction of an escalator tunnel at The Angel Islington Station, London Underground, UK (completed in 1990). The tunnel was constructed at an angle of 30 ° to the horizontal and used a pilot bore (3.7 m in diameter), which was then enlarged to the final diameter of 7.5 m (courtesy and copyright London Underground)

London Clay (lining systems are described in section 4.3). Pilot bores can provide increased stability to the face and also offer an opportunity to verify the ground conditions along the tunnel route. The timber boards shown in Figure 5.2 are used to help maintain face stability. This construction method can be adapted for other lining techniques such as sprayed concrete lining methods as described in section 4.3.2.

5.3 Partial face boring machine (roadheader)

As the name suggests, a partial face boring machine works on discrete areas of the face rather than excavating the whole face in one go. In a similar manner to a full face machine the cutterhead uses a rotating motion and is pressed against the face (Figure 5.3a). Depending on the direction of the rotation of the cutterhead, it is possible to differentiate between a partial cutting machine with a cutterhead which rotates along the axis of the cutter machine boom (axial) and one in which the cutter rotates at right angles to the cutter machine boom (transverse). Figure 5.3b shows the different rotation options for the cutterhead.

A partial cutting machine with an axial cutterhead has to be heavier for the same power of the head motor in order to absorb the reaction forces of the forward drive compared to one with a transverse cutterhead. However, it is generally easier for these machines to produce a smooth circumference

a) Schematic of a partial face boring machine

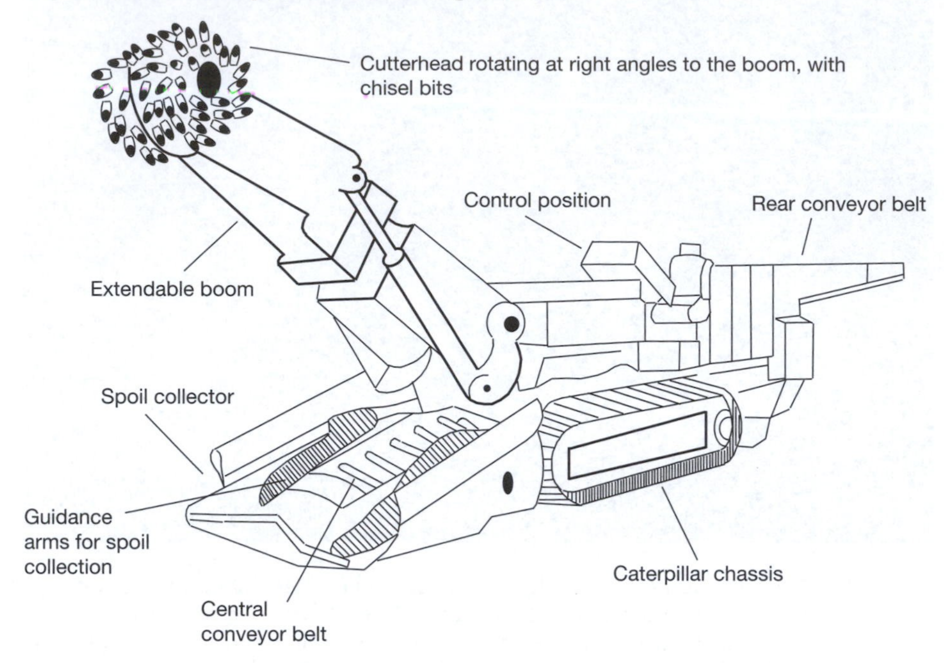

b) Advancement method of a partial face boring machine with:

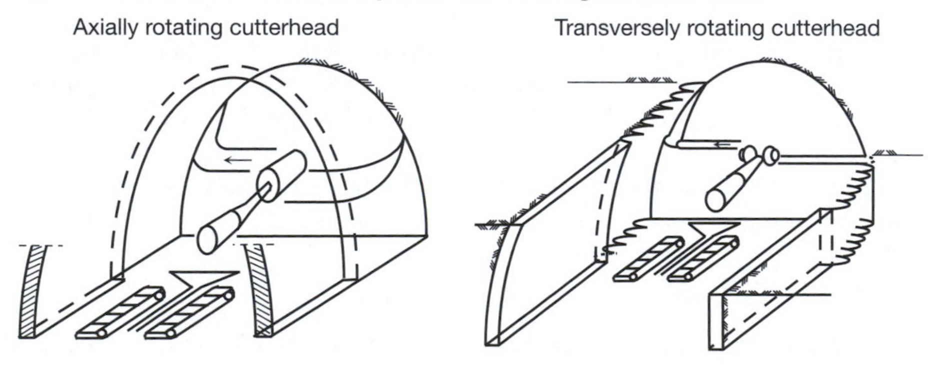

Figure 5.3 Partial face boring machine (roadheader)

to the excavated area from a single standing position. For a partial cutting machine with a transverse cutterhead, the smoothing of the tunnel wall is more time consuming, as the tunnel wall receives a 'wavy' pattern due to the shape of the cutterhead (Figure 5.3b). In this case the machine either has to change position, or the cutter arm has to be moved hydraulically in the longitudinal direction so that the 'wave crests' can be smoothed in the second cutting pass.

Apart from the possibility of the boom being extendable (not essential), it should also be able to move horizontally and be hinged. The cutter arm is often mounted on a caterpillar chassis. In ground with only a short stand-up time the boom can also be connected to a shield frame (section 5.4).

The cutterhead is covered with cutting tools, for example chisels or button bits, which crush the ground. The dust that is produced requires extensive ventilation and dust extraction. However, it is impossible to completely capture all the dust produced. For this reason using a partial tunnelling machine in a ground containing quartz is generally not possible as quartz dust can cause cancer. Another limiting factor for the use of a partial cutting machine is a uniaxial compression strength of the ground of approximately 100 MPa.

As only part of the tunnel face is worked on at any one time, tunnelling using a partial tunnelling machine only produces low levels of vibration. In contrast to a full face machine, a partial cutting face machine is more adaptable to local geology, and offers the possibility of immediately installing support in weaker rock zones. In addition, a partial cutting machine is more versatile with respect to the possible geometry of the tunnel as it can be easily used again on different construction sections or contracts. Therefore it is not necessary to write off the machine at the end of a contract.

Difficulties can occur when cutting below the chassis level of the machine as this is where the spoil removal system is located. Therefore limitations do exist with respect to the geometry, for example for profiles with a high curvature at the invert. In general, the spoil is picked up at the invert using conveyor belts and transported behind the machine where it is passed on to mucking trucks.

An example of a partial face boring machine with an axial cutterhead is shown in Figure 5.4.

Figure 5.4 Partial face boring machine with an axial cutterhead (roadheader) (courtesy of Aker Wirth)

5.4 Tunnelling shields

In its simplest form a tunnelling shield is a steel (or concrete: see jacked box tunnels in section 5.10) frame with a cutting edge on the forward face. For circular tunnels this is usually a circular steel shell under the protection of which the ground is excavated and the tunnel support is erected. A shield also includes back-up infrastructure to erect the tunnel support (lining) and to remove the excavated spoil. Figure 5.5 shows a schematic of a tunnelling shield, in this case with mechanical excavation at the face and a segmental tunnel lining for supporting the ground erected in the shield tail. It should be noted that the principles of Brunel's original shield used for constructing the London Thames Tunnel in 1825 still apply today (see section 1.3).

There are two main types of tunnelling shield, one with partial and one with full face excavation. Depending on the stand-up time of the ground and water flow into the tunnel, different face support techniques exist. Table 5.1 shows the possible combinations of excavation and face support techniques. Due to their complexity, shields with full face excavation, i.e. tunnel boring machines, TBMs, are covered separately in section 5.5.

The support usually adopted with shield tunnelling these days is circular segments. These segments form, when connected together, a closed support ring (see section 4.3.4 on segmental tunnel linings). As the tunnel segments are connected together inside the shield tail, the diameter of the completed tunnel segment ring is smaller than that of the shield, as shown in Figure 5.5. This creates a gap between the ground and the tunnel lining. When the shield is jacked further into the ground the size of this gap is between approximately 50 and 250 mm. In less supporting soft ground it has to be expected that the ground settles by this value. This can result in the softening of the ground and, especially with shallow tunnels, in the settlements reaching the ground surface and having undesirable consequences on surface or near surface structures (this is discussed further in sections 7.1 and 7.2

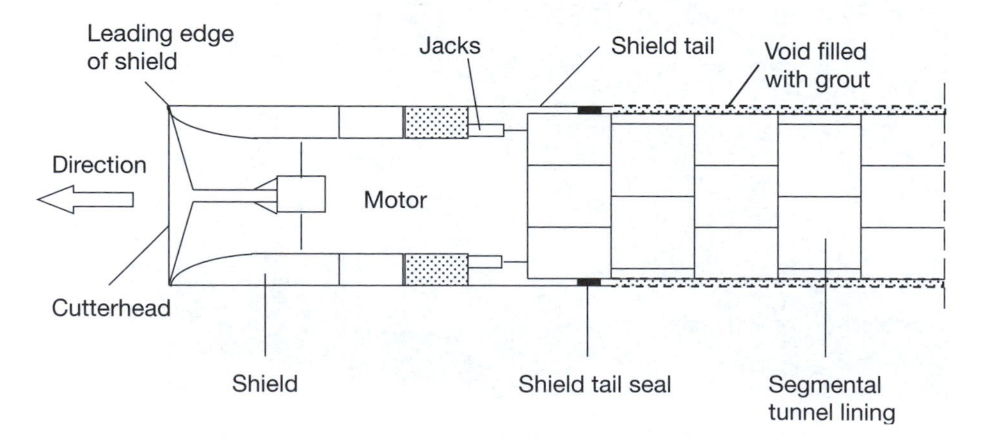

Figure 5.5 Circular tunnel shield with segmental linings

Table 5.1 Possible excavation and face support methods for shields (after Stein 2005)

Shield with partial excavation (described later in this section)	Working face with natural support	Completely unsupported open face
	Working face with partial support	Face divided into sections to aid stability using horizontal benches and/or breast plates
	Working face with compressed air support	Face is pressurized with compressed air to help prevent water ingress and aid stability
Shield with full face excavation (Tunnel boring machines, TBMs, see section 5.5)	Working face with natural support	Very open cutterhead
	Working face with mechanical support	Cutterhead has limited openings (can be closed by face-breasting plates)
	Working face with compressed air support	Face is pressurized with compressed air to help prevent water ingress and aid stability
	Working face with fluid support	Slurry shield machines
	Working face with earth pressure support	Earth pressure balance machines

respectively). In order to avoid these settlements, the gap is generally injected with mortar. At the same time the injection results in a direct bond between the lining and the ground. There are two different methods of injecting the mortar: through the tunnel lining and through the tail of the tunnel shield. The supply of the injection lines through the tail of the tunnel shield has, in contrast to the supply of the injection mortar through the tunnel lining, the advantage that the ring gap can be injected without any time delay and is continuous over the whole circumference. In the case of a supply through the tunnel lining segments there is always a delay until the shield has moved forwards and the injection sockets are outside the shield tail.

Tunnelling shields do not have an 'engine' to propel themselves forwards, but push themselves forward using hydraulic jacks. In order to create the necessary force to push the tunnel shield forwards, jacks are placed around the circumference of the shield. These jacks push against the last erected tunnel segment ring and also push the shield against the tunnel face in the direction of the tunnel construction (Figure 4.24). The tunnel segments transfer the jacking forces into the ground using friction. Of course this principle does not work at the start of the tunnel construction and therefore in the starting shaft a reaction frame is necessary to take the jacking forces (Figure 5.11). The jacks can be operated individually or in groups, allowing the shield to be steered in order to make adjustments in line and level and to be driven in a curve if required. When the shield has advanced by the width of a tunnel segment ring, the jacks are retracted leaving enough room in the tail of the shield to erect the next tunnel segment ring.

There are many possible ways of excavating the ground at the face within the shield. Manual excavation, i.e. by 'hand', is only considered for very special applications, e.g. very short advances, due to the low advance rate. This type of tunnelling is called the manual shield technique. Figure 5.6a shows a manual shield as used on a pipe jacking project (see section 5.11 for a description of pipe jacking) in a predominantly stiff clay stratum and Figure 5.6b shows an example of a 9.17 m diameter open face shield used after the tunnel collapse at Heathrow, UK in 1994, where excavation on multiple levels was required. The top two levels were excavated manually and the lower level used mechanical excavation.

As shown in Figure 5.6b, shields can be split into vertical and horizontal compartments (note the similarities between this shield and Brunel's Thames tunnel shield described in section 1.3). This shield indicates the use of working platforms to aid excavation and stability of the face. Open shields can also have face-breasting plates attached to these working platforms (see Figure 5.10) to aid stability. The theory is that each area between the working platforms can be excavated separately, and safely, as the ground will form a 'plug' in the opening, and hence form a stable face. The stability of the ground, i.e. whether it collapses into the tunnel shield or not, is based on its shear strength. The size of the openings is therefore designed with this in mind, i.e. in weaker, less stable ground the benches must be

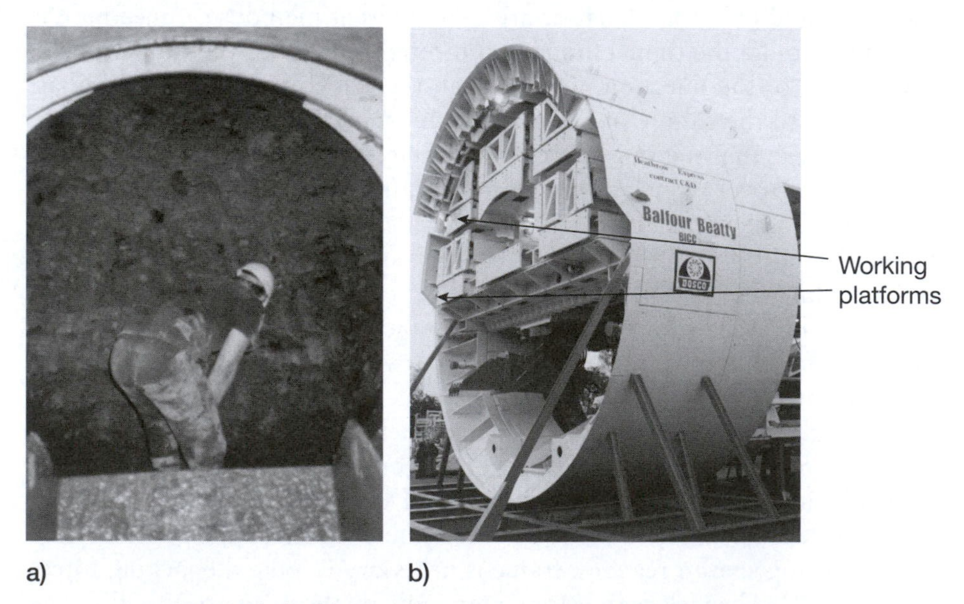

Figure 5.6 a) Manual shield tunnelling as part of a pipe jacking operation, and b) a 9.17 m diameter open shield (courtesy of Dosco Overseas Engineering Ltd)

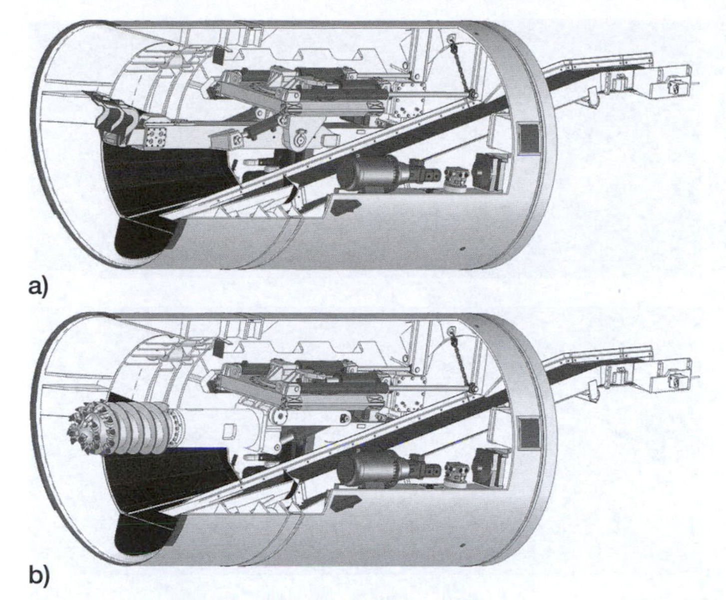

Figure 5.7 Diagrams showing boom-in-shield tunnelling machines, a) boom with a shovel bucket and b) boom with an axial cutterhead (courtesy of Herrenknecht)

closer together. The use of handheld pneumatic excavation tools and low pressure compressed air is limited by current health and safety regulations, for example due to the danger of hand-arm vibration syndrome.

In partially-mechanized shields, an excavator or a partial cutterhead works on the face (for partial face boring machines (roadheaders) see section 5.3). Figure 5.7 shows diagrams of two boom-in-shield machines, Figure 5.7a shows a boom with a shovel bucket to excavate the face and Figure 5.7b shows a boom cutting machine with an axial cutterhead. These are typical techniques used for shields with partial excavation.

Partially-mechanized shields are likely to continue to find a place in the tunnelling industry, particularly at station tunnels or tunnels where the cost of full face tunnel boring machines cannot be justified (Court 2006). If a full face cutter wheel is used for excavating the ground, then this is classed as a tunnel boring machine (TBM) and these are described in the section 5.5.

EXAMPLES OF SHIELDS WITH PARTIAL EXCAVATION

After showing the principles of shields with partial excavation, the following section focuses on some examples in order for the reader to gain an appreciation of the different machines used, including some of their technical specifications.

Figure 5.8 shows an example of a boom-in-shield tunnelling machine as used for constructing the Piccadilly Line Extension (PiccEx) Tunnel to Heathrow Terminal 5 in the UK. The shield had an external diameter of 4.81 m and an overall length of 4.57 m, with the overall length including back-up being 56.5 m. There were 16 rams to provide the thrust to

Figure 5.8 Refurbished Dosco boom-in-shield tunnellling machine as used for constructing the Piccadilly Line Extension (PiccEx) Tunnel to Heathrow Terminal 5 in the UK

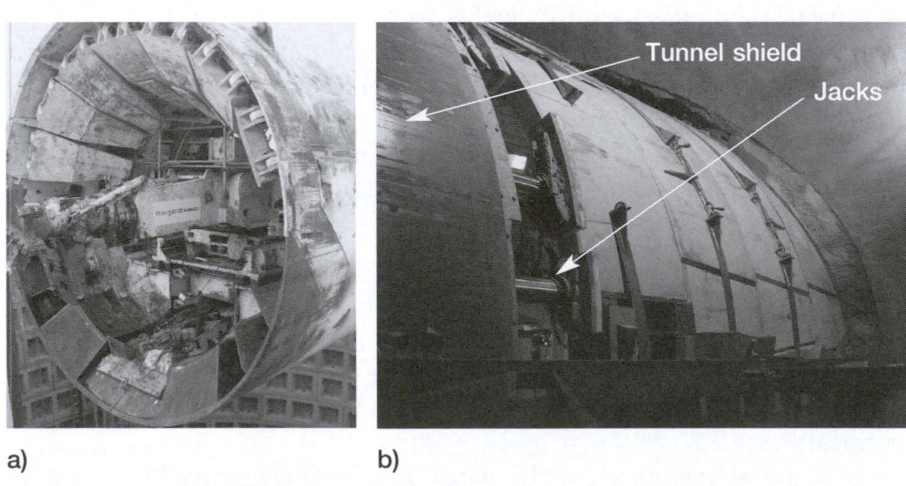

Figure 5.9 a) The boom-in-shield tunnelling machine on removal after completion of the Heathrow Express Extension Tunnel at Terminal 5 in the UK (Manufacturer: Dosco), b) the same shield crossing an existing emergengy exit where external strapping was required to support the erected lining

the shield and the axial cutterhead on the boom was rotated at either 27 or 50 rpm. The PiccEx Tunnel consisted of approximately 1.6 km long twin tunnels of 4.5 m internal diameter. Spoil was removed by belt conveyor and a segmental lining was used. This machine achieved advance rates averaging 26.9 m per day (1st drive Eastbound) and 18.7 m per day (2nd drive Westbound) through a very consistent clay (London Clay) (Williams 2008). (Further details of this project can be found in section 8.2.)

Figure 5.9 shows the boom-in-shield tunnelling machine on removal after completing the Heathrow Express Extension Tunnel at Heathrow Airport Terminal 5 in the UK. This shield had hydraulically operated face-breasting plates at the crown of the shield that could be used to increase the face stability if required. This shield had an external diameter of 6.125 m and an overall length including back-up of 42.8 m. The Heathrow Express Extension Tunnel consisted of approximately 1.7 km long twin bore tunnels of internal diameter 5.675 m (Williams 2008). An interesting aspect of this project was that the shield had to cross an emergency exit area built in advance of the shield tunnel (Figure 5.9b). The distance the shield had to cross was approximately 8 m. The emergency exit was constructed in shotcrete and openings with reduced support were left to help the shield break through the sprayed concrete. Figure 5.9b shows the rear of the shield and the jacks pushing off the segmental linings. However, the segmental linings, although installed in the usual manner, had no support from the surrounding ground and had to be stabilized using steel straps.

An example of a larger diameter boom-in-shield tunnelling machine is shown in Figure 5.10 and illustrates the use of working platforms and face-breasting plates to aid face stability. This 11.77 m diameter shield was used

Figure 5.10 Larger diameter (11.77 m) boom-in-shield tunnelling machine as used on the Dublin Port Tunnel, Ireland (courtesy of Herrenknecht)

as part of the Dublin Port Tunnel, Ireland, and was used to excavate two relatively short lenghs of tunnel, approximatley 300 m long, through glacial tills. Excavation was carried out using booms with shovel buckets at three levels within the shield. An auger mechanism was used to remove the excavated spoil from the shield.

5.5 Tunnel boring machines

5.5.1 Introduction

As mentioned in the previous section, tunnelling shields can be used with partial excavation as well as with full face excavation. The latter is known as a tunnel boring machine or TBM. TBMs exist in many different diameters, ranging from microtunnel boring machines with diameters smaller than 1 m (see section 5.11) to machines for large tunnels, whose diameters are greater than 15 m (the largest TBMs are now in excess of 19 m). TBMs are available for many different geological conditions. This means that, for example, the type of tunnel face support required and the excavation procedure as well as numerous other technical requirements can be solved in many different ways. Every tunnel is different and hence there are often frequent technical advancements in this field. The following sections give an overview of the essential characteristics of TBMs. It should be realized, however, that hardly any of the following information can be assumed to have general validity as there are nearly always exceptions and special cases. Although TBMs are often designed for specific projects, i.e. with a specific diameter and in order to cope with certain ground conditions, these days refurbished machines are becoming more common and projects are actually designed around the machines available. An example of this is when the diameter of the new project is chosen to suit the old machine, with just the cutterhead being redesigned for the specific ground conditions expected.

Most tunnel boring machines, however, do have a 'shield' in common as described in section 5.4. An exception to this is the 'Gripper' type TBM for hard rock and this is discussed in section 5.5.2.1. Although it should be noted that even a Gripper TBM can have a small shield around the cutter wheel to avoid it catching on any rock as the ground deforms under high pressures.

One of the general requirements for the use of a TBM is a consistent geology along the route of the tunnel as the different cutting tools are only suitable for a small variation in material characteristics. A universal machine for all types of ground and soil conditions does not exist (although TBMs with multiple modes of operation such as Mixshields are being developed, see section 5.5.3.4). The combination of different cutting tools on the cutterhead can increase the application of machines to a greater range of ground conditions.

Figure 5.11 Example of a reaction frame used to start off a TBM

Although TBMs can have different mechanisms for moving through the ground, most have to start outside and hence need a reaction frame to start the drive. An example of a reaction frame is shown in Figure 5.11.

In addition, all TBMs using segmental lining for the tunnel support need to have a tail seal at the rear end of the shield. As the tunnel segments are erected within the tunnel shield, there is a gap between the segments and the excavated ground. In order to achieve a rigid connection between the ground and the tunnel lining, thus preventing the ground from moving, the gap is injected with cement slurry. The challenge then is to keep the groundwater, soil and cement slurry out of the tunnel shield. This is achieved by using tail seals. The following points present three examples of construction options for this seal, all of which are still in use.

SHIELD TAIL SEAL USING METAL SHEETS

Early shield tail seal designs consisted of a sprung sheet metal that, due to its sprung mechanism, lay on top of the tunnel segment (Figure 5.12a). This type of seal is problematic as the sheet metal seal can rip off when the cement slurry has hardened. Furthermore, seal problems occur when the outer edge of neighbouring tunnel segments are not flush, but have a step.

SHIELD TAIL SEAL USING A NEOPRENE TUBE

Shield tail seals made from neoprene (Figure 5.12b) are definitely a better solution compared to the sheet metal seal as they are much more flexible.

Furthermore, emergency seals can be added without any problems. The seals cannot be pulled off even when the cement slurry hardens after a break in the tunnel advance.

SHIELD TAIL SEAL MADE FROM STEEL BRUSHES

Currently, seals made of steel brushes are the state-of-the-art (Figure 5.12c). These are positioned in several consecutive rows. The gaps between the brushes are filled with proprietary tail-seal grease (TSG) to seal against water and annulus grout ingress (BASF 2009b). The brush seals are very flexible and can cope with unevenness, especially at the tunnel segment edges.

5.5.2 Tunnel boring machines in hard rock

There are various mechanized full face tunnelling techniques for hard rock and these depend on the quality of the ground. However, the final decision of which machine type to use always depends on controlling the stability of the ground during construction and the expected quantity of water ingress. It is these factors which dictate the final choice of TBM to be used in hard rock, either a Gripper TBM or a shielded TBM. These machines are discussed briefly in the following sections, although further details of hard rock TBMs can be found in Maidl *et al.* (2008).

Hard rock TBMs comprise four key sections, which make up the complete machine (Maidl *et al.* 2008). These are the *boring section*, consisting of the cutterhead, the *thrust and clamping section*, which is responsible for advancing the machine, the *muck removal section*, which takes care of collecting and removing the excavated material, and the *support section*, where the tunnel support is erected.

5.5.2.1 Gripper tunnel boring machine

In stable rock conditions with low water ingress the gripper technique can be utilized. Figure 5.13 shows the principle of the gripper machine technique. The gripper machine locks itself against the ground laterally using ‘gripper shoes’ to establish the required face pressure (Figure 5.14). When the machine is locked, the tunnel is advanced by using hydraulic jacks which move the cutterhead forwards by approximately 0.7 to 1.2 m. Thereafter the cutterhead stops. In order for the tail of the machine to move forwards, auxillary supports are erected behind the cutterhead and at the rear of the machine and the gripper shoes are released. Now the hydraulic jacks that advanced the cutterhead can be retracted while the tail of the shield is pulled forwards. A new working cycle can begin when the gripper shoes of the machine are once more engaged.

Figure 5.15 shows an example of the functions associated with a Gripper tunnel boring machine (TBM) and the lengths for each element, which together make the total length of these machines over 200 m.

Unlike shielded TBMs, where tunnel support, e.g. segmental lining, is fixed and does not change during tunnel construction, the tunnel support system, when using a Gripper TBM, can vary depending on the ground quality. The appropriate rock support devices can be installed immediately behind the cutterhead. These devices can include anchors, steel arches and

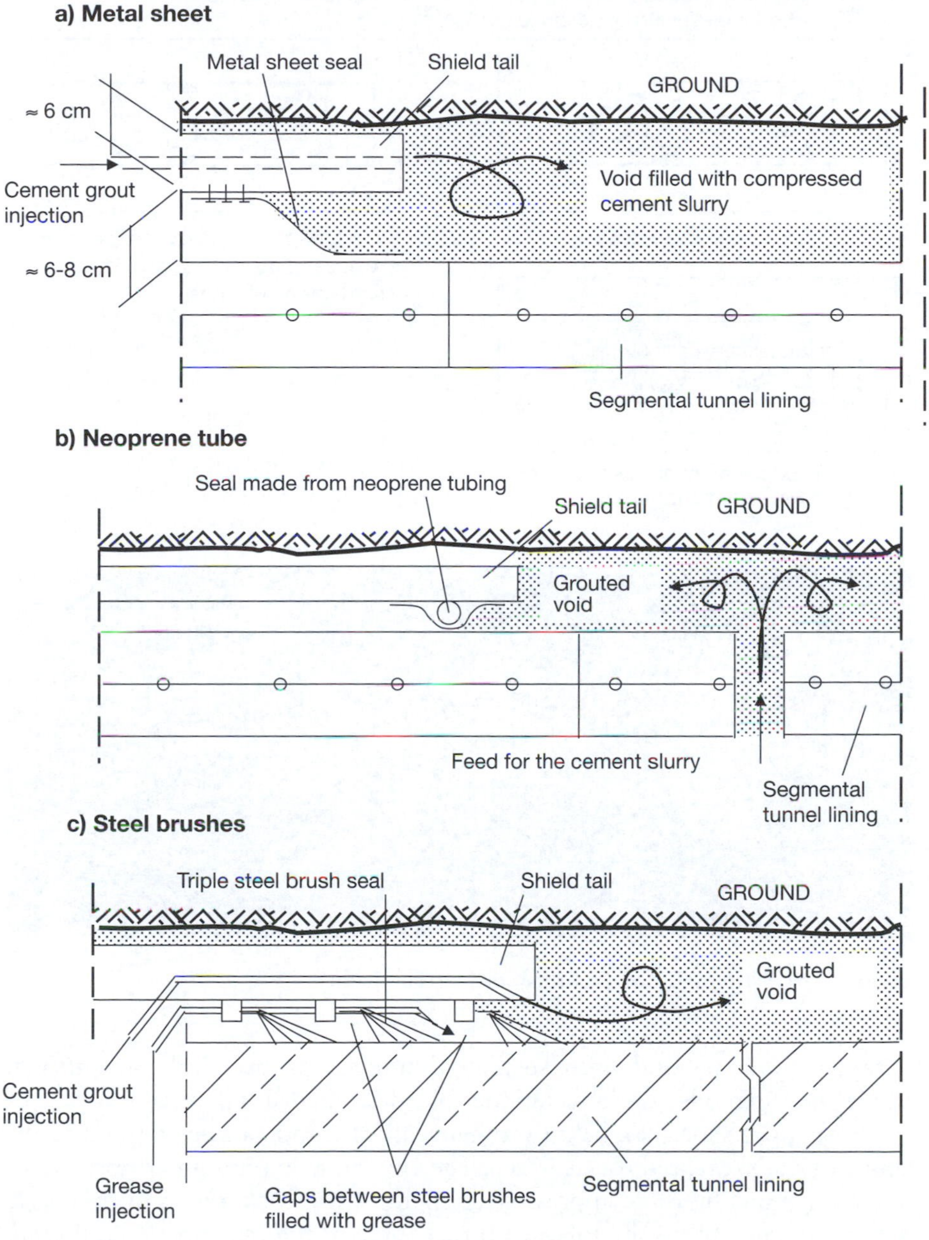

a) Metal sheet

Metal sheet seal
Shield tail
GROUND
≈ 6 cm
Cement grout injection
Void filled with compressed cement slurry
≈ 6-8 cm
Segmental tunnel lining

b) Neoprene tube

Seal made from neoprene tubing
Shield tail
GROUND
Grouted void
Feed for the cement slurry
Segmental tunnel lining

c) Steel brushes

Triple steel brush seal
Shield tail
GROUND
Grouted void
Cement grout injection
Grease injection
Gaps between steel brushes filled with grease
Segmental tunnel lining

Figure 5.12 Different shield tail sealing systems

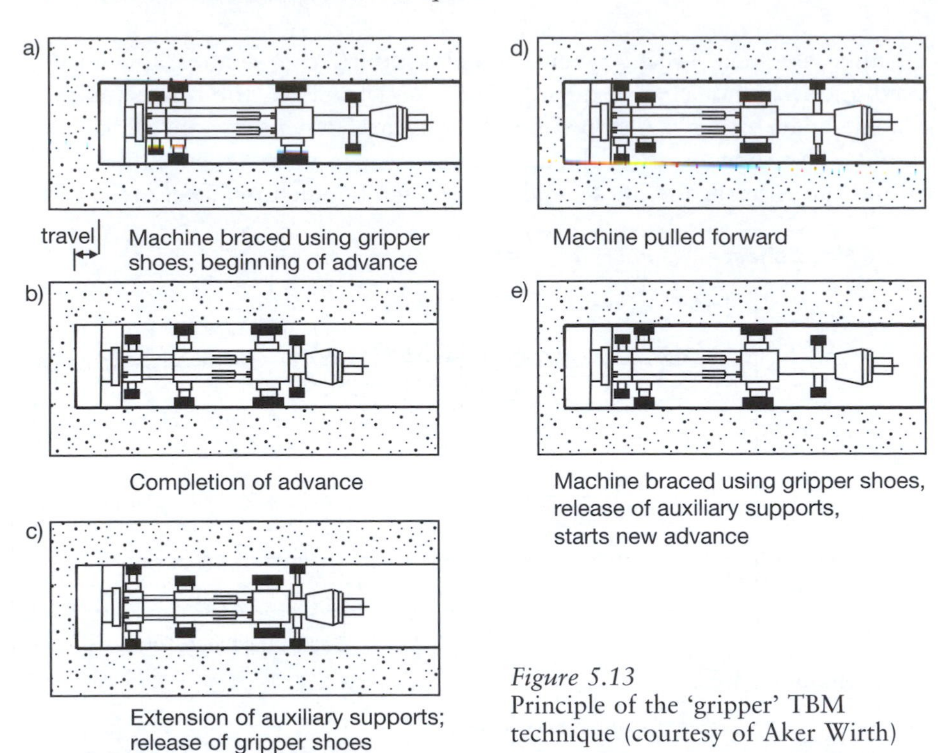

Figure 5.13
Principle of the 'gripper' TBM technique (courtesy of Aker Wirth)

Figure 5.14 Gripper TBM (courtesy of Herrenknecht)

sprayed concrete, and even segmental linings, as used in conventional tunnelling methods (these techniques are described in Chapter 4).

The gripper shoes can be moved hydraulically and can be adapted to the shape of the excavated rock surface. The allowable maximum gripper forces are determined by the compressive strength of the rock and is in the range of two to three times the forward thrust force of the machine (Maidl *et al.* 2008).

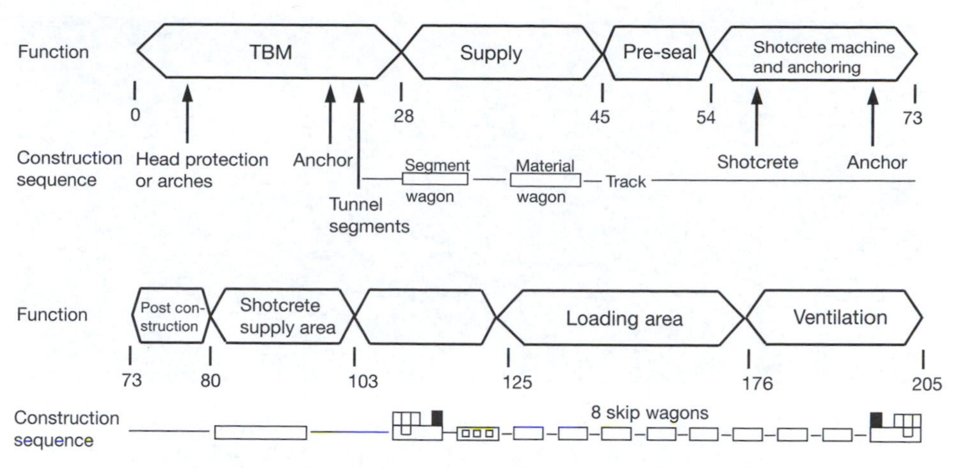

Figure 5.15 Example of the functions associated with a Gripper TBM and typical lengths in metres, with tunnel support using a combination of invert segments and sprayed concrete (after BetonKalender 2005, used with permission from Rowa Tunnelling Logistics)

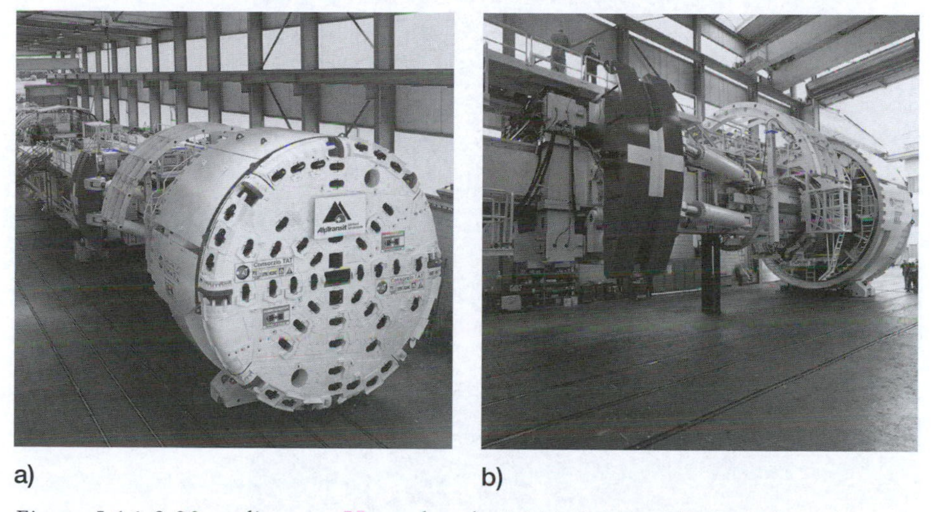

Figure 5.16 8.83 m diameter Herrenknecht S-210 Gripper TBM used on the Gotthard Base Tunnel in Switzerland, a) Cutterhead, b) Rear view showing gripper shoe (courtesy of Herrenknecht)

EXAMPLES OF GRIPPER TUNNEL BORING MACHINES

Figure 5.16a shows an example of an 8.83 m diameter Gripper TBM used on the Gotthard Base Tunnel in Switzerland. The total length of this machine was 400 m and it weighed 3,000 tonnes. The geology that this machine had to cope with was gneiss, granite and schist/shale. Figure 5.16b shows the rear view of the machine and one of the gripper shoes can clearly be seen. This machine used a two-shoe side clamping arrangement, but other arrangements are possible including an X-type clamping where four

Figure 5.17 Gripper machine with multiple 'X' formation gripper shoes (diameter 8.8 m, length 5.7 km, through Gneiss in Quinling China) (courtesy of Aker Wirth)

Figure 5.18 a) and b) 5.03 m diameter Herrenknecht S-351 Gripper TBM used at Glendoe, UK (courtesy of Herrenknecht)

gripper shoes are used in an 'X' formation. This formation is demonstrated in Figure 5.17, which shows an 8.8 m diameter Gripper TBM used in China to construct a 5.7 km long tunnel through gneiss.

Another example of a Gripper TBM is shown in Figure 5.18a. This 5.03 m diameter machine was used on an 8.1 km long tunnel as part of the hydroelectric power scheme at Glendoe in the UK. Figure 5.18b shows a detailed photograph of the cutterhead and the disc cutters on completion of the drive.

5.5.2.2 Shield tunnel boring machines

In contrast to Gripper TBMs, shield TBMs have an extended shield over the front section of the machine. This shield has the function of supporting the ground and protecting the personnel, thus allowing safe erection of the tunnel lining. There are two basic types of shield TBMs for hard rock available; the single-shield and double-shield.

SINGLE-SHIELD TUNNEL BORING MACHINE

The single-shield TBM in hard rock is mainly used in unstable conditions where there is a risk of ground collapse. With these machines, the pushing forces are maintained axially against the installed lining segments. One of the advantages of a single-shield machine is that it can be converted to a closed mode if high groundwater ingress is likely to be encountered (further details on multi-mode TBMs in soft ground are provided in section 5.5.3.4). Figure 5.19(a) shows a 7.6 m diameter single-shield TBM which was used to construct a 5.5 km long tunnel as part of a hydroelectric scheme in Laos.

DOUBLE-SHIELD TUNNEL BORING MACHINE

The double-shield machine (or telescopic shield) combines the ideas of the gripper and single-shield techniques and can therefore be applied to a variety of geological conditions. The double-shield machine consists of a front shield with cutterhead, as well as a gripper section with gripper shoes, a tail shield and auxiliary thrust jacks. Both parts of the machine are connected by a section called the telescopic shield. The operating principle is based on the gripper shoes pressing against the tunnel wall while excavation and segment installation are performed at the same time (the segment installation takes place at the rear of the whole machine). The system adds some flexibility to allow the machine to work either in gripper mode or as a shield TBM. Figure 5.19(b) shows a 5.5 m diameter double-shield TBM which was used to construct a tunnel as part of a hydroelectric scheme in Vorarlberg, Austria. Figure 5.19(c) shows a 3.48 m diameter double-shield TBM which was used to construct a power cable tunnel in Seoul, Korea.

Figure 5.19
a) Single-shield TBM used in Laos (7.6 m diameter, 5.5 km length, variable ground consisting of sandstone, siltstone and mudstone), courtesy of The Robbins Company, b) double-shield TBM used in Austria (5.5 m diameter, ground conditions gneiss and amphibolites), courtesy of The Robbins Company, c) double-shield TBM used in Seoul, Korea (3.48 m diameter, 2.75 km length), courtesy of Lovat

5.5.2.3 General observations for hard rock tunnel boring machines

The cutterhead is commonly arranged with disc cutters which crush or break the rock, although ripper teeth can also be used. The disc cutters are fixed on a rotating bearing and roll in concentric circles on the face when the cutterhead turns. Depending on the type of cutter design, there are disc, toothed or button disc cutters and the rock is loaded either continuously or as a point load when the cutterhead is pressed against the face (Figure 5.20). Through a combination of tensile and compressive forces, a local overstressing of the rock develops which results in its failure. The more gentle excavation using a TBM causes much less damage to the surrounding rock than the use of explosives. Figure 5.21 shows an example of some hard rock disc cutters as used on the Guadarrama Tunnel through granite.

The deterioration of the cutters is mainly dependent on the hardness, fragmentation, abrasivity and ductility of the ground. The cutters should therefore not only be hard-wearing, but also easily replaceable. It is advantageous to be able to change the cutters from inside the machine in the protected area of the cutterhead and not from outside, i.e. in between the unsupported face and the cutterhead. The cutters furthest away from the centre of the cutterhead are most prone to wear (abrasion) as they cover the longest distance at the face. The requirement for high abrasion resistance is also necessary for the cutter bit fixings and bearings, which also have to be designed for the high forces acting on them. In order to minimize deterioration large diameter roller cutters are used. However, cutters with a diameter of more than 17 inches (432 mm) are on the limit of what can

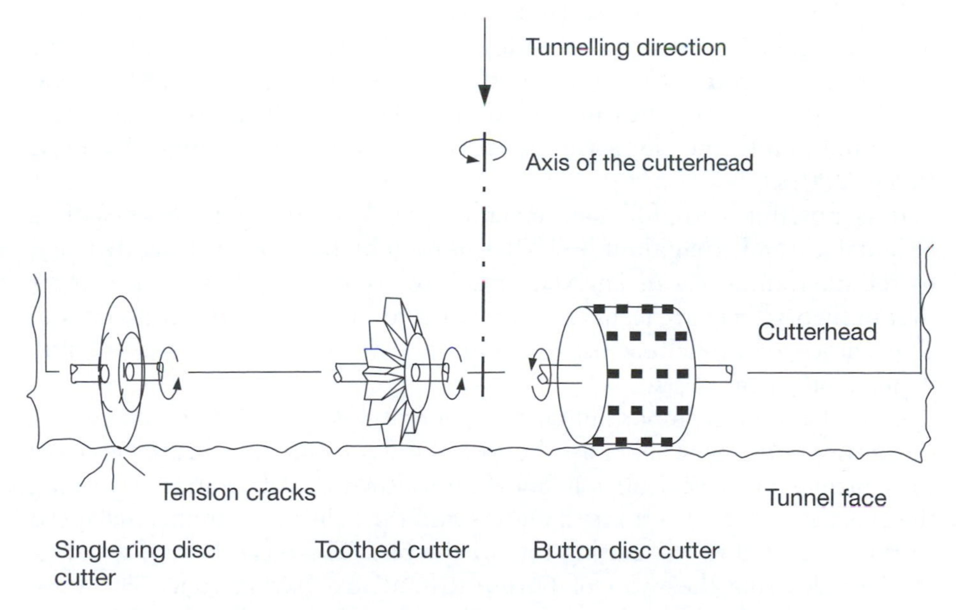

Figure 5.20 Different types of rotating excavation tools for hard rock (disc cutter, toothed cutter and button disc cutter)

Figure 5.21 An example of some hard rock disc cutters (as used in granite). Note the arrangement of the cutters as one moves from the centre of the cutterhead to ensure all the face is excavated (courtesy of Herrenknecht)

be manually handled, and with larger diameters it also becomes increasingly difficult to produce a robust construction of the bearings.

Proprietary agents can also be injected into the face of hard rock TBMs to reduce abrasive wear of the cutting tools. These agents also improve the cooling of the cutterhead, improve the rubber seal durability, and extend the life of the cutter bearing assembly. They also help to reduce dust formation, and hence improve the working environment within the TBM (BASF 2009a).

It is possible with full face machines to drive through rocks with a uniaxial ground strength of 300 MPa or even higher. This is roughly equal to the maximum *in situ* uniaxial tensile strength for ground. This means that in theory there are technically no limits to the use of full face machines. The choice of tunnelling using a full face machine is therefore mainly dependent on the costs.

One of the most important cost considerations for TBMs is the cost of the cutting tools. Not only does the cost of the cutting tools themselves have to be included in the calculation, but also the downtime of the machine during the replacement of deteriorated cutters and the achievable tunnelling speed as this is related to the cutting rate of the bits. The term 'boring speed' is used to describe the ratio of boring distance to boring time. The term tunnelling speed, which ultimately dictates the progress of the construction, is used to describe the ratio between the boring distance to the total time used for all the works in creating a void, including the tunnel support works. An example of speed that can be achieved using a hard rock TBM was up to 106 m/day using a 9.69 m diameter Herrenknecht double-shield TBM during the construction of the Cabrera Railway Tunnel near Valencia in Spain in 2008. The tunnel lining consisted of concrete segments and the geology consisted of hard limestone. The tunnel comprised a twin-tube construction with each tube length being approximately 5.97 km.

An advantage of TBMs when compared to drill and blast techniques (section 5.6) is that the excavation, mucking and stabilizing can be done in parallel with the machine advance. Hence the tunnelling speed is, in general, significantly higher. At least, this is true for ground that is stable or partially stable, i.e. a medium to long-term stand-up time, which, therefore, from the point of view of stability, either requires no support (note: spalling and flaking materials) or only support at a certain time after excavation. It gets difficult for TBMs in a ground with little stand-up time when immediate or even advancing support is necessary and this has to be erected in the area of the cutterhead. In certain conditions the machine has to be stopped during stabilization of the ground. Cramped conditions make the stabilization of the void difficult and the machine should also not be damaged with, for example, sprayed concrete. However, ground with limited stand-up time no longer falls in the category 'rock', which also means it is no longer in the primary application of an open face TBM.

Nevertheless, during the advancement of a tunnel, changing ground conditions have to be expected and contingency plans must be in place. Traversing fault zones is manageable with TBMs, although with a potentially much reduced advancing speed. Very strong and frequently changing rock conditions can limit the use of a TBM, which is why a homogeneous and consistent ground is preferred. As a result, the investment required in the site investigation is larger than that required for blasting methods as the knowledge of the ground should be as comprehensive as possible (no gaps in data) because any modification of the machine during tunnel construction is limited. The largest problem would be if a change to a completely different machine during tunnel construction was required as the additional investment would be substantial (the approximate cost per m diameter for a new machine is on average €1,000,000 to €1,250,000 at 2009 prices) and this is never included in any contingency budget. Furthermore, the delivery time for a new machine of approximately one year and the recovery of the old machine leads to additional unnecessary delays.

With a full face tunnelling machine one is limited to only small variations in a circular profile. If not required from a structural point of view, the desired kinematic envelope of the completed tunnel is often a rectangular shape, e.g. a road tunnel. The circular profile of a TBM thus results in over-excavation, especially in the invert where the spoil is removed and it is filled afterwards with concrete to create the road base.

5.5.3 Tunnel boring machines in soft ground

5.5.3.1 Introduction

Open face TBMs (as discussed in the previous section) are mainly used in ground with a significant strength and stand-up time, but there are certain types of TBM which can be used in soft ground. The different types will be discussed in this section. Generally the main differences are the cutter tools, also known as cutter dressing, on the cutterhead and the face support requirements as the ground stability is generally lower.

EXCAVATION TOOLS

For soft ground conditions different cutter tools from those discussed in the previous section can be used. Some examples of these different cutter tools are described in the following section. Hard rock cutter tools, such as disc cutters, can also be used, however, for example in sandstone or limestone. Normally cutting tools work in soft ground by scraping material from the face. However, there are instances where larger hard inclusions, such as boulders, may occur. These inclusions have to be reduced in size before they can be quarried from the face. This is either done by hand, if the conditions at the face allow access and if the inclusions are sporadic, or by disc cutters. In addition, a stone crusher can be placed behind the cutterhead on slurry machines.

Cutterhead with continuous 'knife' tools The cutterhead consists of radially placed spokes. The long sides of the spokes are made up of continuous cutting edges, which scrape the ground as it rotates. This solution is mainly applied to cohesive soft soil in which there are few larger inclusions.

Cutterhead with scraper teeth, drag bits and round shank cutters In this case the spokes contain pick bits, scraper teeth, or there are round shank cutters placed centrally on the spokes. Drag bits are blunt tools inserted at right angles to the working face and usually have a square shape. Their purpose is to loosen coarser grained soils such as gravel, sand and silt. Fine grained soils, such as clays, tend to be kneaded if picks are used (Stein 2005). Scraping tools (also known as flat bits or cutting teeth) take many forms and often have tungsten carbide inserts. Instead of a continuous cutting edge, they are placed individually and project from the face. Scraper teeth are used mostly in fine grained loose sediments such as clay or clayey silt, but can also be used in coarser grained ground. Round shank cutters are used in harder materials and are pointed cylindrical excavation tools with replaceable tungsten carbide inserts (Stein 2005). A stone crusher may be provided.

Examples of cutter tools Figure 5.22a shows a close-up photograph of the cutter dressing used on a soft ground TBM. This machine (Figure 5.22b) was used to construct the Storm Water Outfall Tunnel (SWOT) at Heathrow Airport Terminal 5, UK. The TBM had an external diameter of 3.345 m and was used to construct a tunnel approximately 4.1 km long with an internal diameter of 3.0 m through London Clay.

Figure 5.23 shows a close-up photograph of the cutter dressing used on a Mixshield TBM. This machine was used in Shanghai, China. The TBM had an external diameter of 15.43 m and was used to construct a road

Figure 5.22 a) The cutter dressing used in soft ground (London Clay), and b) the fully refurbished Lovat TBM (Storm Water Outfall Tunnel, SWOT, at Heathrow Airport Terminal 5, UK)

Figure 5.23
Cutter dressing used on a Mixshield TBM in Shanghai through sand, clay and loose gravel/boulders (courtesy of Herrenknecht)

tunnel approximately 7.47 km long through sand, clay and loose gravel/boulders, up to depths of 65 m and water pressures of up to 6.5 bar.

FACE SUPPORT AND WATER CONTROL

The application of a shield boring technique in less stable soft ground commonly requires the face to be supported. This is in contrast to the open face TBMs which are often used in hard rocks where the ground is able to support itself during excavation. In soft ground, with little or no stand-up time, the ground would simply collapse into the machine and attempts to control the excavation of this material and to prevent large displacements occurring within extensive amounts of the ground around the tunnel heading would be very difficult. In addition, for tunnels constructed below the groundwater table in permeable materials, water flow into the tunnel must be controlled in order to prevent the machine and tunnel from flooding. Groundwater flow towards the tunnel heading also reduces the stability of the ground.

Non-pressurized systems

- Natural support: The natural stability of the face can be enhanced by designing the form of the cutting wheel in such a way that a curved surface develops as the excavation progresses. This enhances the stability of the face by improving the arching effect and hence stress redistribution around the tunnel heading. (Further information on face stability is presented in section 3.3.)

- Mechanical support: Adjustable support plates can be placed in between the spokes to allow an active mechanical support. These are similar to the face-breasting plates used on open face tunnelling shields as described in section 5.4.

Pressurized support systems The principle is that pressure is created at the front of the shield and this supports and stabilizes the tunnel face. In addition this can be used to control water flow into the tunnel. There are two basic types of pressurized closed face tunnelling systems, slurry tunnelling machines (STMs) and EPBMs. There are also machines that operate using compressed air within a bulkhead at the tunnel face inside which an excavator operates (compressed air tunnelling is discussed in section 4.2.9).

STMs were developed specifically for use in coarse grained ('cohesionless') ground that contains little or no silt or clay. EPBMs were developed for use in weak fine grained ('cohesive') soils. However, due to the rare occurrence of pure 'cohesive' and 'cohesionless' soils in practice, there has been an effort to widen the application of both types of machine. Both STMs and EPBMs use a rotating cutterhead to excavate the ground. The cutterhead can contain either picks or discs, or a combination of both.

STMs and EPBMs were developed initially in Japan and Europe. In Japan, STMs were developed in the 1960s with EPBMs introduced in the mid to late 1970s (STMs were actually invented and patented earlier by John Bartlett of Mott Hay and Anderson in the UK, but never developed). In Europe, STMs were in use in the 1970s and EPBMs in the 1980s (BTS/ICE 2005). The term sometimes used for a TBM with a liquid face support is 'hydroshield'. This expression can be traced to a patent filed by a German construction company (Wayss & Freytag) whose shield construction was known by this name. These TBMs are discussed in more detail in the following sections.

5.5.3.2 Slurry tunnelling machines

STMs use a pressurized fluid to stabilize the face during excavation of the ground. There are two systems in use today for maintaining a balanced face pressure. One simply uses the fluid in the pressurized chamber behind the cutterhead (called the plenum) to provide the pressure, and the other uses an air bubble system. The air bubble system has a bubble of compressed air introduced into the plenum in the roof behind the bulkhead and this acts as an accumulator and thereby ensures that a constant pressure is maintained at the face (see 'Mixshield tunnel boring machines' later in this section). The slurry does not only stabilize the face, but also mixes with the excavated material to allow it to be transported out of the machine (Figure 5.24) (Court 2006).

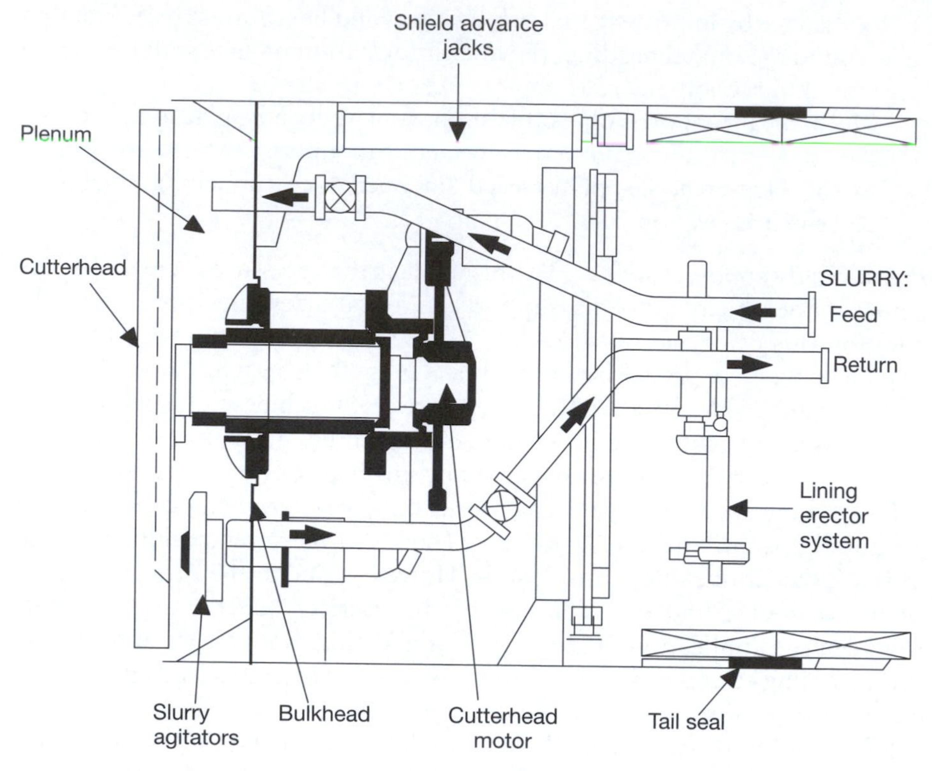

Figure 5.24 Longitudinal section through a slurry TBM (after Whittaker and Frith 1990)

The fluid is pumped to the face, where it mixes with the excavated material. This mixture is then pumped out of the excavation chamber through a slurry line where it is conveyed to a separation plant. To prevent the slurry line from becoming blocked by larger pieces of material it is passed through a crushing unit prior to entering the slurry line. In the separation plant, the excavated material is removed from the transportation fluid using screens, cyclones and, if necessary, centrifuges. This allows most of the fluid to be reused.

If the ground contains a high proportion of fine ('cohesive') particles, water is often an adequate transportation fluid. However, if the ground is predominantly a sand or gravel with few or no cohesive particles then a slurry must be created to aid transportation of the excavated material. This is achieved by introducing bentonite (and possibly polymers) to the water. The name bentonite stems from the placename of Benton in the USA, where this special clay was found for the first time. As a clay water mixture it has thixotropic characteristics: it is like 'jelly', i.e. a solid, at rest but when agitated it is liquid.

It is becoming more common that cutting tools can be changed from 'behind' the cutting wheel. Access under atmospheric pressure conditions

Figure 5.25
a) View taken towards the rear of a Mitsubishi 10.6 m diameter slurry machine used in Japan (note the two slurry pipes at the top right, one for passing the slurry into the machine, which enters at the top of the shield, and one for removing the slurry and excavated material, which comes from the bottom of the shield), and b) the full slurry TBM in the factory showing the cutterhead (courtesy of Mitsubishi Heavy Industries Mechatronics Systems Ltd)

is possible if the main spokes of the cutter wheel are large enough for a person to get inside and change the cutting tools. All works and repairs outside the main spokes and in front of the cutter wheel have to be done via an airlock. The slurry is replaced by low-pressure compressed air during this work.

Figure 5.25 shows an example of a 10.6 m diameter slurry TBM as used in Japan. As can be seen in Figure 5.25a, looking towards the rear of the TBM, slurry machines can be relatively uncluttered and 'clean' due to the fact that the excavated material (spoil) is pumped out without the need for spoil wagons or conveyors.

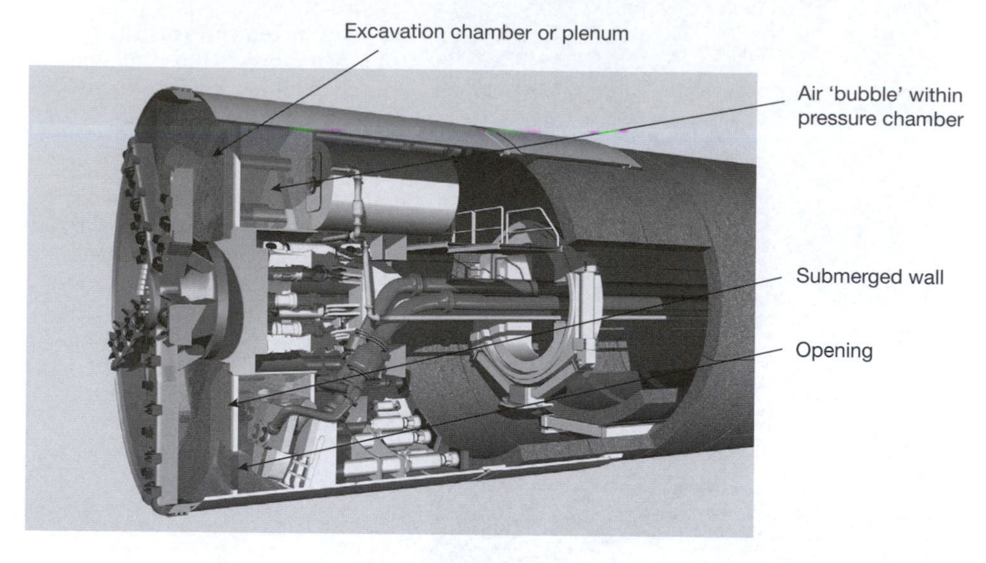

Figure 5.26 Mixshield TBM (courtesy of Herrenknecht)

MIXSHIELD TUNNEL BORING MACHINES

Mixshield TBMs are slurry machines that use a submerged wall/pressure bulkhead combination in order to create an air bubble for face pressure control (Figure 5.26). The submerged wall separates the pressurized front section of the shield into two areas. The area between the submerged wall and the rear pressure wall is called the 'pressure chamber' (Figure 5.26). The area in front of the submerged wall, i.e. immediately behind the cutterhead, is defined as the 'excavation chamber'. The required pressure exchange between the excavation chamber and the pressure chamber occurs via an opening in the bottom of the submerged wall.

The Mixshield design also allows higher face pressures to be dealt with using a closed slurry circuit as the mucking system. Mixshield TBMs are now capable of working under 13 bar pressure. This not only affects the thrust required by the shield, but also the seal systems associated with these TBMs. In addition, there is also the question of access to the face for maintenance and repair activities under these elevated face pressures. Systems that reduce the need to access the face are available, remotely activated standby cutter tools and load detection and wear sensor systems for example, but these cannot entirely remove the possible need to enter the face.

Burger and Wehrmeyer (2008b) compared the torque at the cutterhead for equivalent diameter Mixshield and EPB machines and showed that Mixshield TBMs require significantly lower torques. For example, above approximately 8 m diameter, an EPBM requires approximately double the torque than an equivalent sized Mixshield TBM (approximately 18,000 kNm compared to 9,000 kNm for a 10 m diameter machine). This is because in the case of a

Mixshield TBM the cutterhead is only excavating the ground at the tunnel face into a suspension-filled chamber, whereas the cutterhead of an EPBM not only has to excavate the ground but also has to act as a mixing tool inside the excavation chamber, which is completely filled with spoil.

Burger and Wehrmeyer (2008a) stated that the Mixshield TBM was first used in 1985 on the Herrenknecht S-12 HERA tunnel project in Hamburg, Germany. A list of projects conducted using Herrenknecht Mixshield TBMs can be found in the comprehensive review of this type of machine by Burger and Wehrmeyer (2008a and b). From the projects listed in this review it can be seen that the Mixshield has had a large impact on the development of tunnel construction. There has been a considerable increase in tunnel diameters and hence TBMs. The recent preference for multiple lane transportation tunnels (two and three lane road tunnels) is pushing Mixshield TBMs to beyond 15 m. Other examples include multipurpose tunnels such as the SMART project in Kuala Lumpur, Malaysia, which involves both road and water storage usage.

Examples of Mixshield tunnel boring machines Figures 5.27 and 5.28 show examples of Mixshield TBMs. It is interesting to note that the cutterhead of the Herrenknecht S-108 TBM used to construct the 4th Elbe Tunnel (Figure 5.28) was equipped with a geophysical method to help detect boulders and changes in strata ahead of the machine. The system developed by Amberg and Herrenknecht called SSP (seismic soft-ground probing) was based on seismic reflection techniques and could identify density contrasts in the ground up to 40 m in advance of the TBM.

Figure 5.27
8.160 m diameter, 100 m long Herrenknecht S-192 Mixshield TBM used on the Channel Tunnel Rail Link (CTRL) project in the UK, a) before and b) at breakthrough. This machine drove through alluvium, chalk and flints (courtesy of Herrenknecht)

Figure 5.28
14.2 m diameter Herrenknecht S-108 Mixshield TBM used to construct the 4th Elbe Tunnel in Hamburg, Germany, a) before and b) at breakthrough (length of drive 2.56 km, through sand, glacial drift, silt, gravel and boulders) (courtesy of Herrenknecht)

5.5.3.3 Earth pressure balance machines

EPBMs use the excavated material to support the tunnel face during excavation of the ground. The excavated material enters the plenum in a fluid or plasticized state after having been mixed with a conditioning agent (Figure 5.29). The plasticized spoil is removed from the plenum by using an Archimedean screw (screw conveyor) (Figure 5.30). The screw conveyor is used to remove the excavated material in a very controlled manner so that pressure is maintained in the plenum. At the same time, the pressure at the other end of the screw conveyor is atmospheric, i.e. there is a pressure drop from one end to the other. This means that the plasticized spoil in the screw conveyor needs to form a plug to help maintain this pressure differential. The pressure in the plenum should be high enough to maintain the ground stability and is controlled by a combination of thrust on the cutterhead and the rate of removal of material from the plenum via the

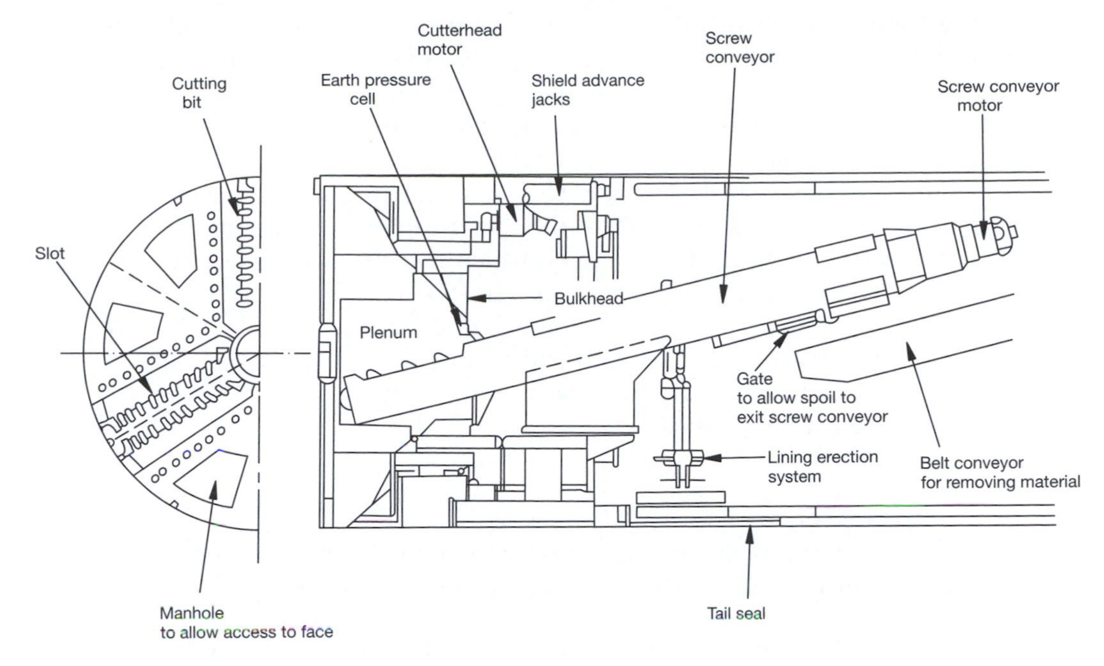

Figure 5.29 Longitudinal section through an earth pressure balance TBM (after Whittaker and Frith 1990)

Archimedean screw. This should be matched against the rate of advance of the tunnel machine. It should be noted that EPBMs have also been used in compressed air mode to support the face instead of using plasticized clay.

Figure 5.30 a) and b) The head of the Archimedean screw, and c) conveyor system used in an EPBM on the Line B metro extension project in Toulouse, France (ground conditions: Toulouse Molasses, i.e. a hard sandy clay with pockets and lenses of very dense sand)

The ideal material for EPBMs is a fine grained ('cohesive') soil with stiff to soft consistency (I_C = 0.5–0.75, consistency index is described in section 2.3.3), which extrudes through the openings of the cutterhead towards the screw conveyor. If the excavated material does not conform to this ideal, it must then be 'conditioned', i.e. artificially altered. This is important as an optimally conditioned material helps maintain pressure in the plenum and hence transmit the pressure to the face to maintain stability. It also helps to control the material in the screw which also allows improved control of the face pressure and settlement control (BTS/ICE 2005). EPB technology has made significant progress over the last couple of decades. This is particularly so in the area of ground conditioning which has enabled EPBMs to be used in coarser grained materials.

Soil can be conditioned using:

- water;
- bentonite, clay or polymer suspension;
- foam (surfactants mixed with water and compressed air);
- foam with polymer (the polymer helps stabilize the foam).

These conditioning agents are injected via ports mounted across the face of the cutterhead. There are also often facilities for the conditioning agent to be injected into the plenum and even into the screw conveyor casing. However, the most effective injection point is through the rotating cutterhead as this ensures that the conditioner is mixed directly with the excavated material. Trial tests to determine the correct conditioning regime are essential. In granular soils the conditioning agent must create a stable plastic consistency in the excavated material that will not degrade until the material has been discharged from the screw conveyor or possibly until it reaches the surface stockpile as it is easier to transport the material in this state (BTS/ICE 2005). The conditioning for soft cohesive soils will generally prevent the material forming into lumps and will therefore assist its flow through the plenum and the screw conveyor.

Slump tests, as used for concrete testing, are used to determine the appropriate conditioning of the excavated soil. An optimum slump for EPBMs is in the range 5 to 15 cm. (BTS/ICE 2005)

EXAMPLES OF EPB MACHINES

Figure 5.31 shows an example of a 9.16 m diameter EPBM used for constructing the Heathrow Airside Road Tunnel, UK.

Figure 5.32 shows an example of an EPB machine used as part of the Channel Tunnel Rail Link (CTRL) project in the UK, which provides a high speed rail line between London and the Channel Tunnel. Two similar machines each measuring 8.16 m in diameter were used to construct two parallel tunnels 4.7 km long through Thanet Sand. The internal diameter of each tunnel on completion was 7.15 m. The cutterheads on these

Figure 5.31 Heathrow Airside Road Tunnel, UK, a) 9.16 m diameter Herrenknecht S-185 EPBM, b) lowering the EPBM into the starting shaft, c) breakthrough, d) the completed concrete segmental lined tunnel (courtesy of Herrenknecht)

machines were equipped with 17 inch (432 mm) disc cutters, bits and teeth and could rotate at 0–3 rpm. The shield for this machine consisted of three elements and articulation was possible between each of these elements. This was important as the machine had to cope with curved sections during the bore. The total length of each of these machines was 112 m and each weighed 1,100 tonnes.

5.5.3.4 Multi-mode tunnel boring machines

In order to overcome some of the problems associated with TBMs with respect to the difficulties of changing ground conditions, efforts to develop TBMs that can operate in a number of different modes have been made, for example slurry and open face, EPB and open face or a combination of all these. These allow tunnels to be constructed in increasingly challenging ground conditions that previously would have been seen as too high risk.

An example of the use of multi-mode TBMs was on the CTRL project in London, UK, where the EPB machines supplied by Kawasaki and Lovat

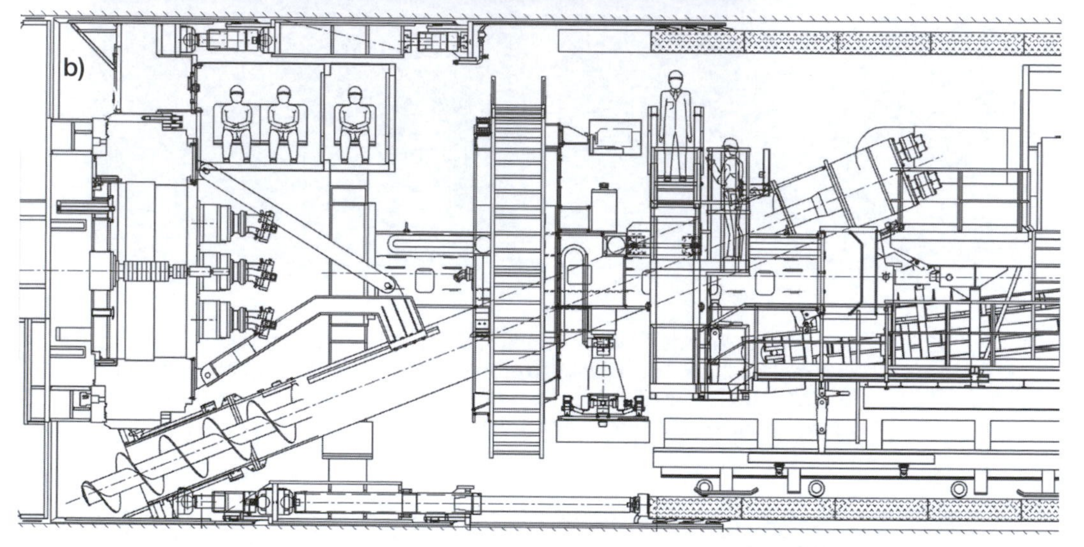

Figure 5.32
a) 8.16 m diameter EPB machine used on the CTRL project in the UK, b) schematic longitudinal section through this machine (courtesy of Aker Wirth)

(contracts 220 and 250 respectively) were designed to work in both 'open' (non-EPB in London Clay) and 'closed' EPB mode in the more variable water-bearing ground with water pressures up to 3 bar (Woods *et al.* 2007).

A Herrenknecht multi-mode machine was used on the 10.1 km-long A86 highway tunnel in Paris, France. The tunnel design was for a two-deck tunnel with three lanes on each level. This 11.565 m external diameter machine tunnelled through the entire spectrum of geological formations under Paris, including marl, chalk, clay, limestone and sands of the Seine river basin, as well as three different groundwater levels (BTS/ICE, 2004). The TBM had to operate in different modes as described by Burger and Wehrmeyer (2008b):

- as a slurry shield with slurry supporting the face;
- as an earth pressure balance (EPB) shield with face support provided by the conditioned spoil;

- in semi-EPB, or compressed air, mode;
- in open mode (spoil discharged via the screw conveyor, but with a non-pressurized excavation chamber).

There will no doubt be further developments in the area of multi-mode TBMs as tunnels are constructed in more challenging and variable ground conditions.

5.5.3.5 Choice of slurry or earth pressure balance tunnel boring machine

Notwithstanding the development of multi-mode TBMs, it is often the case that a single operation machine is still used and therefore a decision has to be made between slurry and EPB machines. This choice is critical and does not only include the ground conditions. The experience of particular contractors, the logistics and configuration of the works, and the requirements to meet the client's specification also play a factor in the choice (BTS/ICE 2005). With respect to the ground conditions it is likely that these will fall between more than one optimum range for each type of TBM and a compromise is required in the choice of machine, unless a multi-mode machine is chosen. However, even if a multi-mode TBM is chosen, the operating parameters for the different modes of operation need to be understood.

PARTICLE SIZE DISTRIBUTION

A general indication of which ground conditions are applicable to STMs (slurry machines) and EPBMs is shown in Figure 5.33.

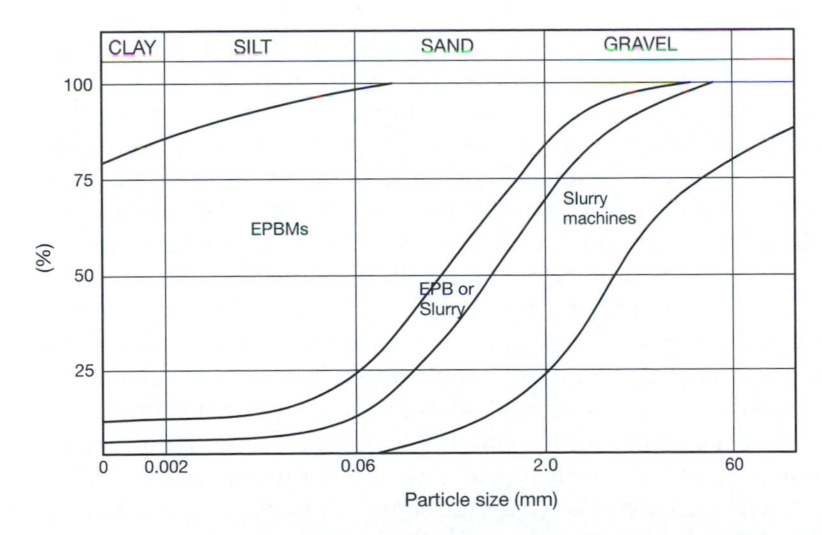

Figure 5.33 Graph of soil particle distribution curves related to the choice of EPB and slurry machines (STM) (after BTS/ICE 2005, used with permission from Thomas Telford Ltd)

According to BTS/ICE (2005), the key to STM operation is the separation plant and therefore loose waterbearing granular soils are ideal for these machines. STMs have problems coping with silts and clays. If the amount of fines (particles smaller than 60 μm or able to pass through a 200 sieve, BSI 2002) is greater than 20%, then the use of an STM becomes difficult, although still possible.

In contrast EPBMs operate more effectively in silty ground, which help to form a plug in the screw conveyor and also aid the control of water inflow. If the fines content in the ground drops below 10% then the operation of EPBMs potentially become more costly due to the need for greater quantities of soil conditioners.

In terms of permeability of the ground, a general rule is that a ground permeability of 10^{-5} m/s is chosen as a point of selection, with STMs applicable to ground of higher permeability and EPBMs to ground of lower permeability (BTS/ICE 2005).

GROUND MOVEMENTS

Both types of machine are able to control ground movements to very tight tolerances as long as they are properly used. Further details on ground movements caused by tunnel construction are provided in section 7.1.

5.6 Drill and blast tunnelling

5.6.1 Introduction

Drill and blast for tunnel construction can be used in geology ranging from hard rock with low strength, e.g. marl, loam, clay, gypsum, chalk, to the hardest rocks, such as granite, gneiss, basalt or quartz. Due to this large range of possible usage, drill and blast can be advantageous for very changeable ground conditions. The drill and blast work and the extent of the tunnel support can be adjusted with every heading advance if required.

In addition, tunnelling by using drill and blast is often preferable to TBM or road header tunnelling if, for example, the tunnel is relatively short so that the high investment costs needed for a tunnelling machine are not economic, or when the ground hardness is very high so that a high wear of the cutter tools leads to an uneconomic application of the machine. Other reasons for using drill and blast for tunnelling can be when a cross sectional profile is required that differs from a circle, or when a very large tunnel profile is required which does not allow the application of a tunnel machine for technical or economic reasons. As with all tunnelling types, tunnelling using drill and blast is principally most economic when it is continuous and similar work processes are used. Therefore when planning a tunnel it is beneficial to split the tunnel into sections where the same advancing schema and the same tunnel support intensity can be used. During drill and blast operations the individual job processes are mainly conducted sequentially, which in general results in a slower tunnelling speed compared with TBM tunnelling.

Ground vibrations need to be considered when using drill and blast for the excavation of a tunnel, especially in urban areas, as these can affect surface and subsurface structures (and humans). It is therefore important to make predictions of the likely effect (possibly by conducting blasting trials as part of the site investigation) and to carry out monitoring during the construction works (New 1990).

When carrying out drill and blast, the order of the processes generally consists of the following cycle: drilling, charging (including adding the detonator), stemming, blasting, ventilation, mucking (spoiling) and supporting. Each of these components is described in the following sections.

Further details on rock blasting and drill and blast tunnelling can be found in Hopler (1998), Persson *et al.* (1994), Holmberg (2000 and 2003).

5.6.2 Drilling

In order to create the drill holes, which are needed to take the charges, rubber tyred drilling carriages are, usually, used as these have relatively high driving speed and are manoeuvrable, i.e. they need to drive to the face and then return to a safe distance (Figure 5.34). Nevertheless, tracked drilling carriages are also an option. From a practical point of view the same drilling carriage should be used when placing the support anchors. Drill carriages carrying two to four hydraulically operated booms are commonly used (Figure 5.35). The speed of drilling is approximately one to five metres per minute. These booms serve to provide exact positioning of the drill hammer. This is essential in order to achieve a successful blast, as the exact position of the boreholes is crucial. A computer navigated drilling vehicle (jumbo) is therefore used as this makes a fast and exact positioning of the booms possible, and it also guarantees a uniform depth of drill hole even for a strongly jointed or uneven face. The drill holes are normally drilled 10% (commonly 20 cm) longer than the desired advance length in order to ensure that the advance length is achieved as the rock is not always completely blasted over the total drill hole length. In very small tunnels, i.e. 4–5 m^2,

Figure 5.34 Example of a rubber tyred drilling carriage 'jumbo' as used on the 2nd tube of the Katschberg Tunnel, Austria (details are provided in section 4.3.3)

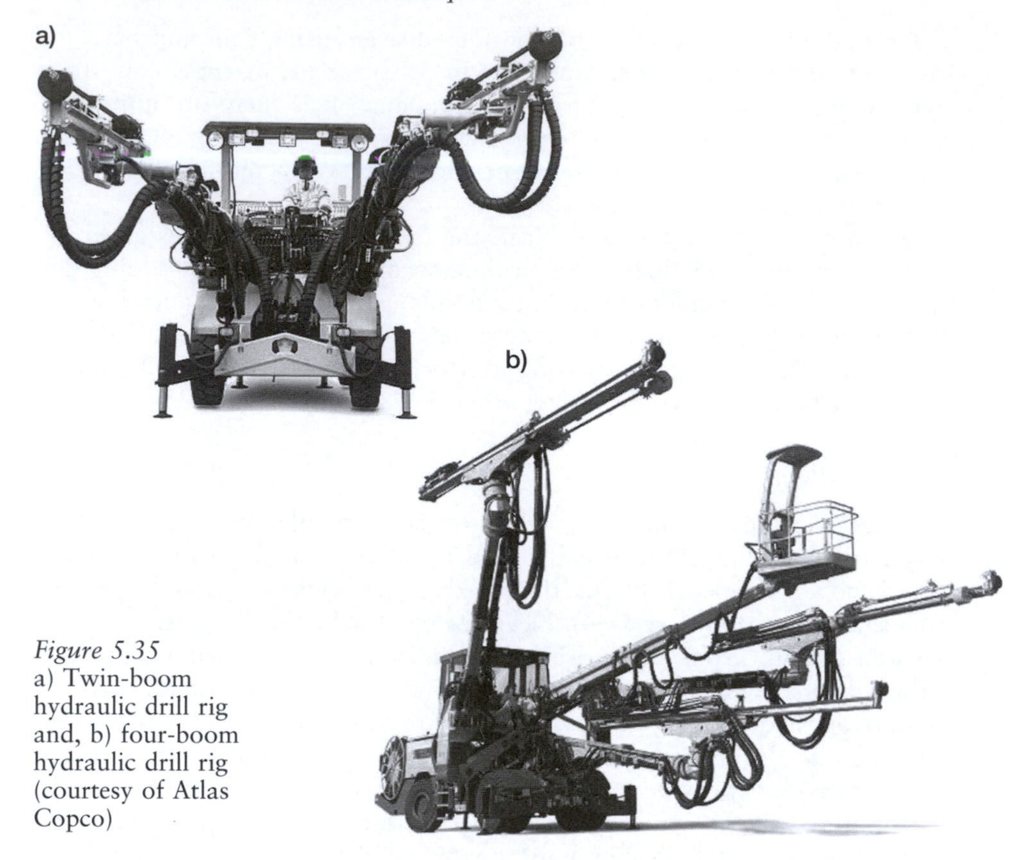

Figure 5.35
a) Twin-boom hydraulic drill rig and, b) four-boom hydraulic drill rig (courtesy of Atlas Copco)

handheld rock drills, sometimes with air legs, may be used. Examples of the drilling operation are shown in Figure 5.36a–c.

The diameter of the drill hole has to be chosen so that the necessary blasting charge can fit into it, and this blasting charge fills the cross section of the drill hole tightly. Possible drill hole diameters range from approximately 30 to 100 mm, with the most common diameter being 40 to 50 mm. The number of drill holes required depends on many factors. Amongst these are the 'blastability' of the ground and the size of the excavation profile. The number of drill holes per square metre reduces as the cross section increases. In general the number of drill holes per square metre does not fall significantly after approximately 30 m^2. For profiles in the range of 60 to 70 m^2 the number of drill holes, depending on the blastability of the rock, is between 1 and 2 per square metre of the excavation area (see Figure 5.45a). In addition, the number of drill holes is influenced by the advance length, the type of charge used (see section 5.6.5.5 on the power of the charge) as well as the shape of the charge, i.e. the diameter of the cartridge. It is very important not to drill into sockets left from the previous round of blasting as they may contain traces of unexploded explosives and hence could be very dangerous.

Figure 5.36
Examples of drilling blasting holes, a) and b) for an emergency cross passage, Katschberg Tunnel, Austria, c) for the Spillvatten Tunnel, Goeteborg, Sweden (an 8 km sewer tunnel, cross sectional area 11 m^2, ground included granitic gneiss)

5.6.3 Charging

Charging is the insertion of the explosives into the drill holes. More details on the different types of explosives are given in section 5.6.5.2. An example of an explosive package (in this case an emulsion charge cartridge) is shown in Figure 5.37. This particular charge comes in package sizes ranging from 40 × 300 mm to 90 × 400 mm.

The explosive material exists as a cartridge, an emulsion or as a powder. Powder explosive can be blown into the borehole using pressurized air. This type of loading is fast and the explosive fills the drill hole completely, resulting in a good explosive effect. Nevertheless the use of powdery explosive is uncommon in tunnelling as loading the mainly horizontal drill holes can be problematic. In addition, powder explosive is not water resistant.

Emulsion explosive can be pumped in depending on its consistency. Similar to powder explosives, the loading is quick and the drill hole can be completely filled. Furthermore, these explosives are generally very water resistant. A special vehicle is necessary to transport the explosive to the face and to pump it. The pump is computer controlled, which guarantees that each hole is filled with the required pre-defined amount of explosive. The vehicle is expensive, but this method is very time efficient and safe. Examples where pumped explosives have been used are on the Katschberg Tunnel, Austria and the Spillvatten Tunnel, Sweden (details of this tunnel are provided in section 5.6.2).

The type of explosive used most commonly for tunnelling is the cartridge type charge as it is easy to handle (see section 5.6.5 on detonation). However, the loading of a cartridge explosive into the drill hole is time consuming because any loose material within the hole has to be carefully removed and the charges have to be guided by hand to the end of the drill hole with the aid of a 'charging' pole. Depending on the requirements of the explosive a load column consists of a number of cartridges which are approximately 12.5 to 70 cm long. Depending on the length of the cartridge they can have a diameter of approximately 2 to 12 cm. A compromise has to be found, as the cartridge should be guided easily into the drilling hole, but needs to fill the drill hole completely to avoid the cartridge being surrounded by air, which lowers the blasting effect. The diameter of the cartridges is usually 5 to 15 mm smaller than the diameter of the drill holes.

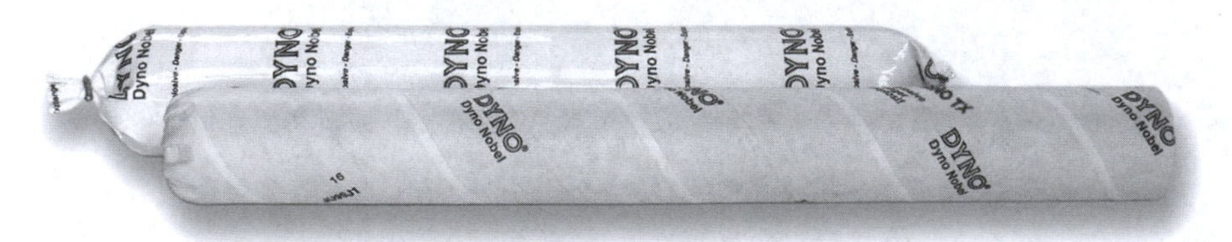

Figure 5.37 Example of a charge package (DYNO® TX, courtesy of Dyno Nobel Inc.)

5.6.4 Stemming

Stemming seals the drill hole and acts to retain the explosive gases within it, thus preventing them escaping into the tunnel. It also allows these gases to act without any energy losses onto the wall of the drill hole. It has been discovered that for the explosive used in tunnelling the stemming does not have an improved effect for a long charged column. The inertia of the molecules within the air column in the drill hole (in relation to the very high detonation speed) is sufficient to act as the stemming. Another reason for using stemming is to prevent any unexploded charge from being discharged out of the drill hole.

Stemming of the drill hole is also used to reduce dust. For this purpose water stemming cartridges, or even better calcium chloride stemming cartridges, can be used. The thin-walled casing of the cartridge bursts during the detonation so that water or calcium chloride powder can bind with the dust. Nevertheless the dust binding effect of the charge is not a primary requirement in tunnelling. Therefore the decision is often taken not to use stemming, in order to save cost and time.

5.6.5 Detonating

5.6.5.1 Detonating effect

At the point of detonation the explosive is compacted and goes through a very rapid and high intensity chemical reaction (up to about 8000 m/s). High temperature and high pressure create a large volume of gas (gas cloud). The fast flowing gas cloud collides with the wall of the drill hole. The force of the impact of this detonation causes the ground to be locally crushed and also leads to the development of cracks. The expanding gas cloud surges into the cracks and bursts open the ground. The gas pressure does the majority of the work during detonation. The following example should demonstrate the extremely high power developed during the detonation of an explosive (Wild 1984). The engine of a car with, for example, a capacity of 1166 cm^3 has a power of 41 kWatts: 1 cm^3 of nitroglycol creates, when detonated, 74×10^6 kWatts. This only occurs, however, for a very short detonation time, in the order of microseconds.

The higher the energy/strength of the explosive the higher the detonation force and the higher the crushing effect of the charge on the surrounding area (Figure 5.38). The energy/strength is the ability of an explosive to break or shatter by shock or impact, as distinct from gas pressure. The rock blasting strength of an explosive is dependent not only on the properties of a particular explosive, but also on the properties of the ground that surrounds the charge. The more energy expended in deforming and crushing the ground close to the drill hole wall, the less energy is left for subsequent fracturing and acceleration of the major part of the blasted rock (Persson *et al.* 1994). Thus in hard rock a high detonation force is required. However,

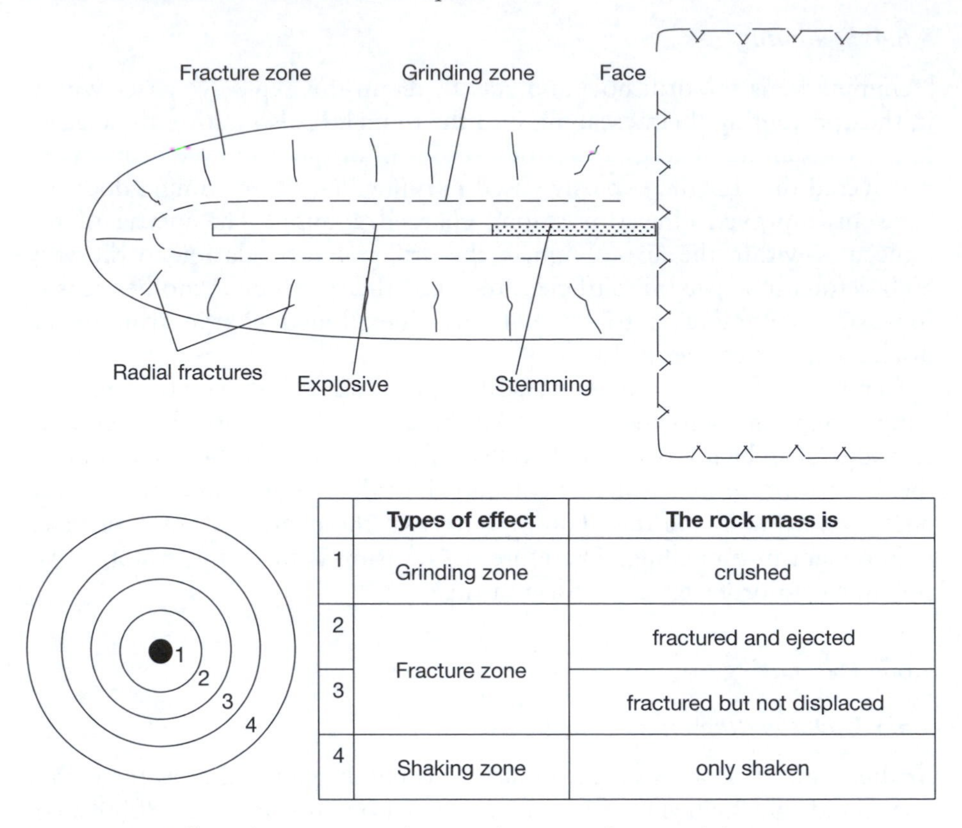

	Types of effect	The rock mass is
1	Grinding zone	crushed
2	Fracture zone	fractured and ejected
3		fractured but not displaced
4	Shaking zone	only shaken

Figure 5.38 Affected zones around an explosion at the tunnel face

in softer rocks a lower detonation force is required, i.e. the explosive relies on the development of cracks and the expansion of the gas cloud, so that the size of the debris fragments created is not too small as this can lead to difficulties with the spoil. Furthermore, large unnecessary crushing can be avoided and the development of dust can be limited.

In contrast to the ignition of fuel, the oxygen required is contained in the explosives so that once set-off, the detonation cannot be influenced or regulated through, for example, the regulation of the oxygen supply.

5.6.5.2 Types of explosive

Examples of some of the different types of explosives used in tunnelling are presented below:

GELATIN-DYNAMITE EXPLOSIVES

As a type of gelatin-dynamite (gelignite) explosive, ammonium nitrate explosives should be mentioned. As with all explosives they consist of a mixture of easily burnable carbon links with oxygen carriers. Commercial explosives of this type contain approximately 50% ammonium nitrate and 18–40% gelatine of nitroglycol, nitrotoluol connections and collodium wool, so that they are malleable and plastic. The available energy increases with an increased gelatine content. These explosives have a high strength and at the same time have a large gas cloud volume, are water resistant and are safe to handle so that they can be transported by road or rail as long as they comply with relevant regulations. 'Safe to handle' means that the explosives are insensitive to mechanical and temperature effects (i.e. impact, friction, heat, frost) and also to unintentional explosion. Such explosives are also called safety explosives.

Gelignite explosives have a detonation speed of approximately 6500 m/s and achieve a gas cloud volume of between approximately 800 and 860 litres/kg. They are used in cartridge form, and due to their water resistance can also be used in wet drill holes.

In summary, gelatin-dynamite explosives have the following characteristics:

- high energy concentration;
- high crushing capacity;
- good detonating transfer;
- high density (sinks well in water-filled drill holes);
- cartridge can be split;
- very good water stability;
- high plume (gas) volume.

EMULSION EXPLOSIVES (DETONATING SLURRIES)

As an alternative to gelignite explosives, explosive slurries can be used. The energy/strength of the explosive slurry is not quite so high (detonation speed of up to approximately 5700 m/s), but the gas cloud volume is 1000 litres/kg of explosive and thus slightly above the value for the gelignite explosives. Emulsion explosives can be pumped depending on their consistency. However, for tunnelling it is mainly used in cartridge form. Explosive slurries are also very water resistant and safe for handling, which is also demonstrated by their ability to be pumped.

An example of a detonator sensitive emulsion in cartridge form is the DYNO® AP cartridge type charge, which ranges in size from 25 × 300 mm to 75 × 400 mm. Its estimated detonation pressure is 63 kbars with a gas volume of 41 moles/kg and a velocity of 4700 m/s.

In summary, emulsion explosives have the following characteristics:

- small percentage of toxic fumes in the detonating plume;
- good water stability;
- solid charges allow easy handling.

In summary, pumped emulsion explosives have the following characteristics:

- no explosive component and hence safe to handle;
- no toxic components;
- drill and blast using large drill holes containing water are easily possible;
- full utilization of the drill hole volume;
- very efficient loading process possible due to the pumping of the emulsion explosive.

POWDER MATERIALS (BLASTING AGENTS)

Ammonium nitrate–carbon carrier explosives (ANC or ANFO) are powder explosives. Approximately 6% diesel oil is added as a fuel to the porous and absorbable ammonium nitrate. This composition renders the explosive, amongst other things, able to run, very safe to handle (it can be blown in) and gives it more of a crushing effect, which means a low detonation speed of approximately 2500 to 3000 m/s and a high gas cloud volume of nearly 1000 litres/kg. Powder explosives are generally not water resistant. They only find limited application in tunnelling.

In summary, powder (ANFO) explosives have the following characteristics:

- for use in dry boreholes;
- initiation is with an amplification charge;
- high plume (vapour) volume;
- good exploitation (use) of the borehole;
- high toxic fumes.

5.6.5.3 Detonators

A requirement of safety explosives is that they do not detonate easily and they should only do so intentionally. Charges are therefore required to set off the explosion, which themselves contain explosive material – these charges are called detonators.

There have been significant advances in the type of detonator since the invention of the mercury blasting cap by Alfred Nobel in 1865. In the 1880s the *electric detonator* was invented. With this type of detonator, the electrical circuit is closed, the contact-element heats the filament bridge and this ignites the primer material (Figure 5.39a). In the first half of the twentieth century, several developments took place, culminating in the millisecond delay detonator in the 1940s. Electrical detonators have many advantages, such as total control of the initiation time, reduction in the air blast and ground vibration and better blasting results with the delays, which is particularly important in tunnel blasting. However, they have a serious disadvantage with respect to the risk of premature detonation due to extraneous sources of electricity, for example, lightning, static stray currents and radio frequency energy. This severe disadvantage has lead to a reduction in usage of electric detonators in tunnelling and they are not allowed in tunnelling in certain countries, for example Sweden.

At approximately the same time, the *detonator cord* (Figure 5.39b) was developed, which is safe for use in extraneous electricity environments, has no limitation with respect to hole size and is inexpensive. However, it causes

a) Cross section of an electric detonator timer

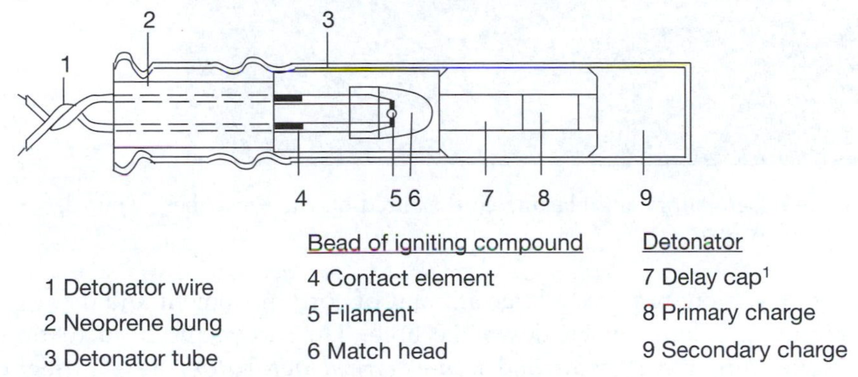

[1] Using non delay blasting (mass blasting) means a simultaneous explosion of charges in all blast holes as contrasted with sequential firing with delay caps.

b) Cross section of a detonator cord

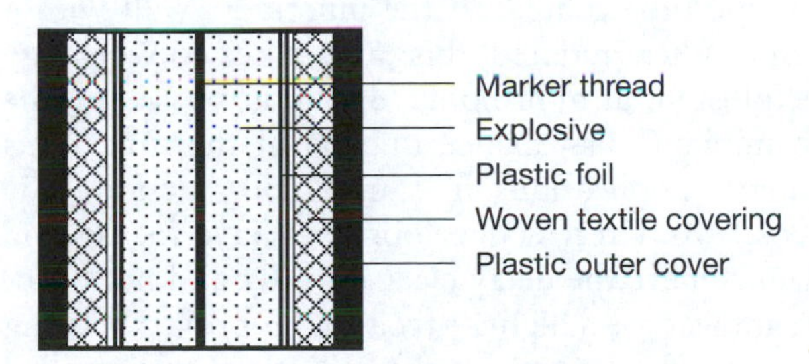

c) Non-electrical detonator

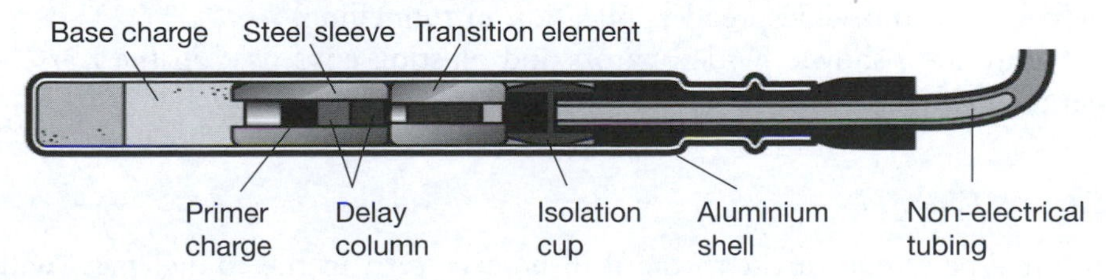

Figure 5.39 Details of detonators

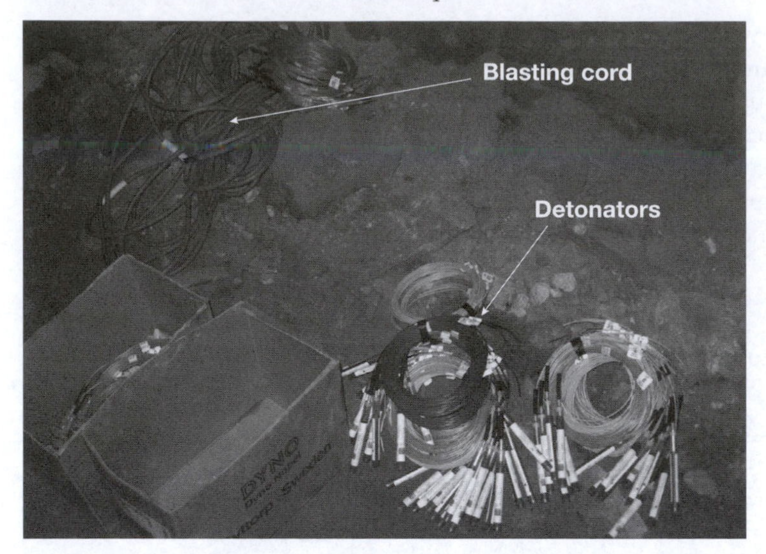

Figure 5.40 Detonators and blasting cord as used on the Katschberg Tunnel, Austria

a 'noisy' initiation, requires a large amount of cord movement and disrupts the stemming column when down the hole. This drove the need to find alternative initiation systems and *non-electrical detonators* were brought to the market in the early 1970s. Instead of electric wires, a hollow plastic shock tube delivers the firing impulse to the detonator, making it immune to most of the hazards associated with stray electrical current (Figure 5.39c). Non-electrical detonators can use a small diameter shock tube, which consists of a three-layer plastic tube coated on the innermost wall with a reactive explosive compound. When initiated, this propagates a low energy signal, similar to a dust explosion, at approximately 2,000 m/s along the length of the tubing with minimal disturbance outside of the tube. This type of detonator finds many applications in tunnel construction. The main disadvantage is its cost. Most recent developments have focused on *electronic detonators*, which achieve the delay electronically and not pyrotechnically. Their main advantages are a higher precision, improved blasting and a reduction in the air/ground vibration as the initiation is controlled by a computer. However, the cost per detonator unit is increased and they require intensive user training. These disadvantages have to be overcome before they find widespread application in tunnelling.

Figure 5.40 shows the detonators and blasting cord used in the Katschberg Tunnel, Austria.

5.6.5.4 Cut types

A cut type is equivalent to the drill hole pattern in the tunnel face, with the drill holes potentially having different lengths and angles. Different cut types are available resulting in different collapse mechanisms. The explosive effect is greater when there is more free surface on which a detonating load can act. Normally in the tunnel there is only one free surface, the face. Therefore the main rule of drill and blast states that in order to achieve the greatest blasting effect the individual charges have to create free surfaces for each other. Hence, they should detonate in a defined spatial and timely sequence.

Generally cuts can be differentiated to be either parallel or angular. In parallel cuts the drill holes run parallel to the tunnelling direction and at right-angles to the face. For angular cuts the drill holes are placed at an angle to the tunnel direction, the existing free surface or the surface that needs excavating.

Figure 5.41 shows an example of a face ready to blast, including the drill pattern used.

The following describes a few examples of collapse arrangements used in tunnelling.

WEDGE CUT

With this angular cut a wedge is detonated out of the centre of the face and after that the remaining part of the advance length is detonated. The wedge can be positioned vertically or horizontally (or at an angle depending on the layering of the ground) as a single or staged wedge. Figure 5.42 shows a wedge cut. The cross sectional area is 92 m^2 and the diameter of the drill holes is 51 mm. The numbers next to the drill holes show the detonation sequence.

Figure 5.41 Example of a face ready to blast in the Heidkopf Tunnel, Germany

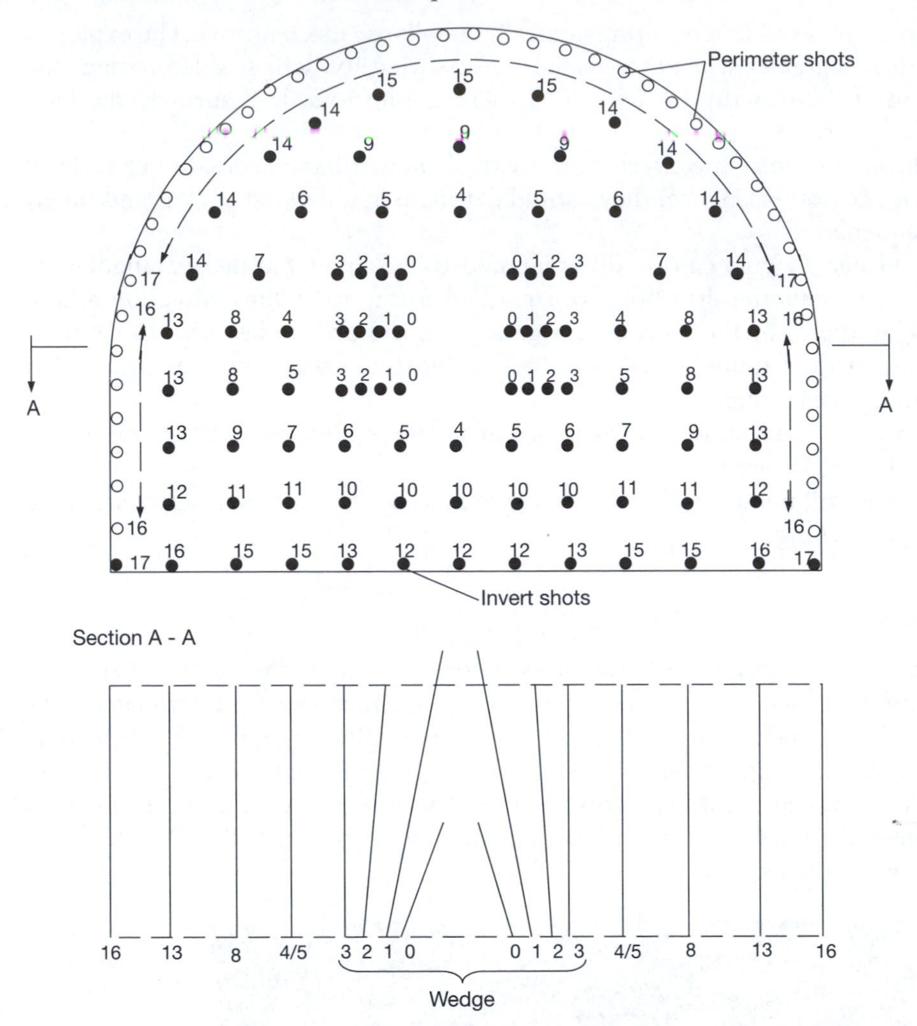

Figure 5.42 Example of a wedge cut drill hole pattern (triple staggered wedge cut with a horizontal wedge arrangement)

If timed detonators are used, the charges detonate in a space of 20 to 30 milliseconds (millisecond effect). This short time sequence is sufficient in order to create free surfaces for the consecutive charges starting from the wedge and to achieve an advantageous blasting effect. As the drill holes are not allowed to run into each other, very precise drilling is required. In tunnelling the detonation charges are generally ignited at the tip of the drill hole furthest away from the face. This results in the greatest detonation power. The detonated ground can be thrown over significant distances, hence a minimum safety distance of 200 to 300 m from the face should be maintained. In order to avoid large pieces of rock being scattered too far, a crushing drill hole, i.e. a drill hole with no explosive, can be placed in the middle of the wedge.

FAN CUT

The fan cut is also an angular cut. For this arrangement several drill hole rows are placed in a fan shape. They have different lengths and are generally positioned against the invert (Figure 5.43a). There are several other basic forms of angular cuts which are randomly applied in tunnelling (for example conical cut). Furthermore the basic forms can be combined (for example a combined wedge and fan cut). Common to all angular cuts is the small amount of explosive material required and the small number of drilling metres, which is particularly true for the fan cut. Generally the drilling works are more complicated for angular cuts as the angular positions of the drill holes have to be precise and the drill carriage has to be frequently repositioned. If the drill hammer on a drill carriage cannot be placed at the desired position or if the advance length does not allow the required angle due to the cramped conditions in the tunnel, a shorter advance length has to be accepted.

PARALLEL DRILL HOLE CUT METHOD

The characteristic for this cut is that the drill holes are the same length and obviously parallel to each other. The positioning and the distance of the drill holes in the middle are important for a successful detonation result. They should be arranged symmetrically and mirrored and, depending on the type of explosive and ground, not lie further apart than approximately 30 to 50 cm (Figure 5.43b). The parallel drill hole method is applicable for ground conditions of lower hardness and toughness.

Due to significantly simplified drilling works, parallel cut arrangements are often used in tunnelling. The drill holes can be drilled from one position using the drill carriage and are the same length for most parallel cut arrangements. There is no interference between the drill hammers for a multi-armed drill carriage due to angling. However, it is important that the drill holes are exactly parallel in order to ensure the success of the explosion. The drill carriages have to be constructed so that the drill hammers can be positioned quickly and reliably in parallel. The blasting effect for the parallel drill hole method relies mainly on a crushing effect. Therefore there are higher requirements for the blasting material, smaller pieces of rock are produced and the debris is thrown greater distances. The last two points can be influenced by a suitable choice of millisecond detonators, and in particular by omitting individual time steps. In order to achieve an economic detonation effect when using a parallel cut it is necessary to use millisecond or half-second detonators and the detonation should start from the deepest end of the drill hole. With small cross sections there is usually no other choice than a parallel cut pattern accompanied by unloaded drill holes, as described in the next section on the 'Burn cut'.

BURN CUT

This is a closely spaced group of boreholes drilled parallel to the direction of advance and perpendicular to the existing face (Figure 5.44). The pattern of boreholes contains both heavily loaded and unloaded holes (Figure 5.44), with the unloaded holes sometimes being drilled at a larger diameter than those loaded with explosive. The empty holes provide a free face for reflection of shock waves. It is important that these holes are accurately drilled and parallel to each other in order to achieve good blasting results.

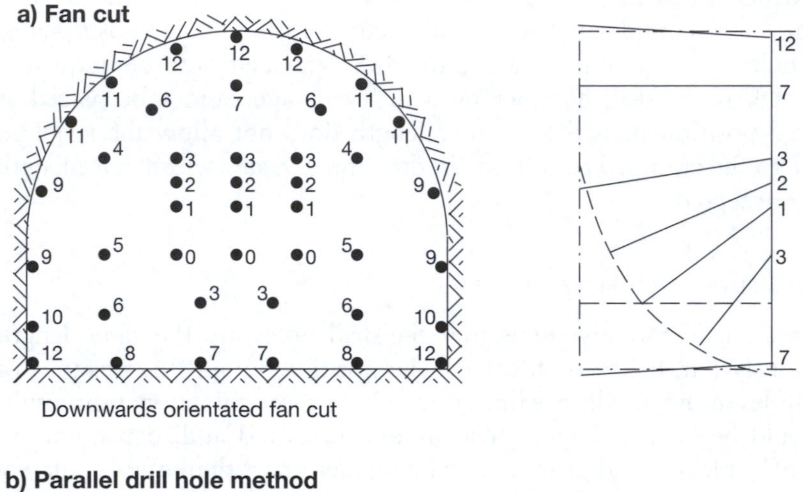

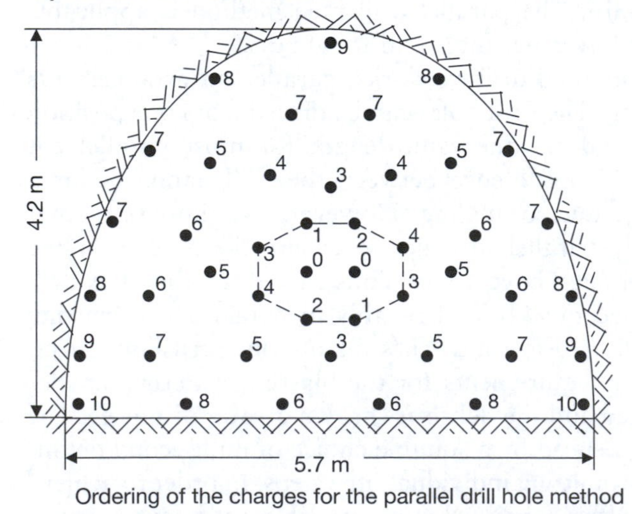

Figure 5.43 Fan cut and parallel drill hole method (the numbers against the drill holes indicate the order of blasting)

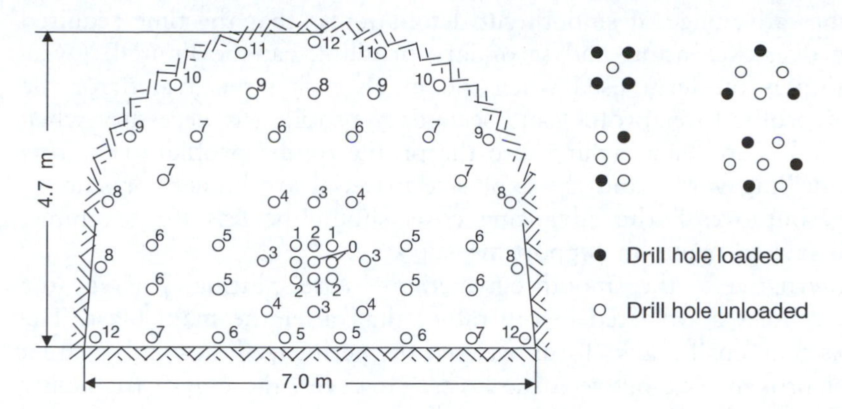

Figure 5.44 Example of the burn cut positioned close to the centre of the face showing examples of the loaded and unloaded drill hole pattern

This arrangement of boreholes is most commonly blasted near or at the centre of the face to break a roughly circular opening. The boreholes that surround this cut area are detonated with millisecond delays and break into this newly created opening. This helps to create a satisfactory advance length or 'pull' for the face.

SMOOTH CUT DETONATION TECHNIQUE

The smooth cut detonation technique, also known as perimeter control blasting or contour blasting, is also a precision detonation technique. For this technique the perimeter holes, which are closest to the edge of the tunnel circumference, are drilled with small spacing and are only loaded with a small amount of explosive. These holes are called trimmers or outer shots and are generally triggered last. The aim is to minimize overbreak and trim the sides of the excavation to the shape and size required. This is particularly important in tunnel blasting as the ground itself is generally part of the permanent structure, and is expected to remain stable and support itself. By creating cracks or opening up joints in the ground by overblasting at the perimeter, the ground may lose all or part of its self supporting properties, thus ultimately requiring additional reinforcement and/or support. Often a blasting fuse is used as a light explosive load. However, in Sweden and other Scandinavian countries small diameter cartridges (11 to 22 mm) are available with a light enough explosive property to prevent damage during smooth blasting. In Figure 5.42 the drill holes numbered 16 and 17 and shown as open circles are the trimmer shots. In addition, the detonation sequence in Figure 5.42 is chosen so that the drill holes 12 to 15 opposite to the trimmer shots are only detonated after 50 to 100 milliseconds. Therefore, the previous and following loads do not build on each other as would generally be the case for the use of millisecond detonators (with detonation steps of 20 to 30 milliseconds).

Another advantage of smooth cut detonation is that the time required for any over-excavation and secondary profiling can be limited. Over-excavation is the term used when too much rock is excavated for the required profile (over-profile) and secondary profiling is necessary when the ground is still encroaching into the profile (under-profile). The costs for the drilling works and the explosive material are higher for smooth blasting, but overall the advancing costs should be less (for example, through savings made on support measures).

An alternative to the smooth cut method is pre-splitting, whereby the perimeter holes are blasted before rather than after the main blast. The theory is that small cracks form between the perimeter holes and create a plane of broken rock between the holes. However, the cracks frequently veer off in the direction of some pre-existing weakness within the ground and so for tunnelling the smooth cut technique is often preferred. It can be used when the face is excavated in benches, where pre-splitting is directed vertically into the top of the bench.

Scaling of the freshly blasted rock profile must always be carried out in order to remove any loose rocks. This scaling can be time consuming as it is usually done by hand using crowbars, and before any spoil is mucked out. Secondary profiling is carried out by drilling a few more holes (1 to 4 additional holes are normally sufficient) and blasting away any under-profile. Scaling must again be done, particularly if the under-profile is in the crown of the tunnel.

5.6.5.5 Explosive material requirements

Figure 5.45b shows the explosive requirements as a function of cross sectional area and advance length for three different types of ground with different strength. However, it is important to note that the type of explosive also influences the required explosion.

The amount of explosive required per cubic metre reduces with increasing cross sectional area due to the relatively smaller interlocking of the face in comparison to a small cross section (also note how the number of drill holes varies with cross sectional area, Figure 5.45a). If the advance length is increased, the interlocking increases, and hence the amount of explosive required increases.

5.6.6 Ventilation

Ventilation is the term used for the artificial ventilation of the tunnel during construction. Air is passed down the tunnel using ducting. It is important to supply air to both the workmen and machinery operating in the tunnel so that the level of oxygen content does not fall below approximately 20% BTS (2008). provides a best practice guide to occupational exposure to nitrogen monoxide, for example produced in the exhaust gasses from diesel and petrol engines, in a tunnel environment.

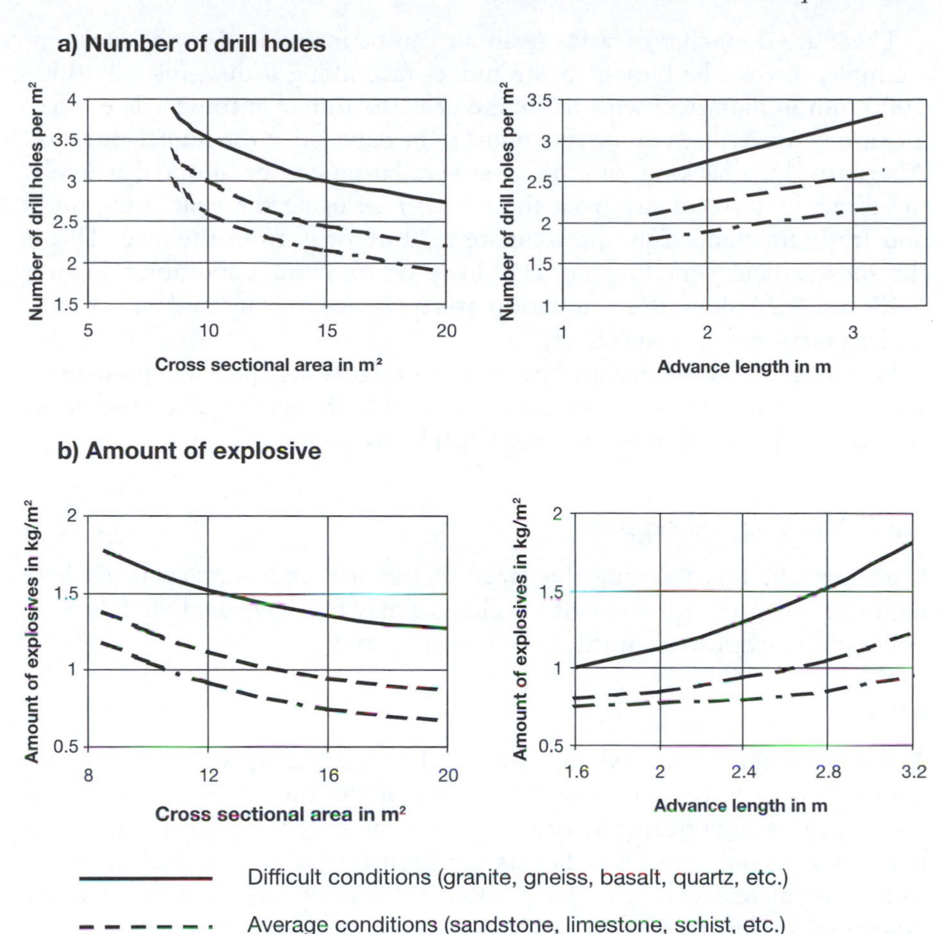

Figure 5.45 Number of drill holes and amount of explosive as a function of cross sectional area and advance length, respectively (after Müller 1978)

With the rock explosives used for tunnelling, the oxygen carrier contains more oxygen than is required in order to achieve detonation. Despite this positive oxygen balance, the amount of oxygen contained in the exploded gas cloud is a long way short of being enough for breathing. The poisonous nature of the gas cloud is merely reduced. Therefore the task of ventilation is not only to provide enough air for breathing, i.e. for the miners during tunnelling, but also to remove the dust and explosive gas cloud and/or sufficiently dilute it. The necessary ventilation time depends on the efficiency of the ventilation, the explosives used and on national laws and regulations. As a general rule, a minimum of 15 minutes artificial ventilation is required before the area of the explosion should be re-entered.

There are a number of ways fresh air can be introduced to the face. For example, air can be blown to the tunnel face along a duct (usually 300–2000 mm in diameter) with its intake near the tunnel entrance. The 'used' air simply moves back along the tunnel to be expelled at the tunnel entrance. Alternatively, a sucking and injection ventilation can be adopted whereby 'used' air is drawn away from the tunnel face using the ventilating ducts and fresh air pumped in approximately 30 m away from the face. This is the most efficient method in very dusty or toxic air conditions. Figures 5.48 and 8.20 show the ventilation system schematically and in a tunnel under construction, respectively.

For long tunnels intermediate shafts designed to provide permanent ventilation can be used. These reduce the length of ducts required and hence the size of the ventilation fans needed (Harris 1983).

5.6.7 Mucking and support

Care must be taken during this stage of the drill and blast cycle as there could be 'misfires', i.e. some of the charges may not have exploded, leaving unexploded explosives in the excavated material.

MUCKING

Mucking is the term used to describe the cleaning up of the excavated material as well as its transportation out of the tunnel. When using full face machines, the transport of the muck is managed using conveyor belts and track-bound vehicles, whereas for tunnelling using explosives track-free removal of muck is preferred. Particularly for large tunnels that are advanced in separate cross sections, it is only possible to use wheeled vehicles. In contrast to track-bound transport, vehicles with rubber tyres have the advantage that they can be utilized for inclines of more than 3%. Furthermore, the need for the labour intensive installation of the track is removed and the investment costs are generally smaller. However, it does require a higher energy input per tonne of excavated material, and as a result of the exhaust fumes from diesel driven vehicles a more elaborate ventilation system is required. The decision on whether the mucking is done by vehicles with rubber tyres or tracks, aided by conveyor belt, or using a combination of these techniques is generally project specific. The advancement performance is mainly dependent on the correct organization of the material flow in the tunnel (logistics). The loading capacity of the vehicles used for example, has to be optimized to the amount of excavated muck and the size of the boulders.

SUPPORT

Depending on the quality of the ground, the next step is the installation of the support for the newly created void. The standard support measures as well as the material and their material properties are described in Chapter 4. Often the support is a combination of steel arches, anchors, steel mesh and sprayed concrete.

5.7 New Austrian Tunnelling Method and sprayed concrete lining

5.7.1 New Austrian Tunnelling Method

The New Austrian Tunnelling Method (NATM) was developed by the Austrians Ladislaus von Rabcewicz, Leopold Müller and Franz Pacher in the 1950s. The name was introduced in 1962 (Rabcewicz 1963) to distinguish it from the 'Austrian Tunnelling Method', today referred to as the 'Old Austrian Tunnelling Method', an old timbering method along with the Belgian, German and English Tunnelling methods (see section 5.2.1).

Although Rabcewicz, Müller and Pacher used techniques and knowledge which were already well known, this was the first time these techniques had been put together in a new and almost revolutionary tunnelling method. Instead of fighting the overburden by a thick lining they acknowledged that the ground, not the lining is the main support of the tunnel. Consequently they reduced the lining thickness radically, down to a mere 20 cm, and they used sprayed concrete instead of the brick lining, which was common at the time. This gave the benefit of a tight and firm coupling between the lining and the ground; whereas the brick lining left a space between the support and the surrounding ground. Furthermore, it was important that the sprayed concrete lining (SCL) was supported by systematic anchoring. Rabcewicz, Müller and Pacher used a flexible approach with respect to the excavation sequence and amount of support. They observed the reaction of the ground as a result of the tunnelling process and used this information to determine the required support and construction sequences. The calculation techniques available at that time could not confirm the stability of this thin lining. Therefore, they used displacement monitoring to prove the efficiency of their support (Schubert 1999).

Müller and Fecker (1978) published 22 principles to fully describe the NATM. The main principles can be summarized as follows.

- The ground offers the main support to the excavated tunnel. The sprayed concrete has only a secondary supporting function (principle 1).
- The original strength of the ground should be preserved. Loosening of the ground deteriorates its strength (principles 2 and 3).
- The support must not be installed too early or too late. It should not be too stiff or too weak (principle 6).
- Force-transfer coupling between the lining and the ground, and the installation of the support at the right time is essential (principles 7 and 9).

- The choice of support and construction sequence is made on the basis of displacement monitoring (principles 8 and 12).
- The method uses a thin sprayed concrete lining (principle 10).
- Increasing the support is not achieved by a thicker lining, but by the use of girders, steel reinforcement and anchors (principle 11).
- The tunnel is to be seen as a composite system consisting of the ground, and the support and stabilizing measures, e.g. sprayed concrete, anchors, steel ribs and similar (principle 13).
- Installation of a structurally acting invert closure (if necessary) will give the ground arch the structural function of a closed tube (principles 14 and 15).
- Full face excavation should be used whenever possible (principle 16).
- The excavation sequence is important for the overall stability (principle 17).
- Maintaining a rounded shape for the tunnel profile (principle 18).
- The inner lining should also be thin. In addition, it should have a force-transfer coupling with the sprayed concrete, but no friction-transfer coupling (principle 19), for example by using a plastic membrane containing air bubbles between the outer and inner lining.

Some explanation of these principles is required. When excavating an underground void, the existing (primary) equilibrium condition of the ground is transformed through a series of stages in which stress redistribution occurs, resulting in a new, stable (secondary) equilibrium condition. NATM has the aim of directing these processes and letting them occur in an economic and technically safe way. In order to achieve this, the level of deformation in the ground should be (principle 6):

1 on the one hand kept small so that the ground does not lose more of its initial stability than unavoidable and;
2 on the other hand must be large enough in order to activate the support of the ground as a closed arch and to optimize the usage of the support measures and the excavation.

In order to prevent a reduction in strength of the ground, it is important to avoid over-excavation. If using drill and blast it is important to blast carefully to maintain an accurate tunnel profile, and to protect the ground outside this profile, and so keep any loosening of the ground to a minimum. This results in a suitable tunnel lining and guarantees its desired slenderness (principles 2 and 3).

Continuous monitoring and visual observation of the ground and the support measures is an integral component of the method (principles 8 and 12). This serves as a proof of stability, a check on the design calculations and the final dimensioning of the support measures during construction, the optimization of the lining thickness as well as the optimization of the construction methods and sequence. The monitoring also takes into account the geological information.

The thickness of the sprayed concrete lining, if used purely to seal the excavation, is normally between 5 and 15 cm. However, if the lining is to act as a structural element, then the thickness is of the order of 15 to 30 cm. The thickness of the lining can of course be greater, but this is not considered to be a 'thin' lining according to principle 10.

If the long-term stability of the sprayed concrete is likely to deteriorate due to aggressive water conditions and/or if there is a requirement for a 100% watertight lining, an inner lining is necessary (principle 19).

Many of the principles are undoubtedly key elements of most of the tunnelling methods used today. Nevertheless, the principles have been heavily discussed since the strict adherence to some of them reduces the flexibility of the method. Some proved not to be of practical use (principle 6), and some did not consider the ground conditions (principles 12, 15, 16) or the size of the tunnel (principle 16) (Schubert 1999). The principles were supported by the official definition of NATM which was published in 1980 by the Austrian National Committee 'Hohlraumbau' (translated from the German):

> The New Austrian Tunnelling Method (NATM) follows a concept which makes the ground (rock or soil) surrounding the void a supporting construction element through the activation of a ground supporting arch.

Unfortunately, this definition did not prove to be unique to NATM since nearly every excavation method tries to preserve the supporting ability of the ground. Rokahr (1995) gave a definition which summarizes the original intention of Rabcewicz, Müller and Pacher and makes it easy to distinguish the NATM from other tunnelling methods: 'NATM is a support method to stabilize the tunnel perimeter by means of sprayed concrete, anchors and other support, and uses monitoring to control stability.'

According to this definition NATM involves:

- support by sprayed concrete;
- support by systematic anchoring if necessary;
- using measurements to control the effectiveness of the support;
- a flexible approach to support measures, i.e. increasing or decreasing the support according to the geological conditions.

It is not NATM when:

- excavation is by TBM;
- support is by segmental lining (steel, SGI, concrete);
- there is no support at all;
- the full overburden is supported;
- no flexible approach is adopted for the support.

A tunnelling method which does not match these NATM features, but uses sprayed concrete as a support should be more generally called 'sprayed concrete lining' (SCL – which is widely used as a synonym for any tunnelling method using sprayed concrete as a support, see section 5.7.2).

In order to use NATM, the ground has to be capable of supporting itself over the length of each advance section, which means that the ground must have a stand-up time. Depending on the philosophy of NATM, the limit of this construction technique is reached when the stand-up time of the ground has to be improved by artificial measures, such as freezing or grout injection. Strictly speaking, NATM assumes that the ground has sufficient stand-up time itself for the construction cycle. The tunnel advance can be achieved using blasting, a partial face boring machine or simply using an excavator, depending on the ground conditions. Generally, the advancement is spatially and timely staggered in the crown heading, bench heading and invert heading (Figures 5.46 and 5.48). Figure 5.47 shows an example of a NATM crown and bench heading.

The length of the advance in the crown can, for example, lie between 0.8 and 4.0 m. As a rule of thumb the length of the advance in the bench is approximately twice the value of the crown.

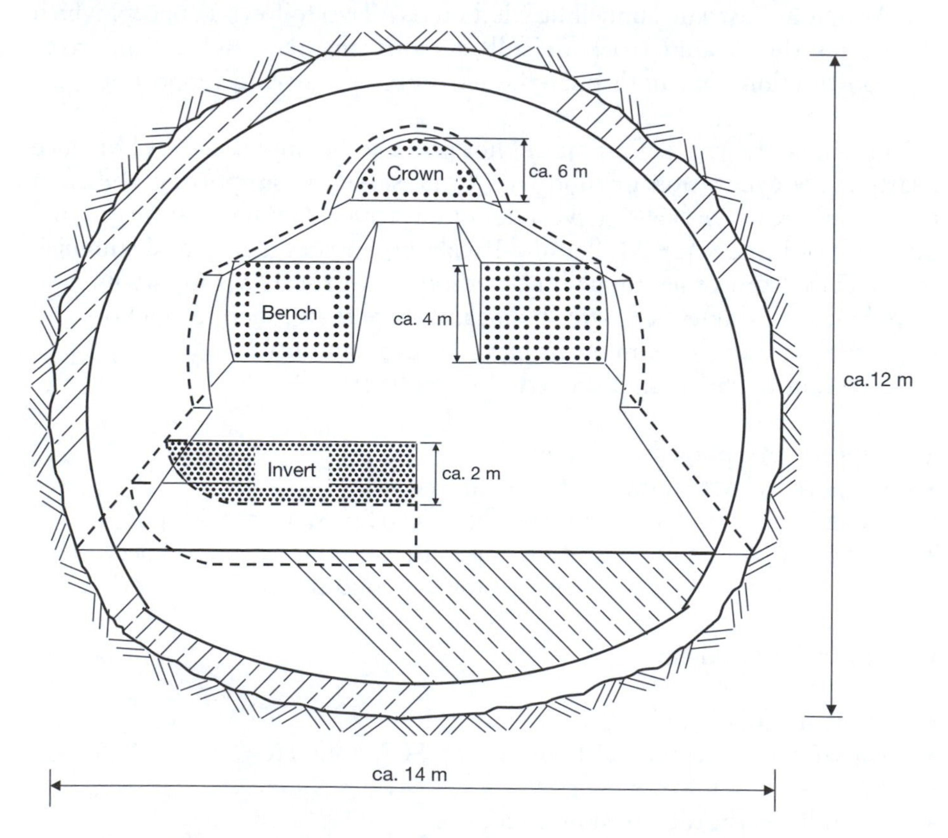

Figure 5.46 Example cross section through a tunnel constructed using NATM

Figure 5.47 Example of a NATM crown and bench excavation during the construction of the Katschberg Tunnel, Austria

The construction process is as follows (see also Figure 5.48):

1 Excavation.
2 Sealing the exposed ground if necessary.
3 Mucking (Figure 5.49).
4 Installation of lattice girders and the first layer of reinforcing bars or mesh reinforcement, and application of sprayed concrete. Depending on the quality of the ground the support might be installed first before the spoil is removed.
5 Potential installation of a second layer of reinforcement and application of more sprayed concrete.
6 If required, installation of anchors, and, if necessary, tightening of anchors a day later and shotcreting of anchor heads.
7 Construction of inner lining.

5.7.2 Sprayed concrete lining

In 1994 three NATM tunnels collapsed at Heathrow airport. The reasons were extensively discussed (HSE 2000, Anon 2000, Rokahr and Mussger 2001). Following the collapse, 'NATM' became disreputable. The HSE started an investigation throughout the UK into the 'NATM' technique used

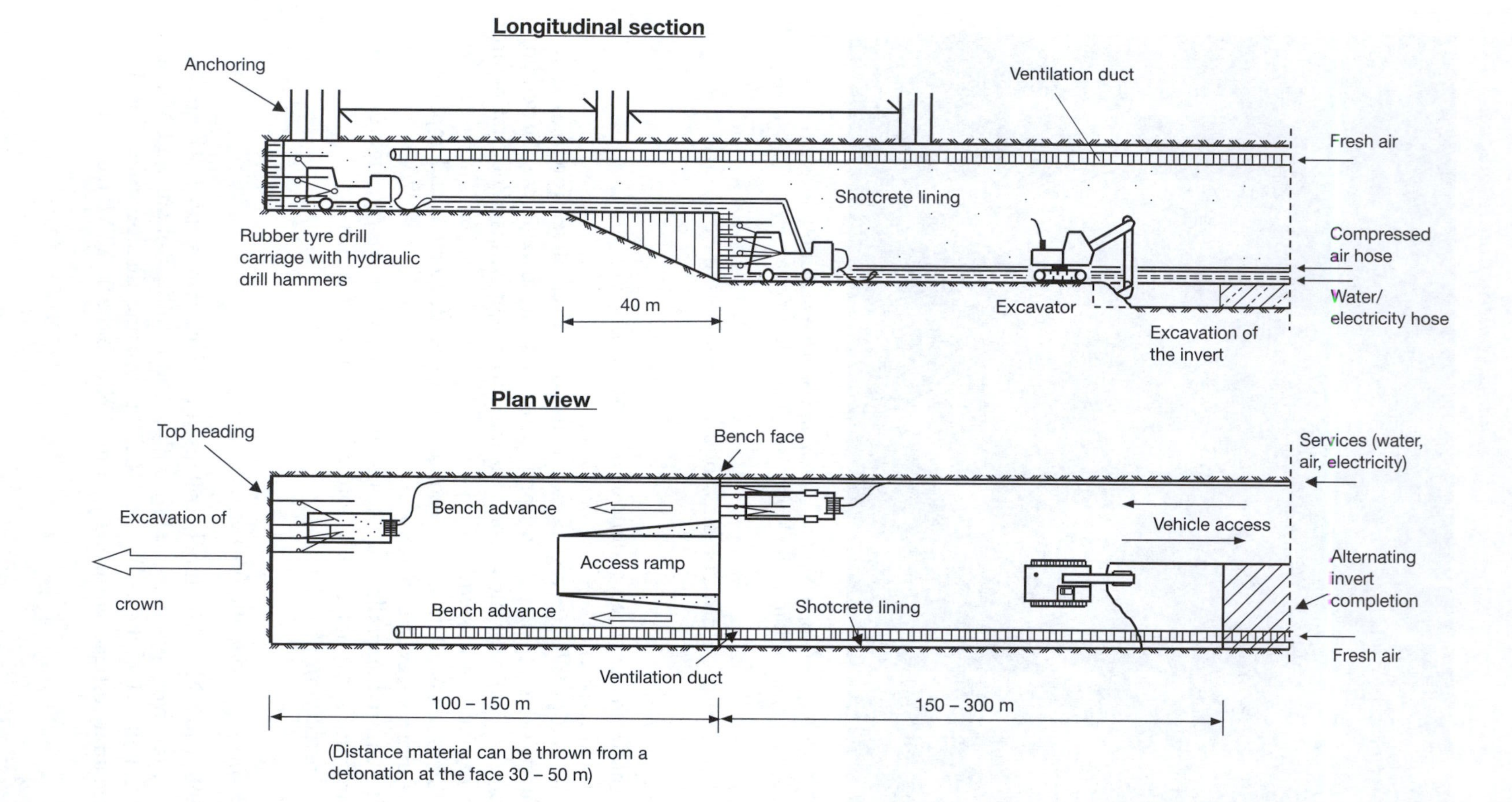

Figure 5.48 Construction activities using NATM for large cross section tunnels in stable ground conditions (no mandatory ring closure)

Figure 5.49 Example of mucking equipment used during the construction of one of the emergency cross passages of the Katschberg Tunnel, Austria

at that time, and as a result published the guideline *Safety of New Austrian Tunnelling Method (NATM) tunnels* (HSE 1996). At the same time the ICE published *Sprayed concrete linings (NATM) for tunnels in soft ground* (ICE 1996). This lead to the use of the term sprayed concrete lining (SCL).

Although the words 'sprayed concrete lining' do not include a construction method (they only refer to the support), the term 'SCL' is, together with 'NATM', used widely nowadays in many countries as a synonym for any tunnelling method using sprayed concrete as a regular support. Unlike NATM, SCL is not related to a specific construction method or to specific principles. SCL is the more general term rather than NATM. However, HSE (1996a) and ICE (1996) give recommendations on how to safely construct tunnels using a sprayed concrete lining as the support. The following gives a short overview.

SCL as used in soft ground in the UK uses many construction techniques related to NATM. It uses sprayed concrete as a primary support, followed by the installation of a permanent lining at some later date. There are attempts to remove the need for the inner lining, for example the LaserShell™ technique (section 5.7.3), but there are issues related to watertightness at the joints in the sprayed concrete lining.

SCL uses an incremental excavation sequence and sprayed concrete as a primary support with, or without, weldmesh, fibres, lattice arches, dowels, anchors and bolts (as described in section 4.2.4). The primary support details are determined in advance of the construction by the designer and are validated during construction by instrumentation and monitoring. The control of convergence and settlement is principally achieved by limiting the length of advance per stage and by providing early invert closure (ICE 1996).

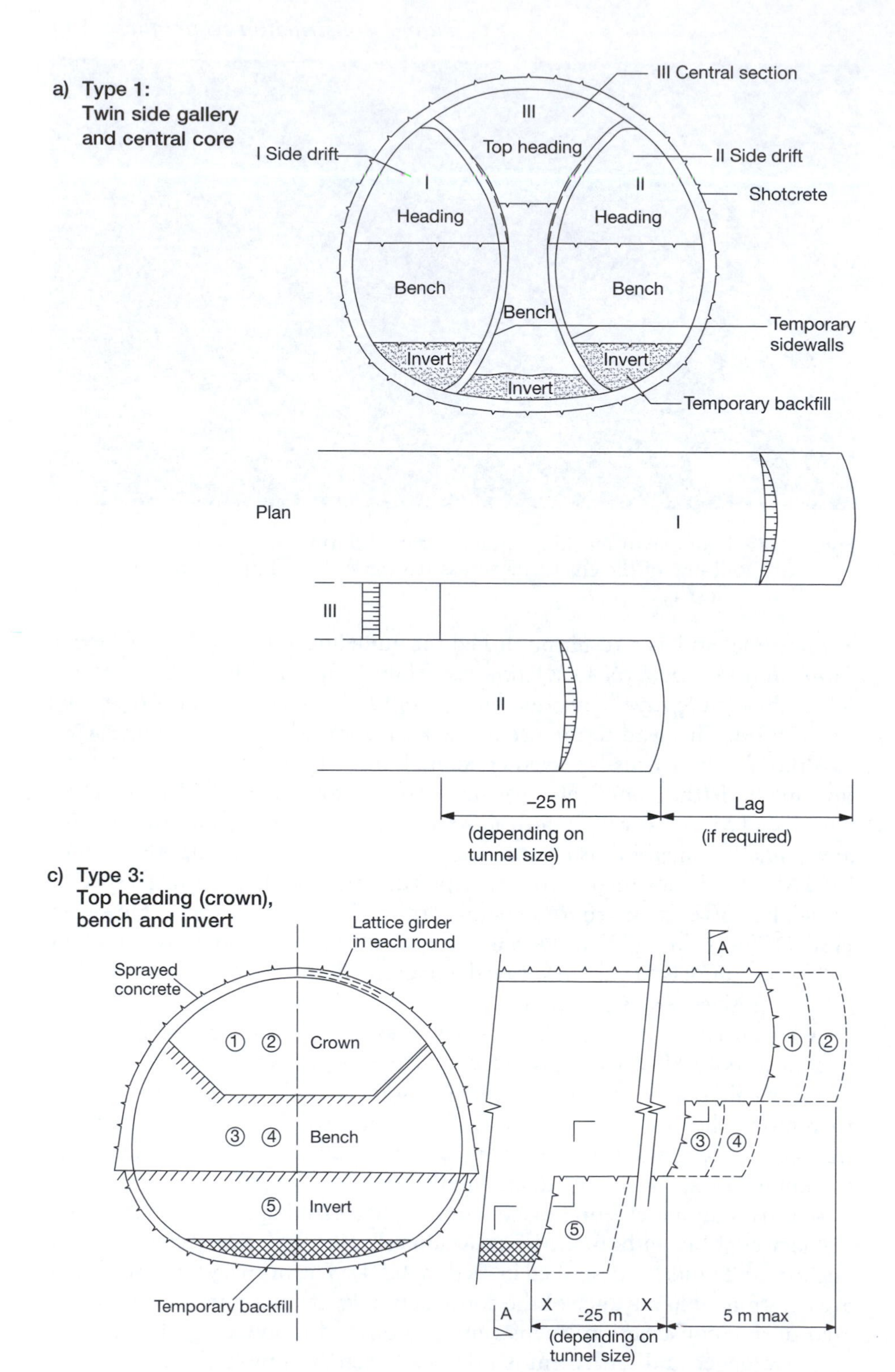

Figure 5.50 Possible excavation sequences for SCL (and NATM) (after ICE 1996, used with permission from Thomas Telford Ltd)

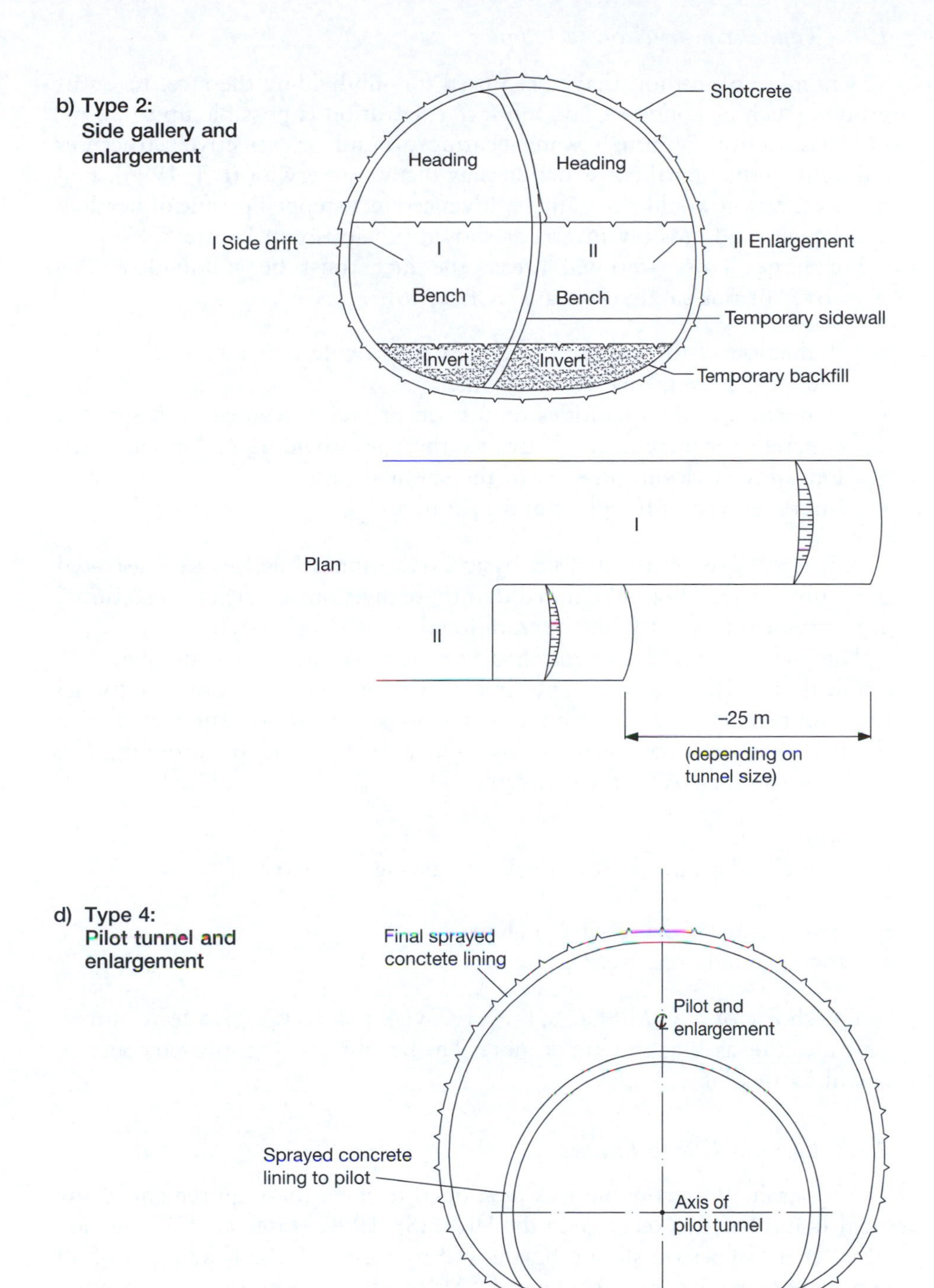

Figure 5.50 (continued)

A number of methods have been used for subdividing the face. In a stiff ground, such as London Clay, full face excavation is possible up to 30 m^2 in cross section, although water-bearing ground or 'sensitive' structures adjacent to the tunnel construction may dictate otherwise (ICE 1996). Full face excavation would normally be advanced in a stepped profile of heading and bench (and possibly invert) as shown previously in Figure 5.46.

For larger cross sectional areas, the face must be subdivided. The objectives of this subdivision are (ICE 1996):

- Reduction of the exposed face area to provide better control of face stability, convergence and settlement.
- Reduction of the quantities of excavation, reinforcement and sprayed concrete per increment of advance thereby providing earlier support.
- Early invert closure in each of the subdivisions.
- Improved access for plant and operatives.

ICE (1996) recommends four basic excavation sequences as illustrated in Figure 5.50. It should be noted that these divisions of the cross sectional area can also be used with the traditional NATM approach.

The use of sprayed concrete has been developed far beyond the early applications. There is nearly no limit to the ground conditions or tunnel size and geometry, which cannot be safely excavated and supported with the use of sprayed concrete (except high water table in soft ground). The range of use for NATM/SCL covers:

- short tunnels;
- non-circular tunnels or tunnels of varying geometry;
- caverns;
- heterogeneous or varying geology;
- high ground pressure.

The choice of NATM/SCL is, however, very project specific, for example 'short' can be as long as 5 km or more. The benefit of using sprayed concrete is still its flexibility.

5.7.3 LaserShell™ technique

The LaserShell™ technique was developed to meet the requirement of the Health and Safety Executive in the UK (HSE 1996, statement 310), which states that 'no person should be allowed to approach the heading until all exposed ground has been supported'. Although the technique was specifically developed for tunnelling in London Clay, it is equally applicable to other types of soft ground. The method is described in section 8.2.5 as part of the case history on the PiccEx Junction tunnels at the London Heathrow Terminal 5 project. A detailed description of this method is also given in Eddie and Neumann (2003 and 2004) and Jones *et al.* (2008).

5.8 Cut-and-cover tunnels

5.8.1 Introduction

The cut-and-cover method for constructing tunnels offers an alternative approach to underground construction techniques. This method involves constructing the tunnel structure in a braced or anchored, trench-type excavation ('cut') and this is subsequently backfilled ('covered'). Pipelines, such as sewers, vehicular tunnels and metro tunnels are often constructed using this technique. In locations with no important constraints on, or close to, the ground surface, using the cut-and-cover method is often cheaper and more practical for shallow tunnels (10–15 m) compared to underground tunnelling methods. However, depths of 30 m are quite common for metro tunnels, which have been helped by advances in construction techniques, such as the use of diaphragm walls (section 5.8.4). (Kuesel and King 1996)

Using the cut-and-cover method can also reduce the risk compared to underground construction, particularly with respect to the health and safety aspects of working underground. The approaches to immersed or bored tunnel sections will often be constructed as cut-and-cover tunnels. It is also common for metro stations to be constructed as cut-and-cover tunnels if there is suitable access from the ground surface. However, in urban areas cut-and-cover tunnel construction can be very disruptive as access to the ground surface over extended areas and for long periods of time is difficult. Even finding suitable sites for access shafts for bored tunnels can be extremely problematic and this can dictate the alignment of a tunnel.

5.8.2 Construction methods

Two basic forms of cut-and-cover tunnel construction are available:

BOTTOM-UP METHOD

An excavation is made from the gorund surface and the sides are supported. The tunnel is then constructed within this excavation. The tunnel may be of *in situ* concrete, precast concrete, precast arches and corrugated steel arches. The excavation is then backfilled, and the surface reinstated. This method has the benefit of allowing good access to the construction area, but means that the surface reinstatement happens last, which in congested urban areas may be unacceptable.

TOP-DOWN METHOD

From the ground surface level, the support walls and capping beams (beams constructed on top of the side walls) are constructed. These walls can be constructed using diaphragm walls, contiguous or secant piled walls or another method (these techniques are described later in this section). The roof of the tunnel structure is then constructed close to the ground surface in a shallow excavation. Access openings are left in this roof structure, but otherwise the ground surface can be reinstated. The remaining construction then takes place beneath this roof structure via the access openings. As some of the surface reinstatement can happen relatively early on in the construction process, the potential disruption in urban areas can be minimised. However, the working conditions are more restricted.

Cut-and-cover construction is often used for underground metro stations, for example Canary Wharf Station on the Jubilee Line in London, UK. Modern stations constructed using cut-and-cover techniques can offer more open areas compared to those constructed using bored tunnel techniques. This construction form generally has two levels, which allows economical arrangements for ticket halls, station platforms, passenger access and emergency egress, ventilation and smoke control, and staff and equipment rooms. The original excavation for Canary Wharf Station was 35 m wide, 280 m long and 26 m deep (Figure 5.51). In this case anchored steel sheet piles provided temporary support to create the initial 11 m deep excavation, 8 m below the groundwater table. Diaphragm walls were used below this level. Extensive pumping was utilized in order to keep the excavation dry.

Figure 5.51 Canary Wharf Station during construction on the London Underground Jubilee Line Extension, UK

5.8.3 Design issues

A cut-and-cover structure must be designed to take the various loading conditions it is likely to experience over the life of the structure. Under normal situations these are likely to be: the development of water and earth pressures, dead loads including the weight of the fill cover, surcharge load and live loads. There may also be requirements for earthquake loading considerations.

An example of the possible loadings for the design of cut-and-cover subway structures are shown in Figure 5.52. An indication of how to calculate vertical and horizontal earth pressures and water pressures with depth is provided in section 3.2.

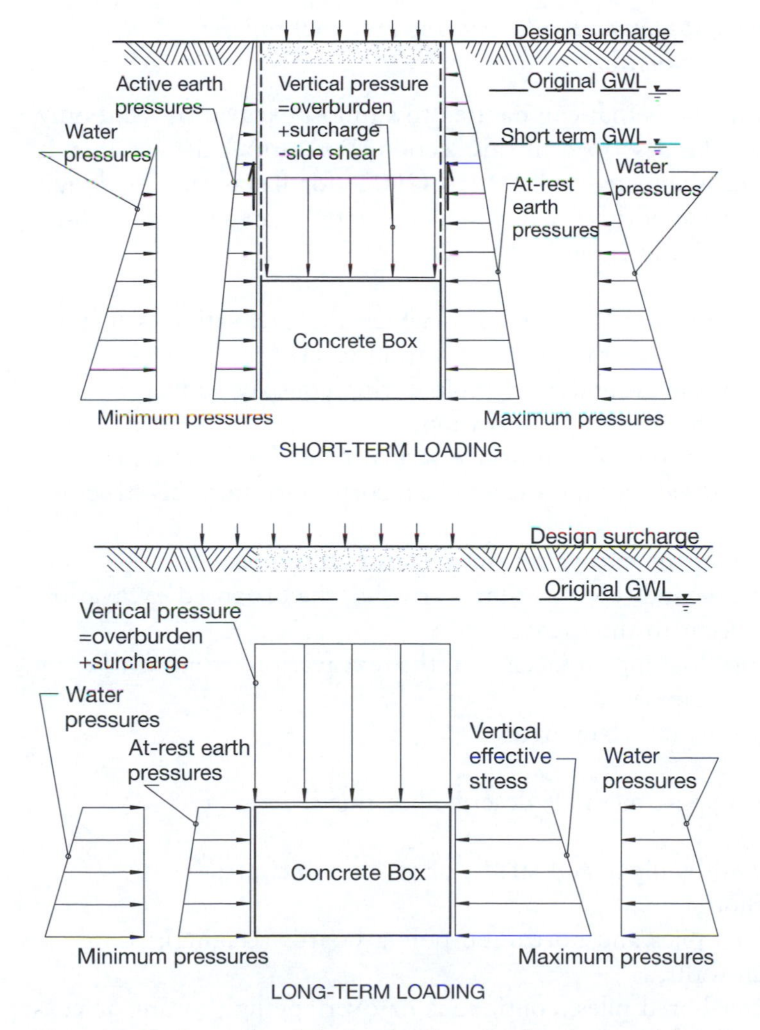

Figure 5.52 Example of design loadings in the short and long-term for a concrete box structure (after Kuesel and King 1996)

When the groundwater table lies above the bottom of the invert or base slab of a subsurface structure, an upward pressure – buoyancy – on the bottom of the base slab equal to the piezometric head at that level must be accounted for. This can be countered by the weight of the tunnel section, which can be increased by thickening the walls, roof and base slab or increasing the width of the base slab. Alternatively, tension piles or tie-down anchors can be provided.

The design of the excavation side supports (shoring systems) must minimize movements and hence subsidence to adjacent structures and services. Any temporary or permanent lowering of the groundwater must also be considered in terms of the effects on adjacent structures and services.

5.8.4 Excavation support methods (shoring systems) for the sides of the excavation

There are many methods that can be used to support excavations and only some of these will be described in this section. For further details on this subject the reader is directed to Macnab (2002) for instance. The design of the excavation support system will depend on many factors, which include (after Kuesel and King 1996):

- the physical nature of the ground in which the excavation is to take place, including below the final excavation level;
- the position of the groundwater table during construction;
- the width and depth of the excavation;
- the configuration of the subsurface structure to be constructed and whether the excavation supports will be incorporated into this structure or not;
- the proximity of the excavation to adjacent structures;
- the number, size and type of utilities crossing the proposed excavation and also adjacent to the excavation;
- the surcharge loading adjacent to the excavation from traffic or construction equipment;
- noise restrictions in urban areas.

The excavation support methods described in this section are:

- sheet piles with walings and struts, or ground anchors;
- ground anchors;
- soldier or king piles and horizontal poling boards (lagging);
- slurry trench walls;
- large diameter bored piles, contiguous or overlapping ('secant' piles).

STEEL SHEET PILING

This provides a relatively simple solution to excavation support in soft ground as long as there are no obstructions in the ground, such as boulders or existing foundations. The sheet piles are driven into the ground to the required depth and they interlock with their neighbouring piles, as shown in Figure 5.53. As the excavation takes place on one side of the sheet piles, walings (horizontal supports positioned along the wall) and struts (supports positioned across the excavation) are placed to brace the excavation and reduce movements of the sheet piles. Struts across the excavation can make access to the excavation difficult and so ground anchors may be used as an alternative (ground anchors are described below). The spacing between struts or anchors, and the number of rows of struts and anchors for the depth of the excavation must be carefully considered at the design stage with respect to minimizing lateral movements in the sheet piles and, therefore resulting surface settlement. In order to reduce the lateral movements in the sheet piles, imposed loads in the struts or pre-stressing in the anchors can be used.

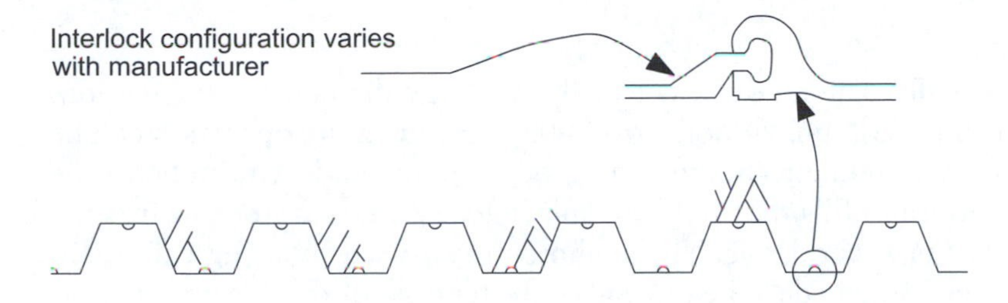

Figure 5.53 Sheet pile excavation support (after Kuesel and King 1996)

GROUND ANCHORS

Ground anchors can be used with sheet piles or other excavation support systems to remove the need for struts. This removes obstructions from within the excavation area. They can also be pre-stressed to help reduce deflections in the earth-retaining wall. Figure 5.54 shows a typical ground anchor arrangement. One of the issues with ground anchors, however, is the fact that they extend far behind the line of the wall and hence potentially outside the limits of the land acquisition for the current project. In this case care must be taken to ensure there is no impact on adjacent basements or other underground structures and that appropriate permissions are obtained from the landowners of the adjacent land. There are a number of anchor types and the choice is based on the ground type.

The construction must ensure that a stable hole is created so that the anchor tendon can be inserted and grouted. The hole can be created by rotary percussive drilling and casing in granular soils or hollow-stem augering in

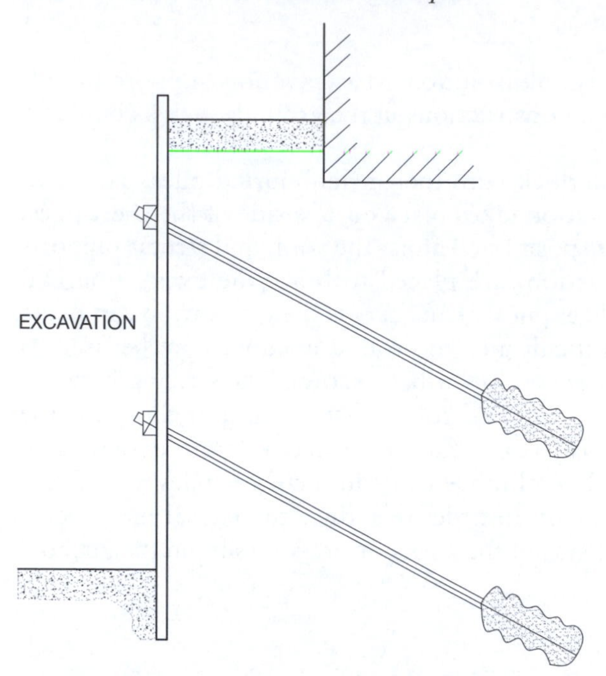

Figure 5.54
General arrangement of an anchored retaining wall (after Megaw and Bartlett 1982, used with permission from John Bartlett)

fine grained soils. The hole size is usually twice the diameter of the tendon, typically 100 to 250 mm (Woodward 2005). The grouting operation can be carried out by a tremie technique, i.e. grout is pumped down a tube to the end of the hole and fills the void by displacing any air or water towards the open end of the anchor hole. The anchors are stressed by using calibrated hydraulic jacks attached to the top end of the tendon, or if it is a bar anchor, a torque wrench, and this acts against the anchor head.

For more information on the design of ground anchors for retaining walls as well as the type of anchors and their choice based on the ground type, the reader is directed to BSI (1989 and 2000).

KING PILES AND POLING BOARDS

King (or soldier) piles are regularly spaced heavy steel sections lowered into a predrilled hole. If bedrock is present at a suitable depth, then the toe of the king pile can be grouted into the rock. The piles require strutting or ground anchors as the excavation proceeds. Between the king piles the ground is supported by horizontal poling boards (wooden planks). Figure 5.55 shows the general arrangement of a king pile and poling board retaining wall. As an alternative to poling boards, the ground between the king piles can be excavated and a slurry trench wall constructed.

DIAPHRAGM WALLS (SLURRY TRENCH WALLS)

Diaphragm walls are reinforced concrete walls, usually cast *in situ* but may be precast panels, constructed in a trench supported by bentonite slurry.

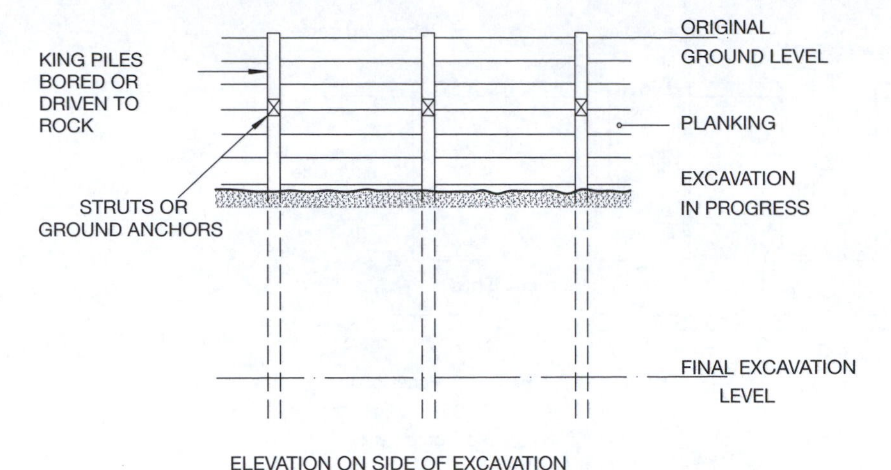

Figure 5.55 General arrangement of a king pile and poling board retaining wall (after Megaw and Bartlett 1982, used with permission from John Bartlett)

They are usually part of the permanent load bearing foundations or perimeter wall (Woodward 2005). Figure 5.56 shows the general principle behind the construction of a cast *in situ* diaphragm wall. The concrete is tremied in at the bottom of the slurry filled panel, displacing the slurry. Typical panel lengths in soft ground are 2 to 3 m, but in stronger ground the panel length can be 6 m. There are a number of excavation techniques available to remove the ground through the slurry. Figure 5.57a shows an example of a grab or clam bucket excavator and Figure 5.57b shows a hydrofraise, which is equipped with two cutter drums with tungsten carbide-tipped cutters rotating in opposite directions.

The support slurry consists of a bentonite slurry and relies on the thixotropic and gel strength properties of the bentonite. Thixotropic means that when the material is not agitated it becomes a solid whereas in an agitated state it is a liquid. As the excavation process is carried out, the bentonite slurry can be screened of the excavated ground and can be reused.

Diaphragm walls can also be constructed as individual panels and used as supporting columns (Chapter 8, Figure 8.8). Diaphragm walls have been constructed to 120 m depth. The main limitation is the construction tolerance with respect to verticality. The hydrofraise offers the most accurate method for constructing very deep diaphragm walls. Further details of diaphragm wall construction can be found in BSI (2008a).

LARGE DIAMETER BORED PILES

An alternative retaining wall construction uses large diameter bored piles at close centres concreted *in situ*. The piles are constructed down to the required foundation level and can be reinforced. The piles may be contiguous or

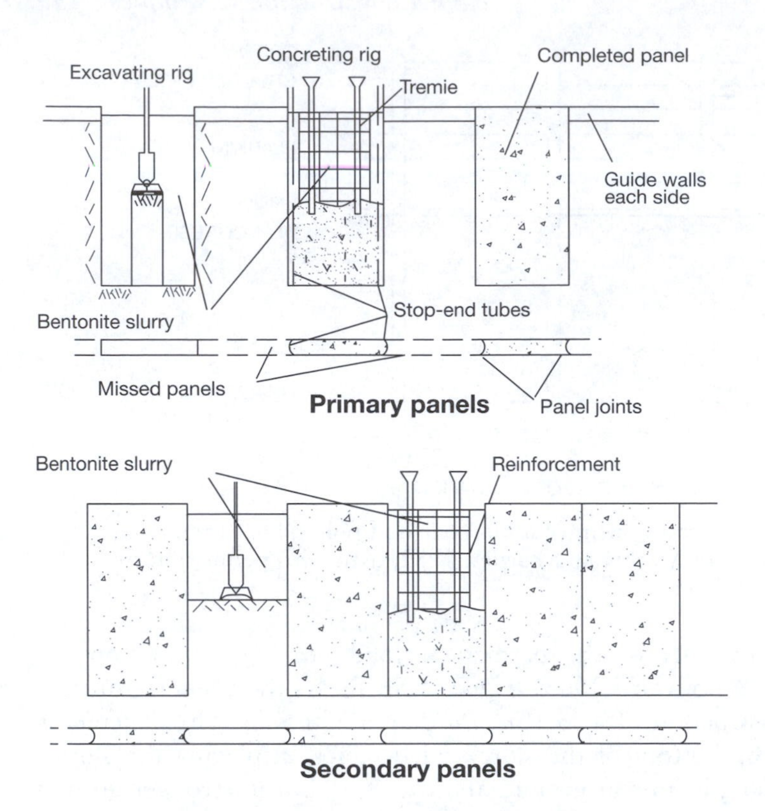

Figure 5.56 Typical construction process for a cast *in situ* diaphragm wall (after Woodward 2005)

Figure 5.57 a) Grab or clam bucket excavator, and b) hydrofraise used for diaphragm wall construction (courtesy of Bachy Soletanche Ltd)

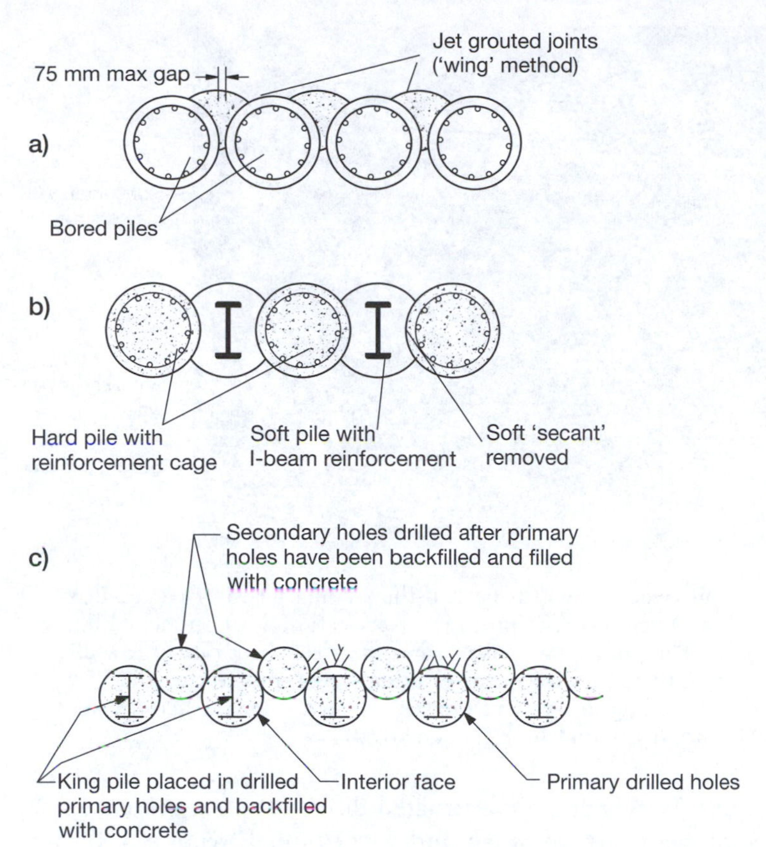

Figure 5.58 General arrangements for bored pile retaining walls, a) contiguous pile wall, b) secant pile wall (after Woodward 2005), c) off-set bored pile wall (after Kuesel and King 1996)

overlapping ('secant' piling). Contiguous pile walls involve successive unconnected piles bored in close proximity in a line (Figure 5.58a). Secant pile walls involve interconnecting piles. The construction sequence uses alternate unreinforced piles. These are then infilled with a reinforced pile which cuts into the unreinforced pile (Figure 5.58b and Figure 5.59). Alternatively the unreinforced and reinforced piles can be constructed slightly off-line from each other (Figure 5.58c). This configuration provides considerably more strength to the wall. Further details on bored pile construction can be found in BSI (2008b).

5.9 Immersed tube tunnels

5.9.1 Introduction

Although a majority of tunnel construction occurs within the ground, for example TBM and NATM tunnelling, there are techniques, most notably

Figure 5.59 Secant pile wall as used as part of the jacking pit for a jacked box tunnel at Owen Street, Tipton, UK (see section 5.10 on jacked box tunnels). The photograph also shows the steel sheet pile headwall with its waling and horizontally raking struts transferring the headwall loading into one of the secant piled sidewalls (courtesy of BAM Nuttall Ltd and John Ropkins Ltd)

immersed tube tunnels, which are constructed differently. An immersed or submerged tube tunnel is a type of cut-and-cover tunnel (section 5.8), but located underwater using pre-fabricated elements constructed in the dry at some distance from the tunnel location and made watertight with temporary bulkheads. These elements are then floated into position, lowered into a dredged trench on the river/sea bed, and joined together. These stages of construction are described in detail in section 5.9.2. One of the key design criteria for this type of tunnel, in contrast to bored tunnels, is the need to ensure adequate stability against uplift.

Immersed tube tunnels are ideal for crossing rivers and estuaries in urban areas. Due to the location just under the river/sea bed, this method can be considerably cheaper than excavating or boring (TBM driving) through the ground under the river/sea bed. The surface infrastructure (roads, railtrack) needs to connect to the tunnel but is constrained by limits on gradients that are suitable for cars or trains. Therefore, the deeper the tunnel, the longer this lead in section needs to be. As bored tunnels are generally constructed at a greater depth, which is necessary for ground stability during construction, they are often longer than immersed tube tunnels and hence more costly. Bored tunnels may also be technically more challenging due to the high water pressures, which can be a problem during construction.

The first immersed tube tunnel was built in the United States for conveying water across the Shirley Gut in Boston Harbour in 1894. The first transportation immersed tube tunnel was the Michigan Central Railroad Tunnel under the Detroit River in the United States which was completed in 1910. Since then there have been many immersed tube tunnels constructed around the world for both service and transportation and some examples are given in section 5.9.7.

5.9.2 Stages of construction for immersed tube tunnels

The main features of the method for immersed tube tunnel construction are (construction 'Stages' refer to Figure 5.60):

- Tunnel elements with a complete cross section and of convenient length (usually 100 to 180 m) are fabricated in a shipyard, in a dry dock or in a casting basin, depending on the type of construction and available facilities (see Figure 5.64).
- These elements, closed by temporary bulkheads, are then floated and towed to their position in the tunnel alignment (Stages 2 and 3). For the moderate water depths normally experienced for immersed tunnel projects, alignment control is achieved via surveying techniques using two towers attached at either end of each tunnel element. The surveying is normally done by theodolite and GPS, and final alignment/realignment towards the previous element can be controlled by hydraulic jacks across the joint. For very large water depths, subsea positioning and survey systems have been applied.
- The elements are sunk (Stage 4) into a pre-dredged trench (Stage 1) by filling temporary water ballast tanks inside the elements, then joined together and watertight connections are formed.
- The tunnel foundation is prepared by either constructing a levelled gravel bed prior to the element immersion, or by placing the elements on temporary foundation pads and subsequently jetting sand into the gap between the elements and the trench bottom (Stage 5).
- The tunnel is then protected by backfilling the excavation and placing rock protection on the top (Stage 6). Sufficient ballast is placed inside and/or on top of the tunnel to provide safety against uplift for the permanent tunnel.
- The tunnel is then sealed and the water ballast pumped out.

The side slope of the tunnel trench should be sufficient to ensure the stability of the trench during the period from dredging and until the backfill has been placed. In harder materials such as rock, these side slopes can be near vertical, but can be as shallow as 1:4 in soft material. Typically, a slope of 1:1.5 is feasible. Dredging is normally done in several stages. These consist of bulk dredging initially, with fine dredging to the final dimensions being conducted a few element lengths ahead of the elements being placed in order to keep the time interval between this operation and the

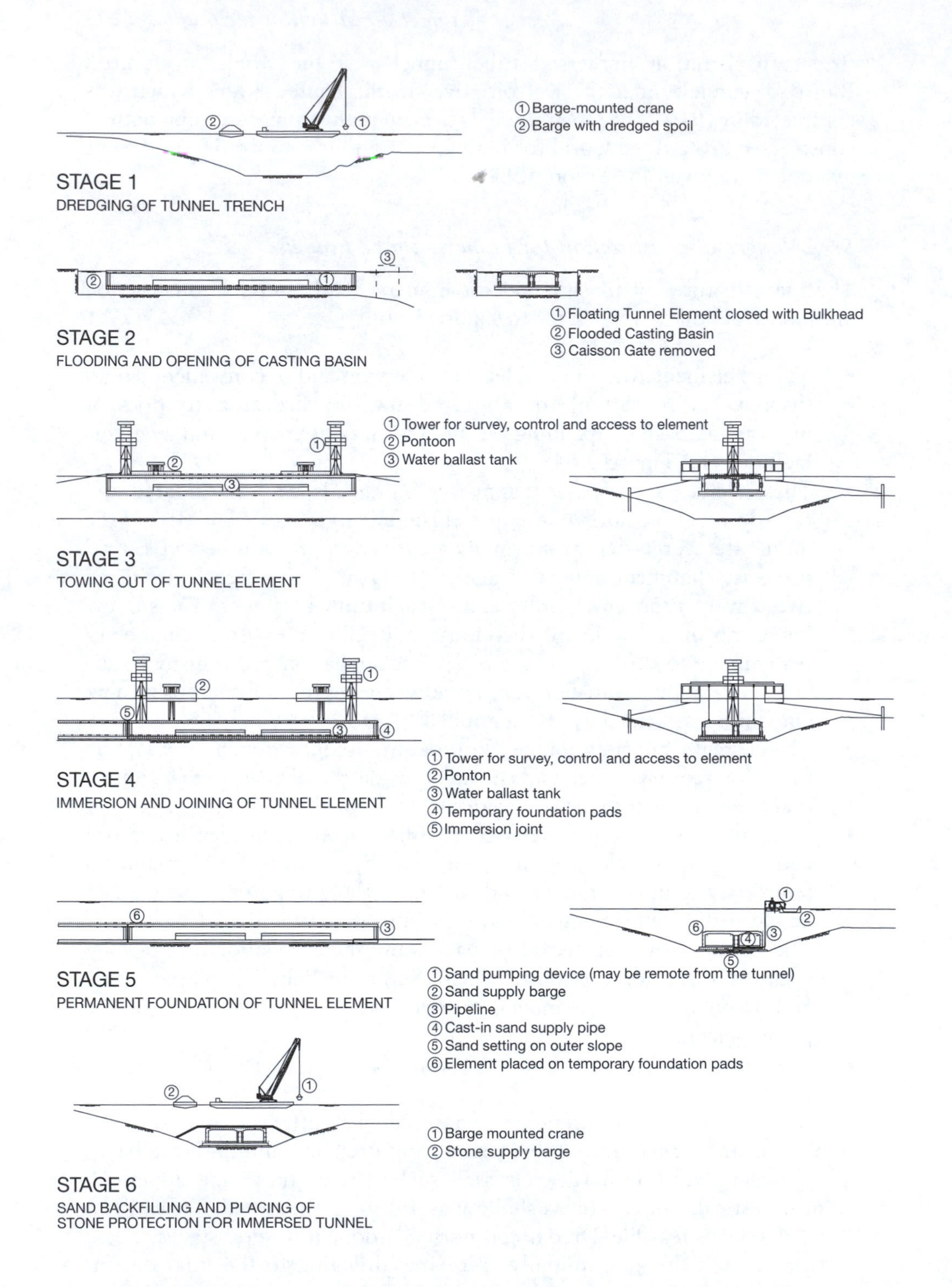

Figure 5.60 Typical sequence for the construction of an immersed tube tunnel (courtesy of COWI A/S)

final element placement to a minimum. Before the foundation course is constructed, the trench is checked for any sediment accumulation and this is removed as required (Kuesel and King 1996).

The top of the tunnel should preferably be at least 1.5 m below the original bottom to allow adequate top protection. If this is not possible, the tunnel may sit proud of the bottom and be protected by backfilling extending, for example approximately 30 m on each side of the tunnel or more, depending on the conditions regarding ship navigation and associated risks of accidental impact. The fill must be protected against erosion by currents with a rock blanket or similar means.

Another consideration is that there must be sufficient time during the tidal cycle when the current is small enough, preferably less that 1 m/s over a 2 hour period, to permit lowering of the elements.

Immersed tube tunnels are very different from bored tunnels and this is illustrated by the following design and construction aspects that need to be considered at the early design stage (after Ingerslev 1990):

- topography, geology, hydrology and meteorology;
- alignment criteria;
- conceptual design as steel or concrete tunnel, including a method of waterproofing by either watertight concrete or an external membrane;
- steel fabrication yard in case of a steel tunnel and dry dock location and lay-out in case of a concrete tunnel (see section 5.9.3);
- schedule for construction as well as for design, including constraints on the construction sequence;
- choice of location with adequate facilities for outfitting of elements;
- access to the interior after placing in case of very large water depths where standard solutions are not applicable;
- towing method and route;
- survey methods during sinking and placing;
- navigation, fairway requirements, and the method of sinking (including plant, equipment, and temporary ballast) and of joining the tunnel elements. These considerations combined may limit the practical tunnel element lengths;
- foundation method and the associated stability for the tunnel elements, both immediately after sinking and in the longer term. For example, temporary supports and screeded gravel foundation and sand flowing (see section 5.9.4);
- methods, and cost, of excavation and disposal of the tunnel trench materials which might include gravels, sands, silts, clays or even rock and boulders. It could be that a longer shallower tunnel on a different alignment may be cheaper than a shorter deeper tunnel, especially if rock excavation is involved as this can be expensive below water;
- source of, type of and method of placing backfill around the tunnel after positioning. Special backfill may be required for seismic conditions;

- protection against accidental dropped or dragged anchors, and ship grounding;
- construction methods and sequence for the immersed tunnel approach structures. This would also include excavation support and methods for dewatering, cut-and-cover tunnels, approach ramps, and ventilation buildings;
- drainage of the interior, including sumps at low points of the alignment and at portals;
- electrical, mechanical, operational and maintenance aspects, as these are often on the critical path and can often delay the opening of tunnels.

5.9.3 Types of immersed tube tunnel

There are two distinct types of immersed tube tunnels, steel shell and concrete. The steel shell type is traditionally preferred in the United States and the concrete type in Europe.

5.9.3.1 Steel shell

Steel shell immersed tube tunnels can be built using either a single or double-shell construction. In a single-shell construction, there is actually an internal concrete lining and an outer steel shell plate that has been stiffened internally. This outer steel shell acts as a permanent watertight membrane, as the formwork for the internal concrete lining and as the structural element to carry the flexural forces before and after placement. In the double-shell construction, there is an additional outer steel shell. The area between the two steel shells provides a convenient space for additional ballast. The steel structure and part of the concrete is normally constructed at a shipyard, either in a dry dock or launched from a slipway. Unlike concrete elements which are sensitive to settlement, shrinkage and creep effects, steel shelled elements have sufficient flexibility and ductility so that these features do not control the design. However, the structural analysis of a steel element must consider each stage of construction separately: fabrication and launching, outfitting and *in situ* loading.

Typical cross sections for steel shell immersed tube tunnels are shown in Figure 5.61.

5.9.3.2 Concrete

Rectangular reinforced concrete elements are generally used for tunnels with four or more traffic lanes, particularly where concrete is more economical than steel, as they are better suited for the rectangular traffic clearance gauge than the semicircular shape of the steel tunnel cross section.

Some typical cross sections for concrete immersed tube tunnels are shown in Figure 5.62.

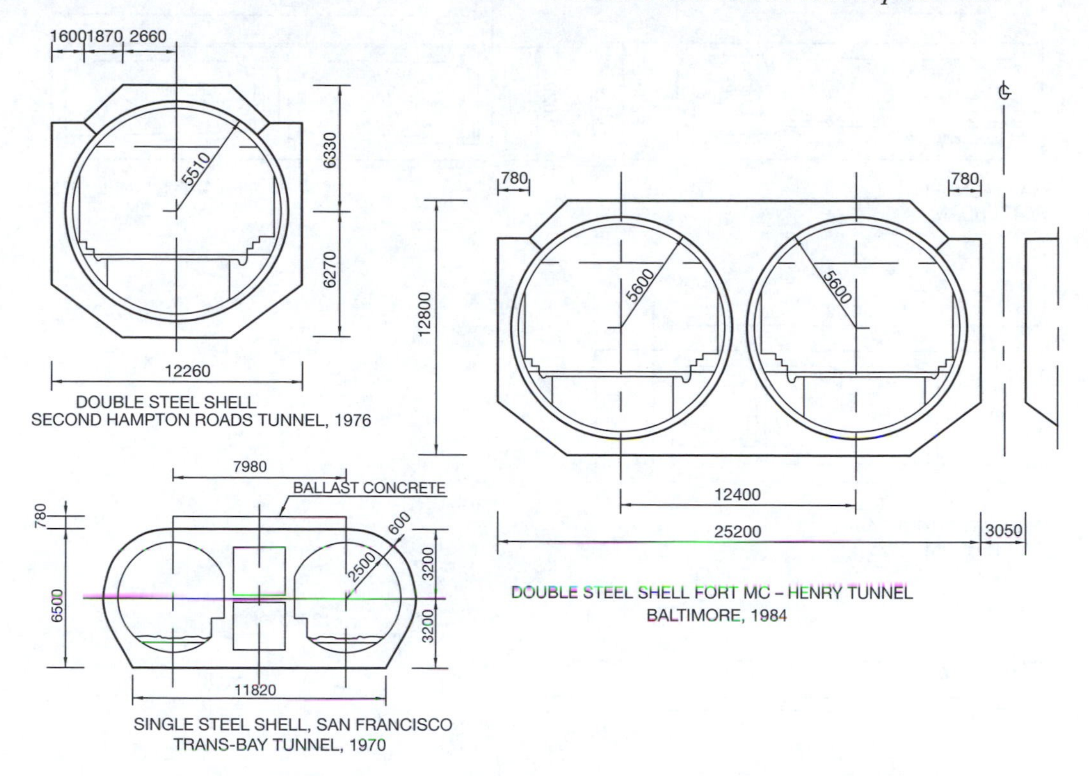

Figure 5.61 Typical cross sections for steel shell immersed tube tunnels (after ITA 1997)

WATER TIGHTNESS AND CRACK CONTROL

Early-age cracking during element construction is the major cause of cracking in concrete structures. This can be alleviated by the use of cooling pipes to control temperature-induced cracks during curing above the construction joints between the walls and top and bottom slabs. Alternatively, temperature cracks can be avoided by continuous casting of the full cross section. Cracking of the structure after placing can be avoided by introducing dilation (expansion) joints every 15 to 20 m along an element to accommodate settlement and temperature movements. Use of low-permeability concrete and quality control of the concrete placement on site during element construction is also very important.

ELEMENT WEIGHT

The concrete tunnel element may not be heavy enough for sinking. In this case it will be buoyant during towing and equipped with internal temporary water ballast tanks. During immersion, the tanks are filled with water and the element position controlled by wires from the element to a number of anchor points and to winches mounted on pontoons on the surface. In case the element is not buoyant, it may be supported by pontoons for floating into position.

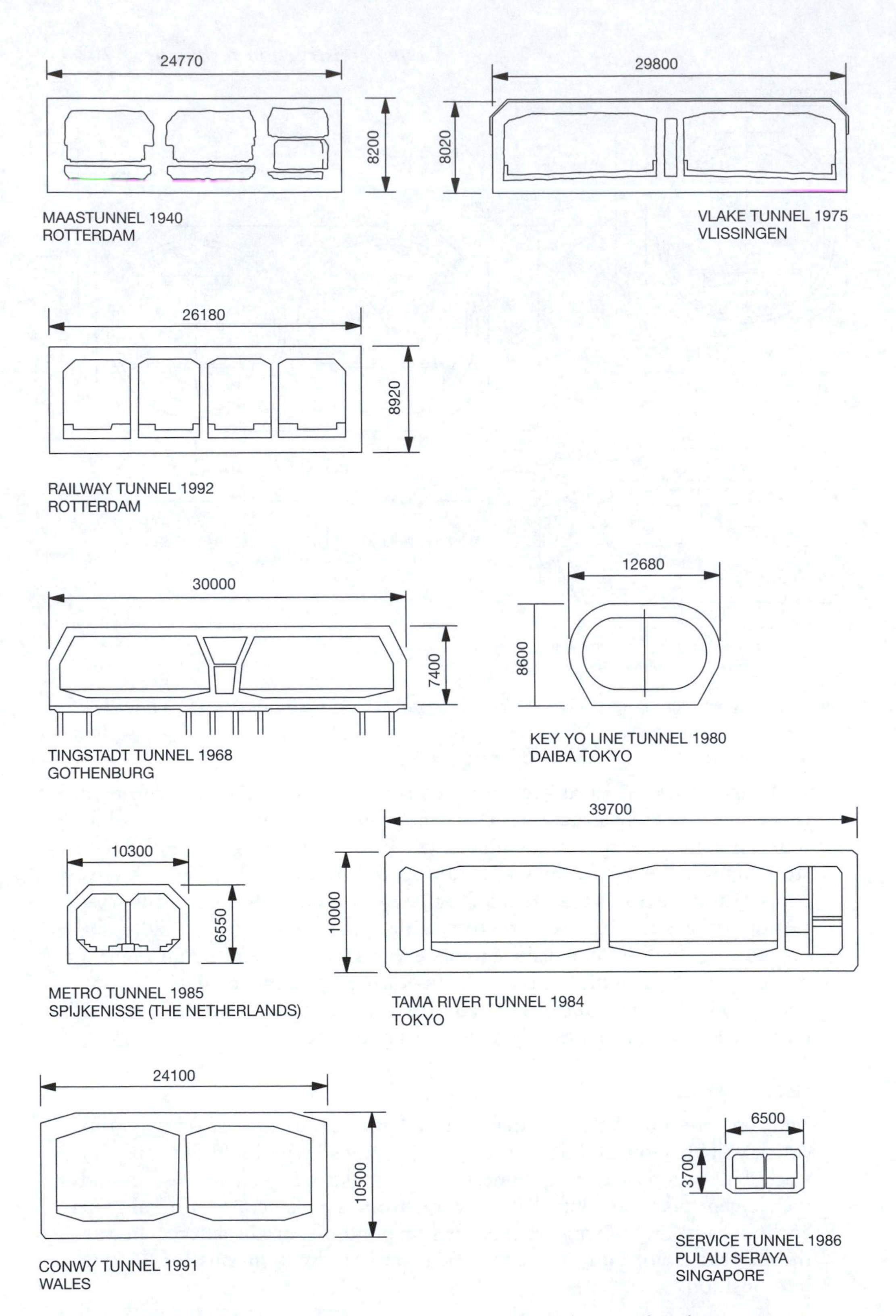

Figure 5.62 Typical cross sections for concrete immersed tube tunnels (after ITA 1997)

5.9.4 Immersed tube tunnel foundations and settlements

Immersed tube tunnels are founded at the base of the trench. Therefore, it is vital that appropriate foundations are constructed to ensure stability and minimize settlements of the tunnel. Two of the commonly used methods for providing a suitable foundation are briefly described in this section, together with the more rarely used method of pile foundations. In addition, brief comments are made with respect to settlement issues.

GRAVEL BED FOUNDATION COURSE

The trench is excavated up to 1.5 m below the bottom of the final tube position. A layer of coarse sand or well graded gravel is then placed in the trench and accurately levelled. The levelling may be done by dragging a heavy 'screed' made up of a grid of steel beams over the surface in successive passes.

SAND FLOW FOUNDATIONS

Sand flowing involves a sand-water mixture being jetted under the tunnel tubes via a pipe system in the bottom slab whilst they are accurately positioned on temporary supports.

PILE FOUNDATIONS OR GROUND IMPROVEMENT

Only in unusual circumstances where the soil beneath the tunnel is too weak to support it and cannot be economically excavated and replaced with better material are immersed tube tunnels placed on pile foundations. Different methods can be considered, including steel piles, stone columns and CDM (cement deep mixing) piles.

SETTLEMENT

Under normal conditions, immersed tube tunnels will not suffer significant soil settlements because their buoyant weight is designed to differ only slightly from that of the soil which it replaces. However, if there are changes in soil types and loading conditions along the tunnel, then differential settlements will take place. These are particularly important for concrete tunnels as they are brittle and more susceptible to cracking compared to steel. Further details on the settlement of immersed tube tunnels from case studies can be found in Grantz (2001).

5.9.5 Joints between tube elements

As the tunnel is made up of individual elements, joining these elements together is critical to ensure a watertight seal. There are many types of jointing system for immersed tunnels, a couple of example jointing methods are as follows.

TREMIE JOINTS

Tremie joints can be used for joining steel elements. This involves joining circular steel collar plates projecting from the tunnel elements. The tunnel elements are jacked together and curved steel sections are inserted externally to the collar plates at the edge of the tunnel elements. The area between the collar plates and the curved steel sections is then filled with tremied concrete ('tremied' concrete is a technique for placing concrete under water and involves pumping the concrete into a submerged space where it displaces the water).

RUBBER GASKET JOINTS

An example of this type of joint is shown in Figure 5.63. The initial seal between the elements is provided by the compression of rubber gaskets (Gina gasket) attached to the face of one tunnel element bearing onto a smooth surface on the adjoining element. Hydraulic coupling jacks extending from one of the elements are attached to the element to be positioned. The jacks pull the elements together and compress the gaskets. This initial compression of the gasket provides a good enough seal for the joint area to be drained from the inside. This brings the full hydrostatic pressure onto the far end of the element, and further compresses the gasket. After dewatering of the tunnel element, the joint area can be entered via doors in the bulkhead and a permanent connection can be made by a gasket mounted from the inside (Omega seal).

Special joints are required for seismic areas. These must typically allow for displacements in any direction of up to 100 mm. Alternatively, the rubber gasket joints shown in Figure 5.63 may be fitted with tension bars to limit the movements.

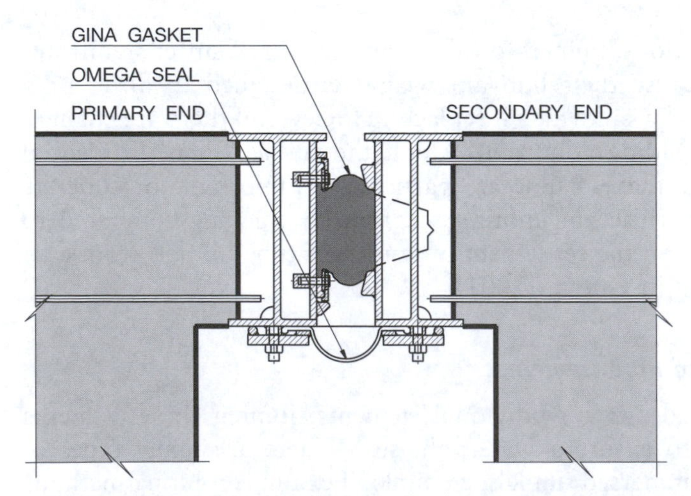

Figure 5.63 Example of rubber gasket joint (courtesy of COWI A/S)

5.9.6 Analysis and design

The analysis and design of immersed tube tunnels differs significantly from tunnels constructed through the ground, for which it is essential to determine the stability of the ground, especially its stand-up time. The engineer has to decide what percentage of overburden acts as a load on the tunnel and how much self supporting capacity the ground has. The tunnel is then designed to cope with the load both during construction and operation. Although the design loads for immersed tube tunnels can also be classified as construction and final in-place loads, they differ significantly in their type. Construction loads include those imposed during the fabrication, launching, towing, outfitting and placing operations. Final in-place loading conditions include normal and accidental/extreme loads. Normal loads include the dead load, water pressure (by far the most dominating load), earth pressure and superimposed live loads that the tunnel is expected to encounter during normal working conditions. Accidental loads include additional loads produced by unexpected events such as earthquakes, floods, anchor dropping, sunken vessels and vessel grounding, and possible explosion loads.

One of the issues related to seismic loading is the effect on the granular material and the possibility of liquefaction of the backfill around and underneath the tunnel. While the backfill at the sides of the tunnel can be loaded to increase its relative density and prevent liquefaction, this is not possible with the sand fill under the tunnel. As the tunnel is lighter than the soil it replaces, its foundation is relatively lightly loaded. To overcome this on the Conwy Tunnel (North Wales, UK), even though it is in a relatively low risk seismic area, a percentage of cement clinker was added to the sand foundation material in order to stabilize it (Stone *et al.* 1990). This had previously been used successfully in areas of much greater seismic activity, such as Japan and the Far East.

The structural analysis is carried out for the cross section and in the longitudinal direction by two separate finite element (FE) models. The cross section FE analysis addresses the ability of the structure to carry the loads by using a frame model to describe the slabs and walls. Compliance with crack width criteria is normally done through dimensioning of the reinforcement. In the case of earthquake loading, the cross section should also be analysed for racking, i.e. distortion due to horizontal forces. The longitudinal FE analysis addresses the ability to carry the imposed loading particularly at the joints due to differential settlements, an uneven foundation, temperature movements and, if relevant, movements due to earthquakes. The calculations will normally consider scenarios with tunnel element installations in summer and in winter in order to consider the axial force in the joints, which depends on the temperature movements of the elements following installation.

Table 5.2 Details of a few recent examples of immersed tube tunnels

Tunnel name	Western Harbour Tunnel, Hong Kong	Ted Williams, Boston, Massachusetts, USA	Medway Tunnel, UK	Øresund Tunnel, Sweden and Denmark	Busan Geoje Tunnel, South Korea	Bosphorus Crossing, Istanbul, Turkey
Construction period	1993–1997	Completed 1994	1992–1996	1993–2000	2003–2011 (expected)	2004–2011 (expected)
Location	Crosses Victoria Harbour	Crosses Boston Harbour	Between Rochester and Chatham, Kent	Between Malmo, Sweden and Copenhagen, Denmark	Between Busan and Geoje island	Connecting the European and the Asian parts of the city
Use	West Kowloon Expressway (Road tunnel)	Road tunnel	Road tunnel under Medway River	Rail and road tunnel	Road tunnel	Rail tunnel
Type	Reinforced concrete	Double steel shell binocular element	Reinforced concrete	Reinforced concrete	Reinforced concrete	Reinforced concrete
Total length (km)	2.20	2.575	–	4.05	3.85	13.6
Length of immersed tube tunnel section (km)	1.36	1.173	0.37	3.52	3.24	1.40
Tunnel element length (m)	113.5	98.30	2 × 126 1 × 118	176	180	135
Tunnel element width (m)	33.4	24.43	25.1	41.7	26.46 (2 × 28.46)	15.3
Tunnel element height (m)	8.57	12.29	9.15	8.50	9.97	8.75
Number of tunnel elements	12	12	3	20	18	11 (19,000 tonnes each)

Figure 5.64 Construction of one of the elements for the Limerick immersed tube tunnel in Ireland (courtesy of COWI A/S)

Figure 5.65 One of the elements for the Limerick immersed tube tunnel being towed into position ready for sinking (courtesy of COWI A/S)

5.9.7 Examples of immersed tube tunnels

Table 5.2 shows details of a few recent immersed tube tunnels and includes information on their construction method and dimensions. Only a small selection of immersed tube tunnels are listed in Table 5.2, as this is intended to provide an indication of the scale of the tunnels associated with this type of tunnel construction. For further examples of immersed tube tunnels, the reader is directed to the report by ITA (1997), which includes details of over 150 immersed tube tunnel constructions from around the world.

Two examples are briefly described here in slightly more detail. The Limerick immersed tube tunnel shown in Figure 5.64 is constructed from five 100 m long tunnel elements, i.e. the total length is 500 m, and is

Figure 5.66 One of the tunnel elements for the Ted Williams Tunnel being towed prior to additional outfitting work (courtesy of Massachusetts Turnpike Authority)

Figure 5.67
The trench was excavated using floating cranes with grab excavators (courtesy of Massachusetts Turnpike Authority)

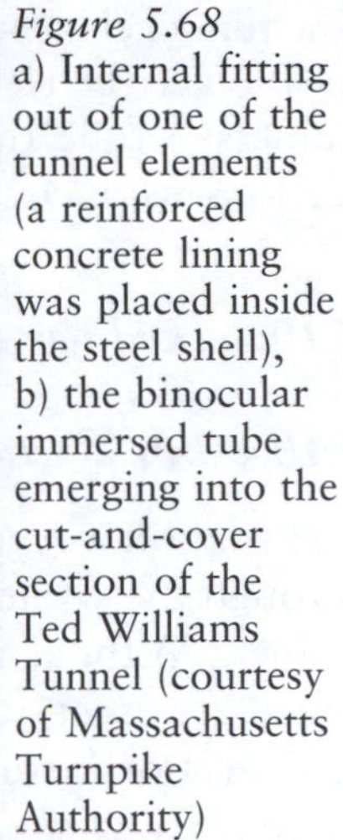

Figure 5.68
a) Internal fitting out of one of the tunnel elements (a reinforced concrete lining was placed inside the steel shell), b) the binocular immersed tube emerging into the cut-and-cover section of the Ted Williams Tunnel (courtesy of Massachusetts Turnpike Authority)

expected to be completed in 2010. The total length of the tunnel, including the cut-and-cover sections at each end is 675 m. The width and height of the tunnel cross section is 22.75 m × 8.45 m respectively. In addition, the bottom slab is extended 1.25 m outside the walls. These 'toes' allow backfill soil above to act as ballast, providing safety against uplift, and is in addition to the ballast concrete cast inside each of the elements. Figure 5.65 shows one of the elements been towed into position ready for sinking.

Figures 5.66 to 5.68 show various aspects of the construction of the Ted Williams immersed tube tunnel, Boston, USA (see Table 5.2 for details). Figure 5.66 shows one of the double steel shell binocular elements being towed. Prior to placement a trench was excavated across the harbour using a floating crane with a grab bucket (Figure 5.67). Approximately 680,500 m^3 of material was excavated to make this trench. Figure 5.68a shows one of the elements being fitted out with a reinforced concrete lining.

The tunnel elements were berthed during the outfitting, and as the outfitting took place the elements sank lower into the water. In order to connect the immersed tube tunnel into the land based tunnels, cut-and-cover tunnel sections were constructed at either end (Figure 5.68b).

5.10 Jacked box tunnelling

5.10.1 Introduction

Jacked box tunnelling involves pushing a precast tunnel section (box) through the ground whilst carrying out excavation at the front face within a shield. Although the technique is not new and has been around for many decades, recent advances in the technique have made it possible to install large section boxes under 'live' highways and railways. The cover depth for these tunnels can be very small, down to a couple of metres (see examples in section 5.10.3). The uses for jacked box tunnelling have traditionally been for pedestrian subways and portal bridge foundations, with later developments including small boxes being jacked one on top of another and filled with concrete to form bridge abutments (Clarkson and Ropkins 1977).

It is important, as with all tunnelling operations, that a project specific comprehensive site investigation is undertaken to determine the ground conditions, in particular their strength and stability characteristics, essential for the system design. Running parallel must be a survey of overlying infrastructure within the zone of influence of the box installation and the identification of any mitigation measures considered necessary.

According to Ropkins and Allenby (2000), the principal benefits of jacked box tunnelling are:

- a non-intrusive construction method;
- minimal disturbance to surface infrastructure;
- traffic flows maintained throughout the construction period;
- traffic flows maintained with only minor restrictions during box installation;
- an efficient structural form incorporating a low bearing pressure foundation;
- a high quality maintenance free structure.

5.10.2 Outline of the method and description of key components

The key components of a jacked box tunnel are a jacking pit, reception pit (as required), jacking base, shield, box, anti-drag system, jacking equipment and face support system (Allenby and Ropkins 2004). The functions of these key components and the stages involved in constructing a jacked box tunnel can be seen in Figure 5.69 and are described below.

STAGE 1

A jacking pit is constructed adjacent to the point where the tunnel is required using traditional excavation and support techniques based on the ground conditions. A reception pit may also be required at the point where the jacked box finishes. Within the jacking pit a jacking base is constructed. The jacking base needs to be slightly longer than the box sections to be used. This jacking base is crucial to the success of the jacked box operation as this provides the reaction for the jacking rig pushing the box forwards. In addition, the jacking base provides a stable base on which to construct the reinforced concrete box and must be smooth and level. It must be accurately constructed to satisfy the launch and installation requirements of the box section. Generally the box is made up of one element, but if space is limited for the jacking pit then smaller multiple box sections can be used. Along with the box construction, a *shield* is built at the front face. The box and shield are generally rectangular in cross section (Figure 5.69a).

The required number of jacking rigs is installed at the rear end of the box reacting against the jacking base. Jacks are used to advance the box section forward and also to provide some steering capabilities as required.

STAGE 2

The box is then jacked into the ground (often this is part of an embankment below a highway or railway) in a carefully controlled and phased sequence. Tunnelling commences by carefully excavating 150 mm of the face and jacking the box forward a corresponding amount, this sequence being repeated many times. One of the main issues with this construction technique is the friction between the outside of the box and the surrounding ground. If these friction forces are not reduced, then the ground, particularly immediately above the box, is likely to be dragged along the box. An *anti-drag system (*ADS) has therefore been developed to cope with this situation and is described in more detail on page 203. As the box is jacked forwards, the friction between the box and the surrounding ground increases, but at the same time the vertical load on the base from the box reduces. In order to maintain the necessary reaction resistance between the jacking base and the ground, the excavated material can be used to provide additional vertical load (kentledge) (Figures 5.69b and c).

STAGE 3

In order to provide stability of the ground at the exit portal, a berm, or portal structure is constructed. Once the box has reached its final position, the interface between the box and the ground is fully grouted. The shield and jacking arrangement are then dismantled and the finishes made to the box, portal structures and approach roads (Figures 5.69a, c and d).

Further details on the face support (shield), jacking equipment and anti-drag system are provided below.

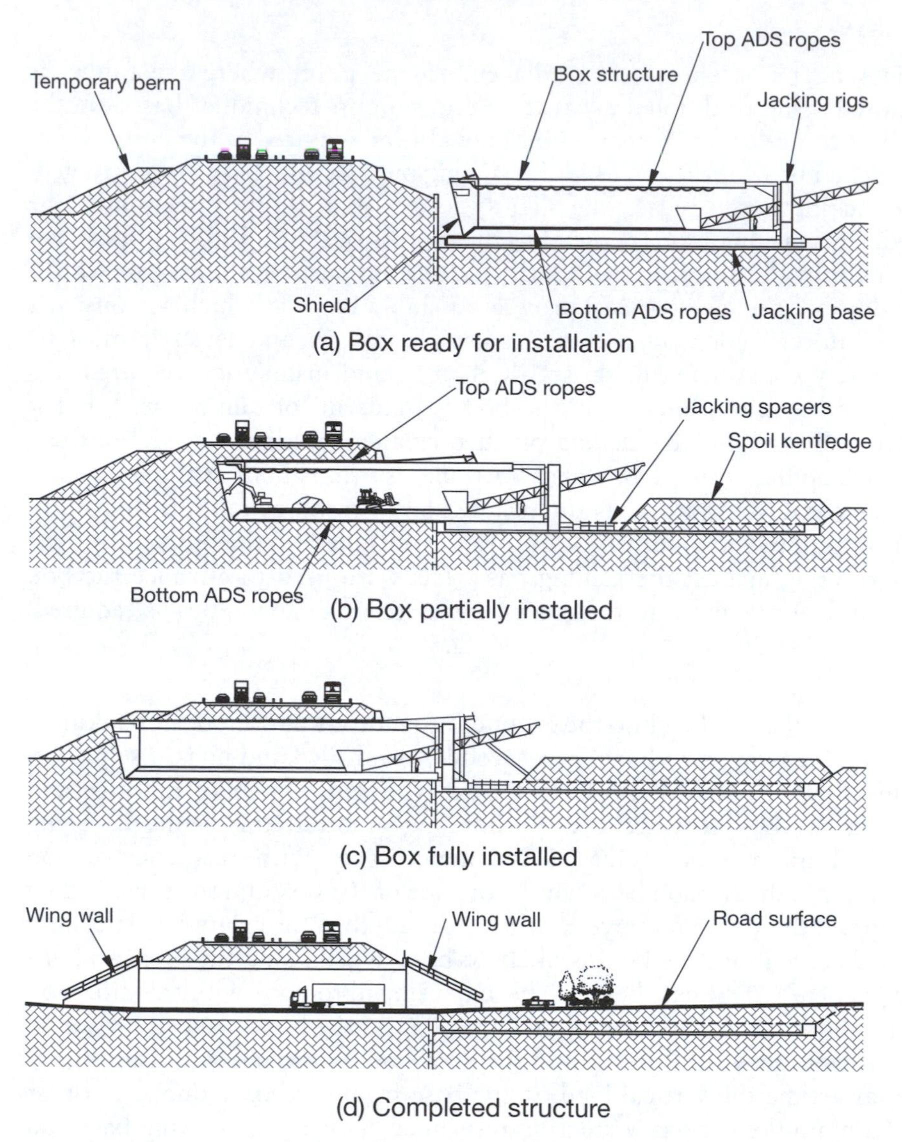

Figure 5.69 Jacked box tunnel installation (Allenby and Ropkins 2007, courtesy of BAM Nuttall Ltd and John Ropkins Ltd)

Face support system This is typically an open face shield generally made up of separate compartments (cells) similar to Brunel's Thames Tunnel shield described in section 1.3, with a cutting edge at the front. The excavation within these compartments is carried out either by traditional hand techniques or excavators, or a combination of each, depending upon the ground conditions, sensitivity of adjacent structures, overlying infrastructure and contractor's preference. Each shield is purpose designed to suit the ground conditions determined from the site investigation and to provide the face support necessary to maintain the integrity of the overlying infrastructure.

Figure 5.70 shows an example of the front face of a jacked box tunnel shield for a project in Lewisham, London and the excavation techniques used on this project. Further details of shields are provided in the jacked box examples in section 5.10.3.

Jacking equipment The required jacking forces are calculated based on the frictional resistance between the box surfaces and the anti-drag systems, the frictional resistance and/or adhesion between the exposed box surfaces and the ground, and from the shield embedment loads. Figure 5.71 shows all the components associated with the jacking process. The hydraulic jacks only have a limited extension and hence spacers between the hydraulic jacks and the cross beam are used to transfer the forces into the jacking

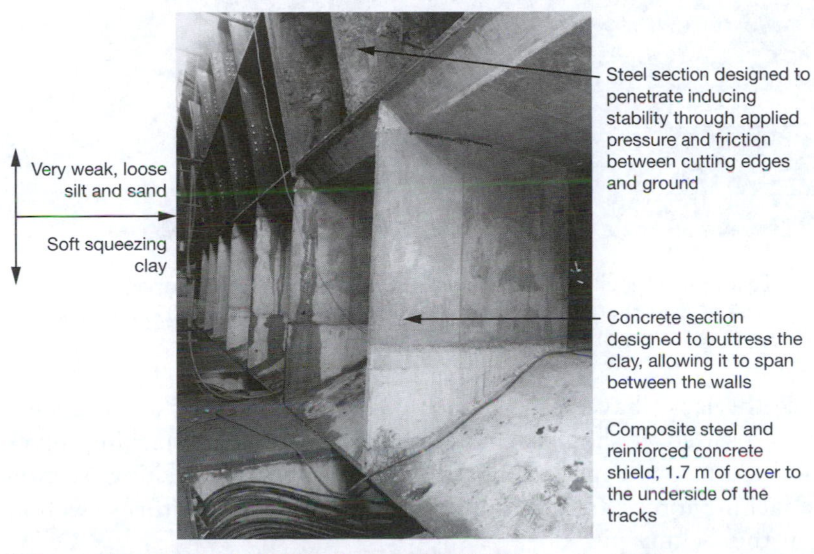

a) Composite steel and reinforced concrete shield

b) Hand mining in a top compartment through very weak, loose sand and silt

c) Rear view of the face during excavation

Figure 5.70 Example shield as used on a jacked box tunnel as part of the Docklands Light Railway, Lewisham, London, UK (box size 17 m wide, 6.2 m high, 48 m long) (courtesy of BAM Nuttall Ltd and John Ropkins Ltd)

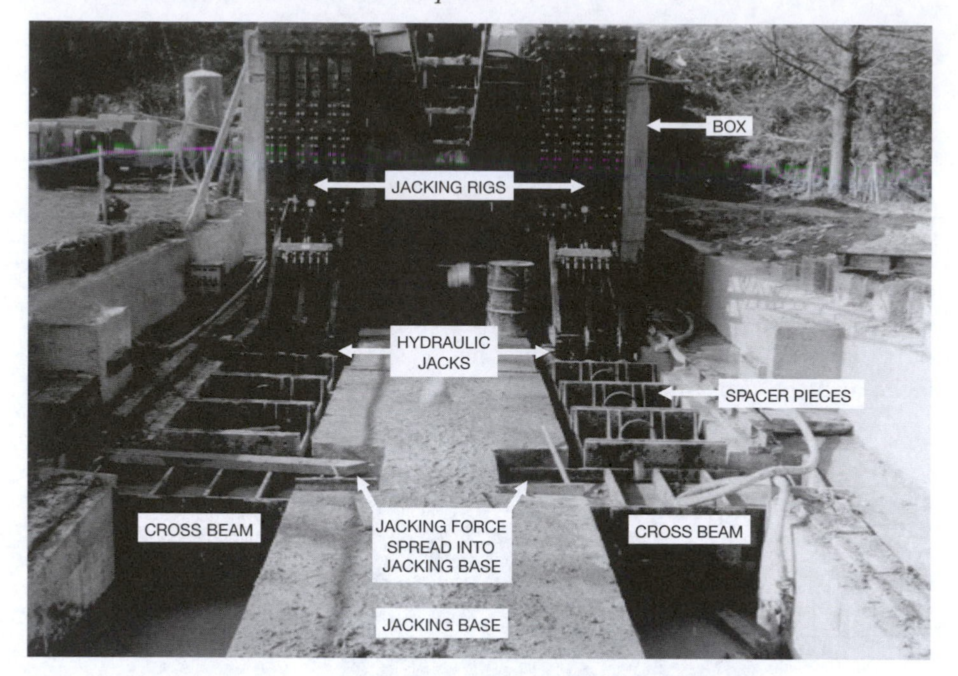

Figure 5.71 Example of jacking equipment and how the reaction forces are passed into the jacking base (Allenby and Ropkins 2007, courtesy of BAM Nuttall Ltd and John Ropkins Ltd)

base. Once the jacks have reached their full extension they are retracted and spacers inserted. It should be noted that in this case the jacking forces are transmitted through the jacking base, whereas in pipe jacking (section 5.11) the jacking forces are generally transmitted through a thrust wall at the rear of the jacking pit.

Anti-drag system A number of anti-drag measures have been developed, including lubrication with bentonite slurry and the use of thin steel sheets, and reinforced rubber 'drag sheets'. Although these systems had some success it was not until 1986 when Ropkins™ developed its highly successful proprietary wire rope anti-drag system that it became possible to effectively control ground drag (Allenby and Ropkins 2007). This proprietary Ropkins™ anti-drag system uses an array of closely spaced wire ropes rolled out from within the front of the tunnel section in the direction of the drive, both at the top and bottom, as the tunnel section is jacked forwards (Figure 5.69). The ropes are stored on cable drums within the box, and one end of each rope is fixed outside the box (Figure 5.72 and also Figure 5.81 in section 5.10.3.2). As the box moves forward the ropes are fed off the cable drums. In this way the ropes effectively form a stationary layer between the ground and the moving box section. The drag forces are absorbed by the ropes and transferred back to the jacking base, effectively isolating the ground from the drag forces. It is important to lubricate

Figure 5.72
Top anti-drag system as used in the proprietary Ropkins™ system (Ropkins and Allenby 2000, courtesy of BAM Nuttall Ltd and John Ropkins Ltd)

the ropes prior to installation and to keep them well lubricated during box installation using strategically positioned injection points. At the end of the jacking process, the anti-drag ropes are left *in situ* because their removal would cause voids and induce additional unnecessary settlement. Grouting of the box/ground interfaces subsequently takes place using cement based grouts commencing at the invert, gradually working up the sidewalls and finally over the roof. Further details of the ADS used on the vehicular under-bridge in the UK are described in the example in section 5.10.3.1.

5.10.3 Examples of jacked box tunnels

Table 5.3 shows a list of jacked box tunnel projects using the Ropkins™ system. The Boston, USA and M1 Junction 15A, UK jacked box tunnel projects are described in more detail in the following sections.

5.10.3.1 Vehicular under-bridge, M1 motorway, J15A, Northamptonshire, UK

The vehicular under-bridge, M1 motorway, J15A was constructed by BAM Nuttall Limited in association with John Ropkins Limited and opened in 2003 (details from Allenby and Ropkins 2004).

This project involved jacking a 45.0 m long, 14.0 m wide, 8.5 m high reinforced concrete monolithic box under a live motorway (Figure 5.73). The ground conditions of the original motorway embankment consist of an engineered clay fill overlying a natural boulder clay. The cover depth for the box section was 2.2 m under the main carriageway and 1.6 m under the hard shoulders.

Table 5.3 Jacked box tunnel projects installed using the Ropkins™ system (Allenby and Ropkins 2007)

Projects	*Size*	*Cover*	*Date*	*Ground conditions*	*Ground treatment*	*Working Jacking Capacity*
Pedestrian and cyclist subway Didcot, Oxfordshire, UK	30 m long 5.9 m wide 3.6 m high	2.0 m	1989	Silt-stone fill overlying soft clay	None	2400 tonnes
Highway tunnels West Thurrock, Essex, UK	30 m long 16.5 m wide 9.5 m high	8.0 m	1991	Chalk with swallow holes loosely filled with sand	None	6000 tonnes
Highway under-bridge Silver Street Station, London, UK	Twin tunnels each 44 m long 12.5 m wide 10.5 m high	7.0 m	1995	Water-bearing gravels above over-consolidated clay containing sand layer with water under artesian pressure	Grouting of water bearing gravels. Dewatering of sand layer	6400 tonnes
Rail tunnel Lewisham Railway Station, London, UK	48.0 m long 17.0 m wide 6.2 m high	1.7 m	1998	Loose silt and sand overlying soft clay	None	4800 tonnes
3 No subways Lewisham Railway Station, London, UK	Up to 32.0 m long 4.4 m wide 3.65 m high	2.0 m	1998	Loose silt and sand overlying soft clay	None	1200 tonnes each
Flood relief culvert Dorney, Berkshire, UK	50.0 m long 23.0 m wide 9.5 m high	6.0 m	1999	Clayey granular fill overlying water bearing sands and gravels, overlying weathered chalk	Artificial ground freezing	7200 tonnes
3 No highway tunnels Boston, Massachusetts, USA	Up to 106.8 m long 24.0 m wide 10.8 m high	6.0 m	2001	Weak water bearing strata with numerous man-made structures, tidally influenced water table	Artificial ground freezing	44989 tons maximum
Highway under-bridge M1 Junction 15A, Northamptonshire, UK	45.0 m long 14.0 m wide 8.5 m high	1.6 m	2002	Pulverized fuel ash and clay fill overlying stiff clay with rock inclusions	None	4800 tonnes

The face was supported via a shield consisting of an open face reinforced concrete cellular structure divided into three working levels, each with seven compartments. The top-level compartments were designed for hand excavation, and the lower and middle level compartments for machine excavation, with an option of hand excavating inside the middle level compartments should the need arise. Substantial steel-plated cutting edges were attached to the shield perimeter, the horizontal working decks and the vertical dividing walls, permitting the face material to be penetrated and buttressed. Figure 5.74 shows a rear view of the shield, annotated with the approximate levels of the ground strata.

The anti-drag systems for this project comprised lubricated wire ropes, as previously described in section 5.10.2, at the top and bottom of the box section. The top anti-drag system ropes were spaced at 26 mm centres and covered the full width of the box section. The ropes were anchored to a steel beam within the shield roof and passed through slots in the shield

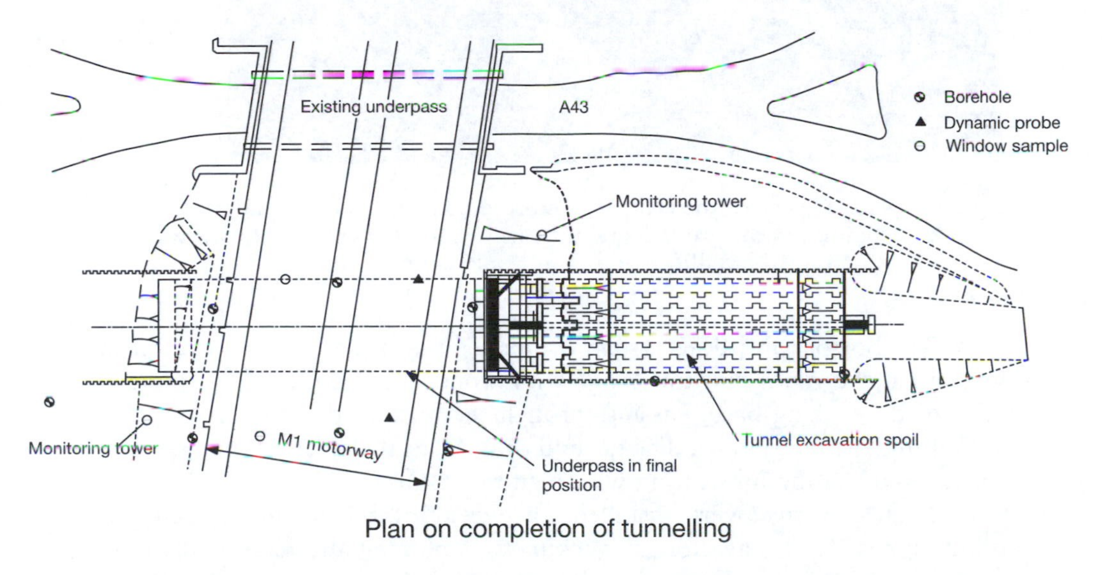

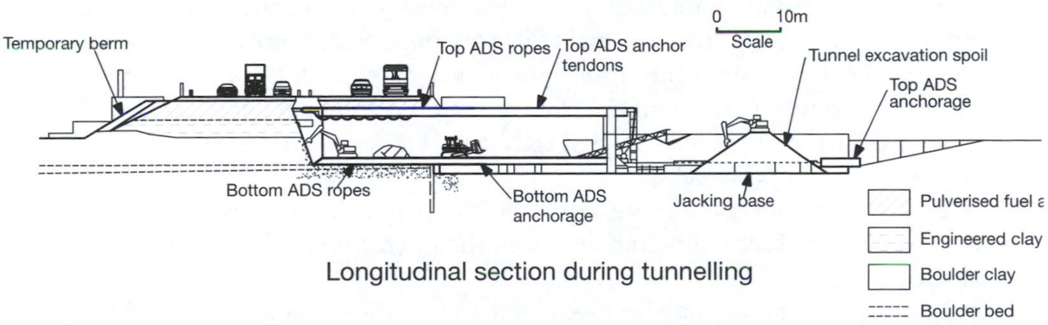

Figure 5.73 Details of the jacked box under-bridge at Junction 15A, M1 motorway, UK (Allenby and Ropkins 2004, courtesy of BAM Nuttall Ltd and John Ropkins Ltd)

Figure 5.74 Rear view of the cellular reinforced concrete shield (annotation indicates the ground strata) (Allenby and Ropkins 2004, courtesy of BAM Nuttall Ltd and John Ropkins Ltd)

roof. The bottom anti-drag system comprised two rope tracks, each 3.5 m wide with the ropes laid touching. These ropes were attached to the lead edge of the jacking base, passing through the base of the box section and stored under a false floor. The ground drag from the sides of the box was further reduced by lubricating with bentonite slurry.

Four jacking rigs were installed to give a total working capacity of 4800 tonnes. The excavation and jacking took place in alternate small increments of 150 mm, while maintaining a minimum penetration into the ground at the front of the shield of 450 mm. Seven miners excavated the top level compartments using pneumatic clay spades and mechanical excavators removed the material from the middle and lower compartments. The installation of the box section took four weeks, working 24 hours per day, seven days per week.

Figure 5.75 shows the box section during the jacking process and Figure 5.76 shows the successful completion of the operation and the front of the shield.

Immediately following the box installation, low-pressure, high volume grouting using a blended PFA/cement grout mix was carried out to the floor, side walls and roof of the box to fill the overcut annulus and surround the anti-drag systems in order to prevent further settlements.

Figure 5.75 Rear view of the box during the jacking operation (Allenby and Ropkins 2004, courtesy of BAM Nuttall Ltd and John Ropkins Ltd)

Figure 5.76 Completion of the jacking process showing the shield (Allenby and Ropkins 2004, courtesy of BAM Nuttall Ltd and John Ropkins Ltd)

Settlements of the motorway were carefully measured using instruments located on 9 m high towers monitoring the carriageway on a 5 m grid and giving displacements in three-dimensional coordinates in real time, as shown in Figure 5.75. With careful jacking and excavation the maximum recorded settlement was 26 mm.

5.10.3.2 I-90 Highway Extension, Boston, Massachusetts, USA

The massive Central Artery highway reconstruction in Boston (the so called 'Big Dig' project) involved major improvements to the road infrastructure and in particular the I-90 highway. As part of this project the I-90 had to pass beneath eleven active railroad tracks, which carry commuter and mainline trains into Boston's busiest rail terminal, South Station. The railroad operation had to remain in service throughout the construction project. In order to achieve this, tunnels were jacked underneath these railroad tracks. The jacked tunnels formed a link between the immersed tube tunnel under the Fort Point Channel and the cut-and-cover sections (Figure 5.77). The cover depth between the top of the tunnel sections and the railroad tracks was approximately 6.1 m (20 ft).

The I-90 Highway Extension in Boston was constructed by the Slattery, Interbeton, J.F. White, Perini Joint Venture with the jacked box tunnelling design, planning and installation services provided by Edmund Nuttall Limited in association with John Ropkins Limited. The jacked box tunnels were constructed in 2001.

Figure 5.77 An overview of the jacked box sites in Boston showing the railroad lines and Fort Point Channel (courtesy of Massachusetts Turnpike Authority)

Figure 5.78 One of the jacked box tunnels during the jacking process, note the cover depth above the tunnel and rail road lines at the ground surface (courtesy of Massachusetts Turnpike Authority)

Three concrete jacking pits were constructed as part of the jacking operations. In each of these jacking pits tunnel box sections approximately 24 m (80 ft) wide and 12 m (40 ft) high were constructed. The lengths of the three tunnels were approximately 50 m (150 ft), 79.2 m (260 ft) and 115 m (380 ft) respectively. The two shorter tunnels were constructed in two section lengths and the longest in three sections. Intermediate jacking stations were positioned between each box section, similar to pipe jacking operations, with the main thrust jacks at the rear of the last box. This allowed each of the boxes to be jacked forwards in sequence thus reducing the thrust required by the main jacks (the intermediate jacks acted as intermediate jacking stations as described in pipe jacking, section 5.11). The sections were jacked at a rate of between one and two metres (three and six feet) per day. Figure 5.78 shows one of the jacked box tunnels during the jacking process.

The ground conditions consisted of Boston Blue Clay, sands, gravels and organic material, and had to be stabilized prior to excavation. In order to increase the stability of the ground during the tunnel jacking operation, and hence reduce the resulting ground movements, ground freezing was utilized on this project (see section 4.2.1 for more information on ground freezing). The freezing process used a closed system of re-circulating brine as the coolant. The brine was passed down vertical freeze tubes into the ground from the ground surface (Figure 5.79).

Figure 5.79 Breakthrough of two of the jacked tunnels, showing the freezing operation at the ground surface (courtesy of Massachusetts Turnpike Authority)

Each tunnel had a rectangular shield divided into six compartments on two levels. The excavation machines were sized to excavate a compartment face typically 4.02 m wide by 5.4 m high (Figure 5.80a). A maximum of four machines were used at any one time in a carefully controlled sequence (Clayton *et al.* 2001). Figure 5.80b shows the Webster Schaeff excavating machines purposely designed for this project. These machines excavated the soil around the freeze pipes, which were then cut away.

The Ropkins™ system proprietary ADS using steel cables was employed on this project to reduce the friction above and below the jacked sections (Figure 5.81).

Figure 5.80
a) Excavating the lower half of a 10 m high face through frozen, waterbearing sands and gravels, and b) Webster Schaeff excavating machines used on the Boston jacked box tunnels (Clayton *et al.* 2001) (courtesy of BAM Nuttall Ltd and John Ropkins Ltd)

Figure 5.81
The inside of one of the tunnel box sections showing the proprietary ADS cable system (courtesy of Massachusetts Turnpike Authority)

5.11 Pipe jacking and microtunnelling

5.11.1 Introduction

Pipe jacking is one of a number of techniques (similar to jacked box tunnelling) for creating smaller diameter tunnels, for example sewers and other conduits, generally up to 3 m in diameter, which attempt to minimize excavation from the ground surface (open-cut). These are often known as trenchless technologies. Although not commanding the glamour of more high profile transportation tunnels, smaller tunnels such as sewers, storm water drains and other conduits such as high voltage cable ducts are an important part of our underground infrastructure. A brief description of the pipe jacking technique is provided in this section. More extensive details on this, and other trenchless techniques, are given in a number of books devoted to this subject, for example Thomson (1995), PJA (1995), FSTT (2004), Najafi and Gokhale (2004) and Stein (2005).

Pipe jacking is often used to install tunnels under highways, railway crossings and canals, i.e. where access to the ground surface is restricted, or where open-cut trenching would create a high level of disruption. In contrast to conventional tunnels where the tunnel lining is constructed directly behind the excavated face, in pipe jacking the complete tunnel lining sections are precast and are pushed into place from a shaft. This forms a string of pipe sections which are all moved through the ground until the desired length of tunnel has been reached.

When constructing pipe jacked tunnels, it used to be the case that individuals working in confined spaces would excavate the face within a shield at the front of the tunnel (see Figure 5.6a in section 5.4 on shield tunnelling). As with larger diameter tunnelling, however, more mechanized methods have been introduced, this being a necessity for tunnels less than 0.9 m diameter (i.e. non-man entry size). For these very small diameter tunnels miniature TBMs have been developed. This small mechanized TBM development led to the term microtunnelling. In its simplest form, microtunnelling is the use of a remotely-controlled, computer assisted, miniature, TBM (EPBM or STM), which is advanced by pipe jacking (Kuesel and King 1996).

There are a number of key issues when using the pipe jacking technique. In particular the fact that the tunnel lining sections are all pushed through the ground. This means that these lining sections have to take considerable axial compressive forces, which consequently dominate their design (Milligan and Norris 1999). In addition, during the installation, large friction forces can develop between the lining sections and the ground and hence the force required to push them through the ground increases as the length of the tunnel increases. These frictional forces can be reduced in a number of ways. The tunnelling shield or machine is made a few centimetres larger in diameter than the installed tunnel sections. This creates a gap into

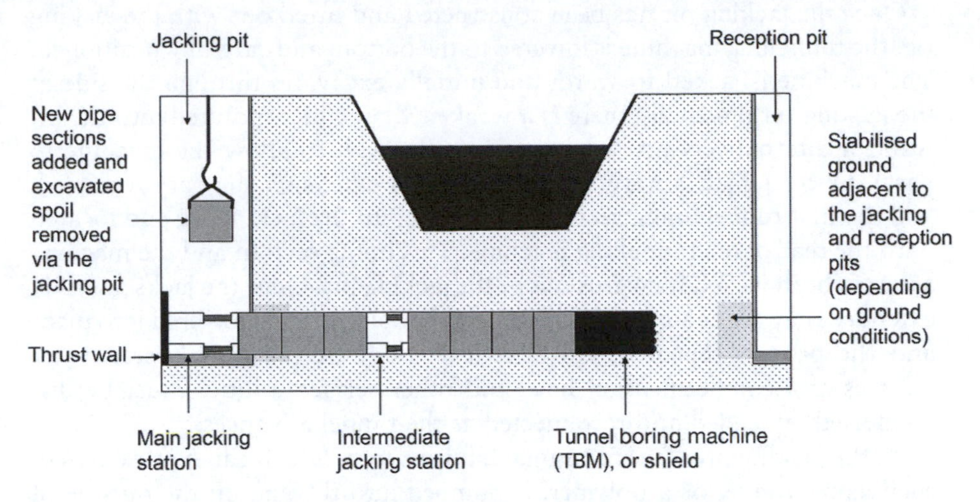

Figure 5.82 Illustration of a pipe jack under a canal (courtesy of A. Royal)

which lubricant can be pumped. The gap and the lubricant help to reduce the friction forces. This gap can close up however, as the ground moves onto the tunnel lining and this can cause displacements in the surrounding ground (Chapman 1999). Although the lubricant helps to stabilize the ground, this gap is often grouted at the end of the pipe jacking process. An alternative way of reducing the jacking forces is to introduce intermediate jacking stations at intervals along the tunnel. This means that lengths of the tunnel in front of the intermediate jacking station can be jacked forwards, followed by others behind, i.e. a reduced length of tunnel is being jacked at any one time, in a caterpillar like operation. These issues are described further in the following sections.

5.11.2 The pipe jacking construction process

Figure 5.82 shows a schematic of the typical components of a pipe jacking or microtunnelling operation. The pipe jacking technique requires two shafts to be constructed to a depth just below the proposed tunnel invert. The jacking pit is where the construction operations happen, with the reception pit really only being required to remove the tunnelling machine at the end of the drive. The jacking pit is therefore essential to the pipe jacking process. The jacking rig is installed at the base of the jacking pit and normally consists of two or four hydraulic jacks (Figure 5.83a and c). Due to the large jacking forces required during the construction operation, the rear wall of the jacking pit, i.e. where the loads are transmitted from the jacking rig, must be designed to take these loads and a thrust wall has to be constructed. The jacking pit must be large enough to enable the tunnelling machine, pipe sections, and excavated material to be handled safely.

Once the jacking pit has been constructed and fitted out with the jacking rig, the tunnelling machine is lowered to the bottom and carefully positioned. The machine is jacked forwards and initially excavates through the side of the jacking pit (normally there is a weakened area incorporated into the pit wall, or alternatively an area is cut into the wall to allow the machine to pass). Once the jacks have reached their full extension they are retracted, and the first tunnel lining section is lowered into the jacking rig and located onto the rear of the tunnelling machine. This lining section and the machine are then both jacked forwards out of the jacking pit. Once the jacks are fully extended they are again retracted and another pipe section lowered into place and the process repeated. All electrical conduits, pressure hoses, water services and communication lines and other service utilities must be disconnected, extended and reconnected as the tunnel advances.

As the jacking process is continued, lubrication, which can be clay-based, such as bentonite, or a polymer, is pumped into the gap on the outside of the tunnel sections to reduce the friction with the ground. At certain intervals during the jacking operation, intermediate jacking stations (also known as interjacks) are lowered into the jacking pit. These consist of two steel sections connected together with hydraulic jacks (Figure 5.83b). These can be used to relieve the forces on the jacks in the jacking pit if the forces get too high, i.e. they are used in addition to the lubrication. They are useful to have along the jacked tunnel as a precaution even if they are not used. Construction rates vary greatly depending on the ground conditions, but are typically between 7.5 and 20 m per day.

For microtunnelling operations the tunnelling machine is controlled remotely with the equipment and controls usually set up at ground level either directly over, or adjacent to, the jacking pit (Figure 5.83d). The shield is normally made up of sections, which can articulate. Steering of the machine is possible using jacks incorporated between these articulated sections. These allow vertical and horizontal adjustment of the tunnel during construction. An example of a control panel for a microtunnelling machine is shown in Figure 5.83e, and this has information on, amongst other things, the extensions of the steering jacks and the laser guidance system. Examples of microtunnelling machines are shown in Figures 5.83g and f.

The tunnel lining sections can be made of different materials, but are commonly made from reinforced concrete, glass fibre reinforced plastic (GRP), steel and vitrified clay. The transmission of axial forces between these pipe sections is very important as any overstressing of the sections locally could cause spalling, i.e. fragments of material breaking off, or more serious cracking and possibly failure. In order to help spread the load at the joints and reduce stress concentrations, strips of material, commonly medium density fibreboard (MDF), are inserted into the joints.

Figure 5.83 a) Jacking forward the tunnel sections from the jacking pit, b) intermediate jacking station (courtesy of Herrenknecht AG/NoDig Media Services), c) jacking arrangement and thrust wall, d) microtunnelling control centre placed over the jacking pit

Figure 5.83 (continued)
e) control panel for a micro-tunnelling machine, f) and g) examples of microtunnelling machines

5.11.3 Maximum drive length for pipe jacking and microtunnelling

Recent recommendations by the Health and Safety Executive in the UK provide guidance on construction methods for a range of tunnel diameters (Table 5.4a), and also limits on the maximum drive (tunnel) lengths between shafts for pipe jacking and microtunnelling operations (Table 5.4b).

5.12 Horizontal directional drilling

Horizontal directional drilling (HDD) is one of a number of techniques that have successfully been used over the last 20 years to install, or refurbish, pipelines with minimal surface disruption and was originally developed in the oil industry. HDD is a versatile tunnelling technology and is commonly used for installing pipelines and cables under rivers with drive lengths of up to 3000 m. Brief details of this technique are given in this section, but further information can be found in Bayer (2005) and also the other references mentioned in section 5.11 on pipe jacking.

The construction procedure for HDD is shown in Figure 5.84. HDD uses surface-mounted drilling rigs to cut a pilot bore between two points (approx. 100 to 200 mm in diameter, depending on the size and length of the final tunnel). The bore is then enlarged with reamers (drill heads pulled back through the original pilot bore) until the required diameter is achieved to allow the installation of the final pipeline, which is then pulled into place by the drilling rig. Normally the bore is enlarged to 1.2 to 1.5 times the diameter of the final installed pipe to reduce the friction on this pipe during the pullback operation. Throughout the process the bore is stabilized and flushed of cuttings (i.e. the excavated material is removed) by using drilling mud that is pumped into the bore from the surface via the drill head or reamer.

HDD requires large volumes of drilling mud (two to four times the volume of the bore, depending upon the soil conditions). This mud exiting the bore can be reused, but must be cleaned of cuttings first. Mud mixing and pumping equipment is required on site to ensure a continuous supply of the drilling fluid. These may form part of the drilling rig, on small projects, or require substantial plant alongside the drilling rigs for larger projects (Figure 5.88). The mud tends to be based on bentonite slurries. However, add-mixtures and polymers may be incorporated to improve performance. In granular soils polymers are essential in order to prevent loss of drilling fluid into the surrounding ground because of the high permeability and also to prevent collapse of the material into the bore.

Figure 5.85 shows an example of a medium sized HDD rig, including the drilling head used for the pilot bore. One of the key issues with HDD is to know the position of the drill head during the pilot bore. This can be monitored using a 'sonde' that is placed just behind the drill head and

Table 5.4 Recommendations for a) construction methods for a range of diameters, and b) maximum drive lengths for pipe jacking and microtunnelling (HSE 2006, used with permission)

Table a) – Nominal internal diameter of pipeline or tunnel linings

<table>
<tr><th>Excavation technique</th><th><0.9 m</th><th>0.9 m</th><th>1.0 m</th><th>1.2 m</th><th>1.35 m</th><th>1.5 m</th><th>1.8 m</th><th>>1.8 m</th></tr>
<tr><td>Pipejack – machine; remote operation from surface</td><td colspan="8">Acceptable
(See Table b)</td></tr>
<tr><td>Pipejack – machine; operator controlled below ground</td><td colspan="3">Not acceptable</td><td colspan="5">Acceptable</td></tr>
<tr><td>Pipejack – hand dig</td><td colspan="3">Not acceptable</td><td colspan="5">Avoid</td></tr>
<tr><td>Tunnel – machine operator controlled + mechanical erector</td><td colspan="5">Not acceptable</td><td colspan="2">Avoid</td><td>Acceptable</td></tr>
<tr><td>Tunnel – hand dig + mechanical erector</td><td colspan="5">Not acceptable</td><td colspan="3">Avoid</td></tr>
<tr><td>Timber heading – hand dig</td><td colspan="3">Not acceptable</td><td colspan="5">Avoid</td></tr>
</table>

Table 5.4 (continued)

Table b) – Indicative drive lengths (e.g. between shafts) and maximum number of drives

<table>
<tr><th>Excavation technique</th><th><0.9 m</th><th>0.9 m</th><th>1.0 m</th><th>1.2 m</th><th>1.35 m</th><th>1.5 m</th><th>1.8 m</th><th>>1.8 m</th></tr>
<tr><td rowspan="2">Pipejack – machine; remote operation from surface</td><td colspan="3">Drive length limited only by capacity of jacking system</td><td colspan="2" rowspan="2">250 m</td><td rowspan="2">400 m</td><td rowspan="2">>500 m(7)</td><td rowspan="2">>500 m(7)</td></tr>
<tr><td colspan="2">Man entry not acceptable</td><td>Avoid man entry</td></tr>
<tr><td>Pipejack – machine; operator controlled below ground</td><td colspan="3">Not acceptable</td><td>125 m</td><td>200 m</td><td>300 m</td><td>500 m</td><td>>500 m(7)</td></tr>
<tr><td>Pipejack – hand dig(6)</td><td colspan="3">Not acceptable</td><td>*25 m – 2 drive lengths</td><td>*50 m – 2 drive lengths</td><td>*75 m – 2 drive lengths</td><td colspan="2">*100 m 1 drive length. Plan to use minidigger if over 2.1 m dia.</td></tr>
<tr><td>Tunnel – machine; operator controlled + mechanical erector</td><td colspan="5">Not acceptable</td><td>*250 m</td><td>*500 m</td><td>>500 m(7)</td></tr>
<tr><td>Tunnel – hand dig + mechanical erector(6)</td><td colspan="5">Not acceptable</td><td>*50 m 1 drive lengths</td><td colspan="2">*100 m 1 drive length. Plan to use minidigger if over 2.1 m dia.</td></tr>
<tr><td>Timber heading – hand dig(6)</td><td colspan="3">Not acceptable</td><td colspan="5">*25 m – 2 drive lengths. Minimum cross section inside frames 1.2m high × 1.0m wide</td></tr>
</table>

Notes for Table 5.4:

1 This guidance should be read in conjunction with BS 6164:2001 (BSI 2001b). It is intended to be used only by those competent to design pipe jacks and tunnels.
2 This guidance for designers has been agreed by HSE and the tunnelling industry (BTS/PJA). It is based on experience of the occupational health and safety risks arising from heavy physical work, including the use of vibrating tools, in a confined space along with the need to be able to evacuate quickly/effect a rescue in a range of reasonably foreseeable situations.
3 Complying with the guidance does not relieve the designer of the duty to consider the risks arising from the foreseeable hazards of pipe jacking/tunnelling including manual handling, noise, heat, vibration and confined space working. Neither does it relieve the designer of the duty to ensure there is potentially adequate space to allow a safe means of access and egress along with adequate working space within the tunnel/pipe jack. The minimum diameter required for construction may in some cases be determined by the criteria above rather than by consideration of the hydraulic requirements for or the intended use of the pipe jack/tunnel.
4 Indicative drive length and the number of drives of that length, have been determined from consideration of access and escape requirements. Again, complying with the guidance does not relieve the designer of the duty to consider the risks arising from the range of foreseeable emergency events which could arise and which could necessitate escape or rescue of those underground.
5 The drive lengths given in Table b) are indicative. Designers should note that for entries not marked * it is acceptable to exceed the indicative drive lengths by up to 25%. However exceeding these lengths by over 25% should be avoided. Exceeding the indicative lengths by over 75% should be considered as not acceptable.
6 All hand dig is categorized as 'not acceptable' or 'avoid' – the lengths given in Table b) for items marked * are indicative and are already in the category 'avoid'.
7 Drive lengths exceeding 1000 m should be considered not acceptable unless the pipe/tunnel is of a sufficiently large cross section to allow the contractor to incorporate an access envelope 0.9 m wide by 2.0 m high within the pipe/tunnel and clear of services including a ventilation duct and a spoil conveyor.
8 For guidance on side connections see relevant PJA publication.

Definitions:

Acceptable – designers should undertake an assessment of the risks normally associated with small size pipe jacking/tunnelling and specify the appropriate mitigation measures.

Avoid – designers should undertake a robust technical assessment and risk assessment to justify their decisions to deviate from 'acceptable' criteria. Designers should identify appropriate risk mitigation measures. They should seek the advice of the Planning Supervisor/Co-ordinator and only proceed if the Planning Supervisor/Co-ordinator is satisfied that due attention has been paid to health and safety in undertaking the design and that appropriate risk mitigation measures have been identified. Contractors being asked to construct a pipe jack/tunnel in this category should also seek advice from the planning supervisor/co-ordinator on the adequacy of their risk mitigation measures.

Not acceptable – designers should not specify the use of pipe jacking/tunnelling of this size and construction method. An alternative design solution should be sought.

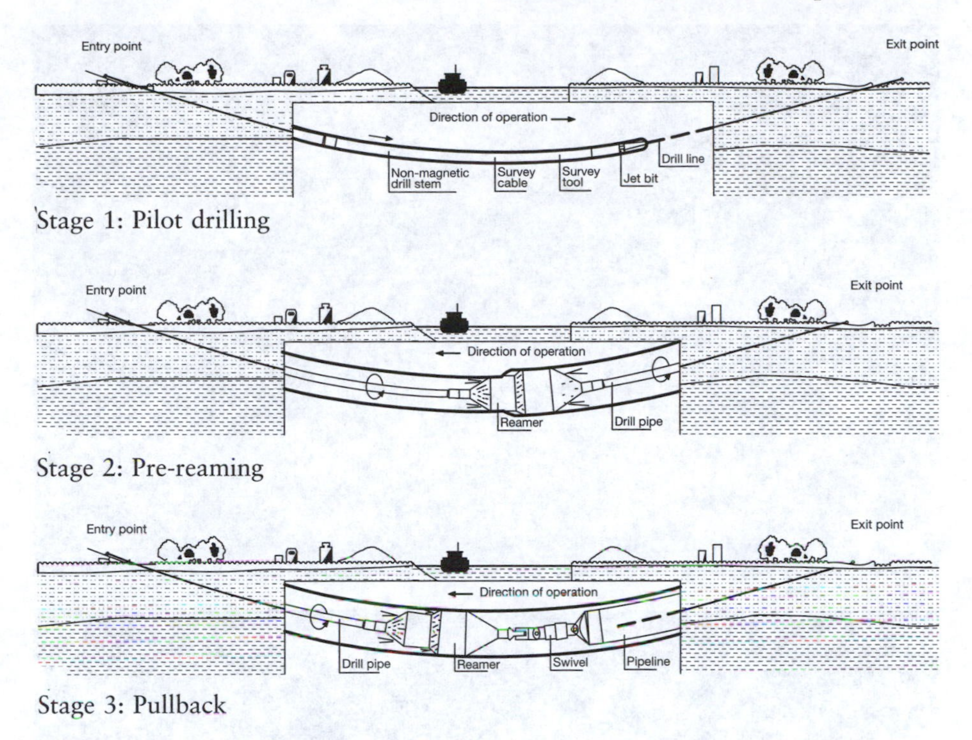

Figure 5.84 Typical HDD installation, including pilot drilling, pre-reaming and pullback operations (the central section of each stage is enlarged for clarity) (courtesy of Herrenknecht)

transmits its position wirelessly. The signal can be received at the ground surface by a person walking the route of the bore with a detector. Alternatively when access to the ground surface is difficult or for very long drive lengths, the signal can be sent via a wire from the sensor directly to the drilling rig along the drill pipe.

HDD can be used to install pipelines through most soil or rock conditions both above and below the water table if the cutting tools, drilling rate, fluid pressure and flow rate are configured correctly. One of the key differences between drilling through soft ground and hard rock, is the drilling head configuration and reaming devices. When drilling in soft soils 'chisel' type drill heads are used (see Figure 5.85b) and the drilling rig provides thrust and rotational forces to the head. In order to drive in a straight line, the rods are continuously rotated as they are pushed forwards. To navigate curves, or to change alignment, the rotation is stopped and the drill rods are just pushed forwards (in this case the forces on the angled drill head causes the head to deviate from its current direction). When drilling in rock, drilling bits are used (Figure 5.86). The drilling rig again provides thrust and rotational forces to the head, but in addition, a mud-

Figure 5.85 a) An example of an HDD rig (note an example of a 'reaming' device beside the rig on the left), b) a close-up of the pilot drill head (note the 'chisel' type drill head with an angled face)

motor behind the drilling bit powers the cutting tool. A mud-motor uses the force of the drilling mud being pumped to the head to turn the drill bits. In order to navigate curves, once again the rotation is stopped and the drill rods pushed forwards. However, in this case a bent-sub arrangement located behind the cutting tool causes the drill head to deviate in direction (Figure 5.87).

Figure 5.86
a) Rock drill bit and
b) rock pre-reamer
(courtesy of Prime Drilling HDD-Technology)

Generally the achievable drive length with HDD is related to the diameter of the installed pipe; the greater the diameter of the pipe the shorter the achievable drive length. Some typical installation values are:

- 1 m diameter pipes can be installed for lengths of up to 1200 m;
- Longer drive lengths can be achieved for smaller diameter pipes (< 1 m);
- 2 m diameter pipes can be installed for lengths of up to 500 m.

HDD is also commonly used for installing smaller diameter pipes (cable ducts) over shorter distances (< 0.5 m diameter over 200 m).

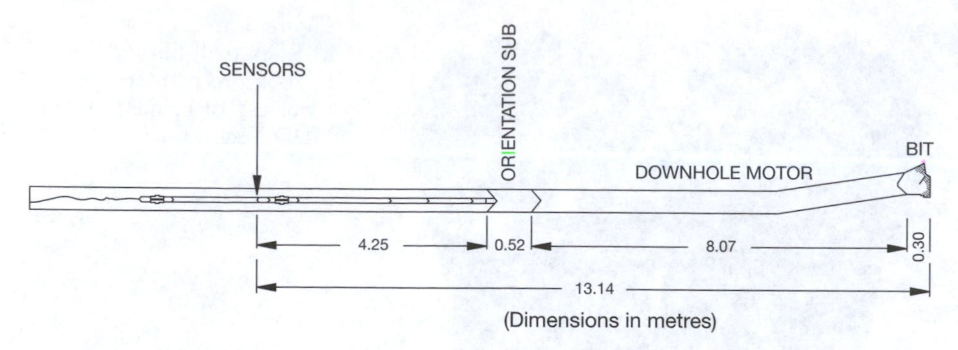

Figure 5.87 Schematic view of a typical bent-sub assembly. The sensors for locating the drill head are spaced behind the motor to reduce magnetic interference (after Riggall 2008)

EXAMPLES OF LARGE HORIZONTAL DIRECTIONAL DRILLING INSTALLATIONS

In terms of longer HDD drives, some typical examples from China and elsewhere during 2007 include (Ma and Najafi 2008): the Wei River Crossing in China involving a drive length of 2,873 m to install 660 mm diameter and 114 mm diameter steel pipes; the Modao Gate Crossing in China involving a drive length of 2,630 m to install a 660 mm diameter steel drainage pipe; and the Elbe River Crossing, Germany involving a drive length of 2,626 m to install a 350 mm diameter HDPE (High Density Polyethylene) pipe.

Figures 5.88 and 5.89 show an example of a larger HDD rig and pullback operations.

Figure 5.88 Example of a large HDD rig and associated facilities, pumps and mud recirculation system (courtesy of Prime Drilling HDD-Technology)

Figure 5.89 Pullback operations showing back reamer and final pipe (courtesy of Prime Drilling HDD-Technology)

6 Health and safety, and risk management in tunnelling

The health and safety of personnel carrying out the construction work as well as the general public is of paramount importance. This chapter introduces this important subject and the topic of risk management.

6.1 The health and safety hazards of tunnel construction

Contribution written by Dr D. R. Lamont C.Eng., FICE., Head of Tunnel and Ground Engineering, Civil Engineering Specialist Team, Health and Safety Executive, UK.

6.1.1 Introduction

Tunnelling is increasingly recognized as an environmentally friendly way of providing road and rail capacity in an increasingly congested world. However, not all tunnels are of large diameter and the many small diameter water, sewage and cable tunnels which are built every year by the utility providers should not be forgotten. Many hazards are common to soft ground and hard rock tunnels.

The risk to health and safety is not confined to those directly undertaking tunnel excavation as members of the public can also be affected. Over the past few years there have been a number of spectacular tunnel collapses around the world which have resulted in both workers and members of the public being killed. With mechanized tunnelling, the risk to the workforce from ground collapse has largely been removed except for those entering the cutterhead for inspection and maintenance purposes. When a collapse does occur with a shield driven tunnel in an urban area, it is probably those on the surface, likely to be the public, who are at greatest risk.

Tunnels are high value assets both in terms of their intrinsic worth and their value within the national infrastructure. Massive social and disruption costs can arise when a tunnel ceases to be available for operational use such as after a major fire. In these circumstances when the tunnel is no longer fit for operational use, it is almost inevitable that attempts will be made to recover and repair the tunnel and this has safety-related implications for those involved in the recovery operations. The two major fires in the Channel Tunnel of 1996 and 2008 illustrate this. Consequently the use of appropriate measures to protect the tunnel lining from fire is one aspect that should be considered at the design and construction stage.

6.1.2 Hazards in tunnelling

All the health and safety hazards of normal civil engineering construction can be found in tunnelling along with a few which are specific to tunnelling. In most cases the risks arising from these hazards present more severe consequences in tunnelling. This increase in severity is due to a number of factors including:

- The degree of uncertainty in the nature and variability of the ground through which the tunnel is being driven.
- The confined space of the tunnel environment particularly in small utility tunnels.
- A safety culture at all levels in the workforce which has until recently been poorly developed.
- A lack of commitment from all parties to the project in addressing occupational health and safety.
- Failure by the industry, to learn from the experiences and mistakes of others.
- Work in compressed air.

Comprehensive guidance on the hazards of tunnelling and on mitigating the risks arising from these hazards can be found in the current version of British Standard 6164 'Code of practice for safety in tunnelling in the construction industry' (BSI 2001b).

6.1.3 Techniques for risk management

The management of health and safety risk is no different to the management of other project risks. The hazards which arise in tunnelling should be identified from experience and by reference to relevant technical publications (Lamont 2006). It is rarely necessary to use formal hazard identification techniques, such as hazard and operability studies. Once the hazards have been identified, the risks which could arise should be assessed in terms of their likelihood of occurrence during the life of the project and their consequence. Risk assessment techniques used in tunnelling extend from the use of simple likelihood/consequence matrices to numerical quantified risk assessment techniques.

The use of sophisticated numerical techniques is only possible where appropriate input data exist and this information is not readily available in tunnelling. It is particularly important to consider the risk of low frequency but high consequence events such as collapse and plan for such events, which in practice happen more frequently than many realize. Risk management is described in more detail in section 6.2.

6.1.4 Legislation, accidents and ill health statistics

Occupational health and safety in tunnelling is normally subject to the same legislation as surface construction. The principal statute in the UK is the Health and Safety at Work Act 1974, which sets out generic goal setting requirements applicable to all work activity. This is supported by the Management of Health and Safety at Work Regulations and industry specific regulations such as the Construction (Design and Management) Regulations. Hazard specific regulations relevant to tunnelling include the Work in Compressed Air Regulations. A comprehensive description of health and safety legislation relevant to construction can be found in Appleby and Lamont (2009). In the author's experience, the UK tunnelling industry has taken occupational health and safety very seriously in recent years and the fatal and major accident rates have reduced significantly. However, although accident statistics are frequently used to measure the effectiveness of the safety management system, they are a poor measure as they are negative indicators, i.e. indicators of failure in the safety system. In addition they are subject to significant error, due to under-reporting of incidents and ill health. In general, few regulatory authorities or contractors publish detailed statistics on tunnelling accidents and ill health. One exception to this is with decompression illness which arises from work in compressed air. Engineers have had an interest in the causes of and cures for decompression illness since compressed air work was first undertaken in the mid-nineteenth century and many have published their thoughts and experiences. The Proceedings of the Institution of Civil Engineers contain a comprehensive selection of relevant papers which were summarized by Lamont (2007).

Most countries have some form of labour inspectorate which is ultimately responsible for occupational health and safety. Some countries also regulate health and safety through statutory social insurance organizations. These link premiums to performance and often provide training along with compensation and rehabilitation services for injured workers. In recent years, the major global re-insurance companies have taken a direct interest in tunnel construction works particularly seeking to reduce their exposure to claims by raising standards of ground risk management in the industry. Initially they worked with the British Tunnelling Society to produce a code of practice for the UK but this document has now been extended to be applicable internationally (ITIG 2006).

6.1.5 Role of the client, designer and contractors

Clients and designers, as well as contractors, have a contribution to make in ensuring good health and safety performance in tunnelling, and their responsibilities are set down in legislation and guidance in many countries. In the UK the Construction (Design and Management) Regulations lay down a statutory framework within which clients, designers and contractors must acquire and share safety related information overseen by a 'CDM coordinator'. Designers have specific duties to consider the health and safety of those affected by the building of a project and those working in the completed structure. The regulations require a 'principal contractor' to be appointed to oversee cooperation and coordination in matters relating to health and safety, between all contractors working on a project.

The client can set a framework for project procurement covering both design and construction, which should include requirements for the health and safety strategy for the project. In addition, the client can ensure that those whom they, directly or indirectly, employ to design and construct the tunnel works, in turn make available adequate resources, including finance and time, in order to address health and safety issues.

Designers can strongly influence health and safety, although in the past they have often shown little inclination to do so. The fundamental safety-critical aspects of tunnel design are diameter, alignment, shaft size/positions and portal location and once these have been fixed, the rest of the design process becomes more one of detail. Examples of how a designer can influence health and safety include choosing an alignment which facilitates the use of a TBM by avoiding a rock/soil interface or routing a tunnel away from, rather than through contaminated land. Where this is not possible, the designer should consider the impact on those building the tunnel and pass on relevant information to the contractor along with advice on risk mitigation measures.

In small utility tunnels the lack of working space contributes directly to the health and safety risk to the workforce, hence the primary factor in determining the minimum tunnel diameter may be the need to provide adequate working space to build the tunnel safely. Designers can use the specification to eliminate techniques or materials which are hazardous and should consider 'buildability' when designing openings in the tunnel and changes of cross section.

It is good practice to include requirements for fire fighting, atmospheric monitoring systems, communication and ventilation systems in the contract documents and to consider how to integrate what is required during tunnel construction with what the client requires in the finished tunnel. It can be argued that during construction these are the contractor's responsibility, as they are part of the temporary works: however, the client pays for them anyway through the contractor's overheads.

Any contractual arrangement which brings together design and construction expertise such as partnering, joint ventures or early contractor involvement can help mitigate risk in the tunnel during construction through sharing experience and improving buildability.

6.1.6 Ground risk

This has the potential to affect the most people in the event of a tunnel collapse. Those affected by a collapse include the client who suffers financial loss, those building the tunnel who are at risk of death or injury and the public who may also be at risk of death or injury. Spectacular tunnel collapses, such as that at Heathrow in 1994 in the UK (HSE 2000), do occur with disappointing regularity, hence engineers should always consider them in their risk assessments and plan their emergency measures accordingly. When they occur, the consequences will most likely be so great that there will be political repercussions in addition to the disruption to the works.

For all tunnel projects, adequate site investigation is essential. The designer must know the geology and hydrogeology in order to adequately address all the risks from the ground. The most comprehensive site investigation possible is required to identify ground parameters, discontinuities, water, gas and contamination (see Chapter 2).

Designers should liaise closely with contractors to ensure the stability of the tunnel under construction. This liaison must go beyond just the stability of the permanent works to include the stability of the tunnel at all stages of construction. In rock tunnels, the stability of the ground through which the tunnel is being driven has to be considered along with the stability of the ground around the tunnel intrados.

Often a primary sprayed concrete lining is classed as temporary works and considered to be the contractor's responsibility, however it is also the primary means by which the tunnel is supported during construction which makes it of fundamental importance for the safety of all those in the tunnel (and also for safeguarding the client's asset). The sequencing of the excavation process, particularly in complex tunnel layouts, can be crucial in ensuring safety. Designers should always ensure that the construction sequence which they envisaged in their design is adhered to.

Contractors should have a proper appreciation of the engineering principles behind the design and ensure they adhere to the design and specification and do not sacrifice quality of materials and workmanship to achieve cost savings and productivity. Quality assurance schemes have a place in tunnelling, but are no substitute for good engineering practice and supervision. It is important to learn from the mistakes of others.

6.1.7 Excavation and lining methods

The method of tunnel excavation can influence safety. In soft ground, most tunnels are now driven by shielded TBMs or a combination of the New Austrian Tunnelling Method (NATM) or sprayed concrete lining (SCL) techniques. In rock, TBMs or drill and blast techniques are normally used (see Chapter 5).

NATM/SCL is an observational method and as such requires considerable engineering input if proper management of the ground risk is to be achieved. Designs must be developed for both the most probable and the most unlikely conditions along with a number of incremental steps in between. Action or trigger limits in terms of relative values, e.g. differential movement between measuring stations, absolute values, e.g. total deflection, and rates of change must be determined. Throughout construction monitoring must be carried out and results compared against the alarm or trigger limits (see section 7.3).

It is very important that contingency plans, whose impact has been predicted in advance, should be in place during construction. It should be possible to put these into effect sufficiently quickly when alarm/trigger limits are exceeded to allow an effective recovery of the situation. In addition emergency plans should be in place which can be put into action when the worst happens and the contingency plans have not been effective.

6.1.8 Tunnel boring machines

Tunnel boring machines have become highly sophisticated but complex machines, and there are many hazards associated with their operation. One of the most hazardous areas of a TBM is the segment build area within which the erector operates. Here, heavy segments are handled whilst visibility for the erector operator can be poor. Miners are expected to place packing between segments as well as bolting up the segments to secure them in position. The risk of serious personal injury is always present.

The power consumption of large TBMs can be between 5 and 10 MW and supply voltages of 11 kV are becoming common. High standards of electrical safety are necessary if electrical accidents are to be prevented. The problems of working in a wet metallic environment, potentially explosive atmospheres, possible oxygen enrichment and compressed air, all add to the complexity of the electrical engineering problems in tunnelling.

TBMs for rock tunnelling are similar in many respects to those for soft ground tunnelling. In addition, self propelled machines such as roadheaders and specialized drill rigs for tunnelling often referred to in the industry as 'jumbos' can be used depending on rock strength. Specially adapted excavators can also be used for certain applications. One result of increased mechanization has been the marked reduction in hand tunnelling and its associated hazards of manual handling, noise, vibration and heat strain.

There are a number of European standards relating to the mechanical and electrical hazards of tunnelling machinery, which meet the requirements of European Directives on machine safety. For example, BSI (2005 and 1997) cover the safety of shielded and unshielded tunnel boring machines respectively. BSI (2002b) sets out requirements for the safety of roadheaders whilst BSI (2002c) relates to airlocks. Manufacturers supplying machinery into the European Community normally certify their machines to meet these requirements. The standards address a wide range of topics such as access to the cutterhead, handling of heavy components, rotation/stability, walkways and access openings, visibility, control systems, hydraulic and electrical systems and fire protection. These standards apply equally over the wide range of machines which are currently manufactured. This range extends from microtunnelling machines of under 1 m diameter to the largest TBMs currently being made of over 15 m diameter. Hence the requirements have to be somewhat general in nature. A separate standard covers airlocks and bulkheads whilst explosion protection of tunnelling machinery is covered by yet another standard.

6.1.9 Tunnel transport

Tunnelling often requires the transport of large numbers of men and considerable quantities of materials over long distances. A railway system is often used in bored tunnels, however wheeled or occasionally tracked plants are used in other tunnels. Much of the plant is of a specialized nature because the restricted space in the tunnel prevents conventional construction plant and vehicles from turning or slewing.

A tunnel is a confined space in which visibility is often poor due to lack of lighting. Consequently, there is a high risk of collision between men and machines which has resulted in a number of fatal and serious injury accidents in recent years. The provision of vehicle and pedestrian routes which are adequately separated and lit, the maintenance of vehicle lights in a serviceable condition and the provision of high-visibility clothing are all important means of mitigating these risks.

Increasingly other methods of removing excavated material from the tunnel are being used. These include slurry systems and conveyors and both give major safety benefits by significantly reducing the number of transport movements required in the tunnel. An added benefit is that neither method utilizes diesel engines, which generate contaminants for the tunnel atmosphere.

6.1.10 Tunnel atmosphere and ventilation

The quality of the tunnel atmosphere is very important and contaminants in the tunnel atmosphere affect everyone working in it. The most common atmospheric hazards and contaminants are oxygen deficiency and the presence of harmful gases such as carbon monoxide, the oxides of nitrogen and carbon dioxide, and potentially explosive gases such as methane, and radon, which is radioactive. Other atmospheric contaminants include dusts containing silica. None of the atmospheric contaminants can reliably be detected without the use of monitoring equipment. In all cases the risks arising from them should be mitigated by ventilation. Waste heat from plant and equipment also builds up in the tunnel atmosphere and has to be controlled by ventilation.

Frequently the tunnel ventilation systems fail to function effectively due to poor design or maintenance. Ducting can be wrongly positioned and thus fail to supply fresh air to the miners or it may not pass the required quantity of air if it has been blocked or joints and leaks in it have not been sealed.

6.1.11 Explosives

In rock tunnelling, extensive use is made of drill and blast techniques. Specialized tunnel drilling equipment capable of drilling a number of holes simultaneously often under computer control is used. The main hazards are dust, noise and vibration and the risks associated with storing and using explosives. The main risks from using explosives include premature detonation and atmospheric contamination from the dust and blast fume released by the blast.

6.1.12 Fire, flood rescue and escape

Among the most significant safety hazards of tunnelling, to which the workforce is exposed, are fire and smoke. In particular it is the rapid spread of smoke through the tunnel system, rather than radiant heat generated by a fire, which can lead to fatalities. As recent fires in the Channel Tunnel have shown, the tunnel lining can also be severely damaged by fire. In most tunnels under construction, the main sources of fuel for a fire are the large quantities of plastic, rubber and other flammable materials found on plant, and equipment, along with the significant quantities of hydraulic fluid and possibly diesel fuel kept underground. Reduced flammability hydraulic fluids are available and should be used in all underground plant along with flame retardant grease around the TBM. All hydraulic systems should be well engineered.

Equally important is the need for effective fixed onboard fire suppression systems on all plant and equipment. These should be supplemented by handheld extinguishers and a fire main with hydrants and hose reels in the complex tunnels. Fixed systems have the advantage of allowing everyone to evacuate the tunnel and not requiring someone to remain in a position of danger to fight a fire. Good housekeeping is another vital precaution in minimizing the build up of flammable rubbish, which typically in tunnelling includes timber, plastic bottles, paper, discarded hoses and cables.

In all tunnels there should be an underground alarm system as well as one or more communication systems. In large or complex tunnels these should be linked into the main tunnel control systems along with a comprehensive fire detection system.

It is normal practice to issue oxygen self-rescuers to everyone going underground. These should be worn on the belt in order to be readily accessible in an emergency.

In every tunnel there should be adequate arrangements for escape and rescue. These can either be based on a team made up from the contractor's own work force or from the local fire and rescue service. Sometimes it is in the contractor's interest to provide the local emergency rescue services with specialized equipment such as long duration breathing apparatus. To facilitate escape and rescue, a clear and well signed walkway should run throughout the length of the TBM and from its outbye end to a place of safety. This place of safety can be on the surface or in a so called 'rescue chamber' underground. In very long tunnels a dedicated emergency train may be required.

6.1.13 Occupational health

Occupational health is seldom allocated the priority it should be, given the number of days lost to ill health. The over-riding principle of occupational health to which all industry should subscribe is that 'no one should arrive home from work less healthy than when they left home to go to work'. The reasons for occupational health provision are two-fold:

- to address ill health due to work;
- to ensure fitness for work.

Most of the occupational health hazards of construction resulting in ill health are also present in tunnelling. They include dermatitis from the use of cementitious materials, serious respiratory problems from exposure to dust, hand-arm vibration syndrome, noise induced hearing loss and severe musculoskeletal injury.

As an example of ensuring fitness for work, no one can work in compressed air without first undergoing a medical examination to ensure their fitness for such work. Thereafter routine medical checks are required at intervals of 3 months or 28 days depending on the pressure to which they are exposed. Fitness for work can also be a safety issue, e.g. checking the eyesight of plant operators or locomotive drivers to ensure their vision is adequate.

People suffering from ill health are often no longer able to work and therefore many cases of occupational ill health go unrecorded ('healthy worker effect'). In an industry where peripatetic workers make up a significant proportion of the workforce, even more cases of ill health than for general construction may be going unreported. As tunnelling workers often have no access to regular health care when living away from home, the provision of occupational health facilities and even basic general health facilities becomes even more important.

6.1.14 Welfare and first aid

The provision of basic welfare in tunnels under construction is improving. Space for basic toilet and washing facilities is limited in small tunnels, but in larger tunnels there is enough space for toilet and washing facilities on the TBM or in the tunnel. A system for cleaning and maintaining the toilets is essential. The poorer the toilet facilities, the greater the need for hand cleaning facilities. Research has shown that significant reductions in the number of cases of minor ill health can be made by providing basic welfare facilities. In addition, messing facilities, with a supply of cold potable water are also required. A means of boiling water and heating food as part of the TBM equipment aids welfare and reduces the risk from improvised electrical installations. First aid provisions must be available to meet the requirements of the project in terms of shift working and remote working.

6.1.15 Work in compressed air

Compressed air working was first introduced in the mid-nineteenth century and has been a useful groundwater control technique ever since. The main occupational illnesses arising from it in tunnelling are decompression sickness and aseptic bone necrosis – collectively referred to as 'Decompression illness'. The normal symptoms of decompression illness are joint pain of varying severity or occasionally neurological symptoms. It is readily treatable by recompression and slow decompression back to atmospheric pressure. Osteonecrosis, which results in the breakdown of joint surfaces and causes disability, is only treatable by surgical replacement of the affected joints.

The era when large soft ground tunnels below the water table, such as the Dartford and Clyde Tunnels, were hand dug by miners working in compressed air has passed. Although such working practices have virtually ceased there is still a small legacy of bone necrosis cases from earlier exposure. TBMs, such as slurry machines and earth pressure balance machines, require the application of compressed air within the cutterhead to facilitate face inspection and cutterhead maintenance. Whilst the number of exposures has probably been cut by over 95% compared to hand digging the tunnel, some work under pressure is still required. There is also a trend for tunnels to be dug ever deeper in more challenging geological conditions, thus increasing the working pressures.

With air-only decompression, the incidence of decompression illness can exceed 2% for some pressure/time combinations. This is unacceptable for the twenty-first century. Many countries including the UK, have now adopted routine oxygen decompression to reduce the incidence of decompression illness. Overall, however, the compressed air tunnelling industry lags far behind the diving industry in its hyperbaric engineering practices. In a very small number of tunnels, saturation techniques as in offshore diving, requiring the use of mixed gas breathing, have been used and working pressures of over 10 bar are being required.

Not only does hyperbaric working present a health hazard, it also presents a safety hazard as fire risk increases directly with increasing atmospheric pressure. Oxygen leakage during decompression raises the oxygen concentration and results in an even higher risk. Work at pressure also leads to increased risk from heat strain and can also exacerbate exposure to contaminants.

The Health and Safety Executive provides extensive guidance on compressed air working in its guidance document HSE (1996b). Further information can also be found in Lamont (2007).

6.1.16 Education, training and competence

The traditional image of tunnel workers, and one which they have been reluctant to change, is of a hard working, hard living, macho culture. Concern for one's safety and health has not been a priority. It is vital if health and safety standards are to be improved, for more resource to be put into raising the general competence of all those in the industry as well as their competence in health and safety matters. There are training initiatives in some countries for tunnel operatives and first line supervisors, and the number of universities offering postgraduate courses in tunnelling is growing. Competent supervision of tunnelling works is vital and front line supervisors play a key role in fostering greater awareness of health and safety issues amongst the workforce. Large tunnelling projects may need to set up their own training facilities, for example CrossRail in the UK set up its own 'Tunnelling Academy' to train the large number of workers required for this project.

All new employees in the industry should undergo comprehensive induction training. Site-specific training, even for experienced employees who are new to a site, is also necessary. Engineers and managers now undertake training in health and safety matters as part of their professional education and continuing professional development. This training extends beyond what is required for personal safety to what is required to ensure the safety of those affected by their professional activities. Many national and international tunnelling organizations provide training materials and courses.

6.1.17 *Concluding remarks*

There is still scope for improvement in standards of health and safety in tunnel construction. Experienced practitioners should share knowledge, guidance and good practice with those entering the industry. Good standards of health and safety require the commitment of resources in terms of time and money but a productive workforce can only be sustained if working conditions are healthy and safe. Respect for people through respect for their health and safety must be our goal in the twenty-first century.

6.2 Risk management in tunnelling projects

6.2.1 *Introduction*

This section provides a brief introduction to the concept of risk management in civil engineering and particular tunnelling projects. This can only be regarded as an overview and the reader is encouraged to consult more detailed reference materials such as Eskesen *et al.* (2004), Clayton (2001), and also the Code of Practice for Risk Management (ITIG 2006). A Technical Guidance Note on Geotechnical Risk Management for Tunnel Works has also been produced by the Geotechnical Engineering Office of the Hong Kong Government (TGN25 2005).

The concept of risks and risk management is not new in construction, and back in 1993 Sir Michael Latham stated when reporting on construction procurement methods for the UK Government, 'No construction project is risk free. Risk can be managed, minimized, shared, transferred, or accepted. It cannot be ignored.' With respect to tunnelling operations, the late Sir Alan Muir Wood stated 'Uncertainty is a feature that is unavoidable in tunnelling. But it can be understood and controlled so that it does not cause damaging risk.'

Therefore the management of risks within a tunnelling project is vital to ensure a successful project. As Sir Alan Muir Wood mentioned, uncertainty in tunnelling is unavoidable. One aspect of uncertainty in tunnelling can be attributed to the ground, which, as described in Chapter 2, is characterized from laboratory and field tests conducted on only a very small proportion of the total ground affected by the tunnelling operation. Generally, the ground parameters are given a range of values and hence the risk management is important, taking into account the best and worst case scenarios as well as values in between. However, it is not only uncertainty, but the hazards that are involved in the overall tunnelling project which need to be considered in any analysis.

In order to evaluate risk, an understanding of the difference between hazard and risk is important, as well as other useful definitions given by Eskesen *et al.* (2004):

- *Hazard* – A situation or condition that has the potential for human injury, damage to property, damage to environment, economic loss or delay to project completion.
- *Risk* – A combination of the frequency of occurrence of a defined hazard and the consequences of the occurrence.
- *Risk acceptance criteria* – A qualitative or quantitative expression defining the maximum risk level that is acceptable or tolerable for a given system.
- *Risk analysis* – A structured process which identifies both the probability and extent of adverse consequences arising from a given activity. Risk analysis includes identification of hazards and descriptions of risks, which may be qualitative or quantitative.
- *Risk assessment* – Integrated analysis of risks inherent to a system or a project and their significance in an appropriate context, i.e. risk analysis plus risk evaluation.
- *Risk elimination* – Action to prevent risk from occurring.
- *Risk evaluation* – Comparison of the results of a risk analysis with risk acceptance criteria or other decision criteria.
- *Risk mitigation measure* – Action to reduce risk by reducing consequences or frequency of occurrence.

The risk management strategy for a project must consider all aspects, from the design life, durability and repair and maintenance of a structure. It is important to realize that the opportunity to minimize risks is highest during the early feasibility stage of a project and this opportunity decreases rapidly once the project moves into the design and construction stages, as illustrated in Figure 6.1. The cost of change also increases substantially the further one gets into the project.

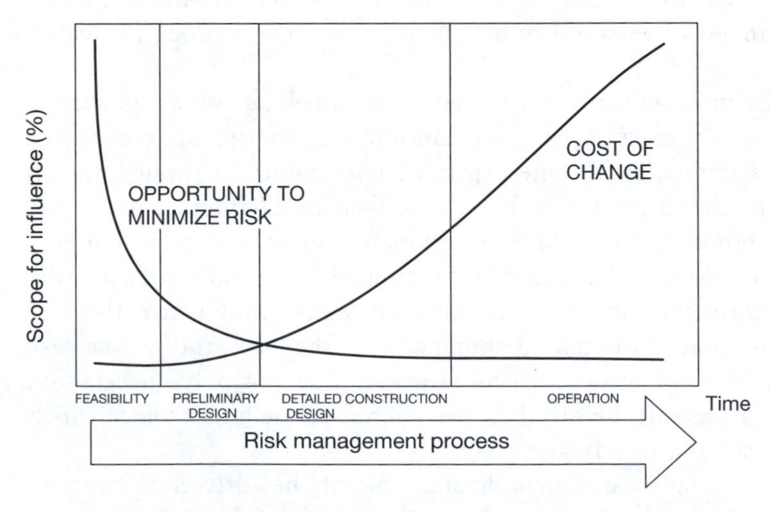

Figure 6.1 Risk management and impact versus time (after Caiden 2008, used with permission)

Although risk management is often regarded as negative, it should also be seen as providing an opportunity for doing things better. Hence, the following elements should form part of a risk management procedure:

- provision of an auditable framework to effectively identify, analyse, evaluate and treat risks on projects;
- ensuring the correct people with the most subject knowledge are involved;
- keeping budgetary and programme creep under control by pre-empting problems;
- ensuring insurability;
- provision of the necessary checks and balances to satisfy financiers or funding agents.

The process of risk management involves a number of steps which can be illustrated as shown in Figure 6.2. Figure 6.2 indicates that risk management is not a linear process and several aspects have to be considered. The key steps that need to be processed are: identification of the risks; assessment of the risks; and addressing the risks. It is important to realize that the risks include political, financial, legal, regulatory, contractual, technical and operational, i.e. are not just restricted to the actual construction operation. When assessing the risks and trying to understand these, it is essential to identify: the potential hazards/impacts; potential consequences; likelihood of occurrence; data/information sources; interested and affected parties; uncertainty, variability and unknowns. Once the risk analysis has been carried out, the risks need to be addressed and either accepted, avoided,

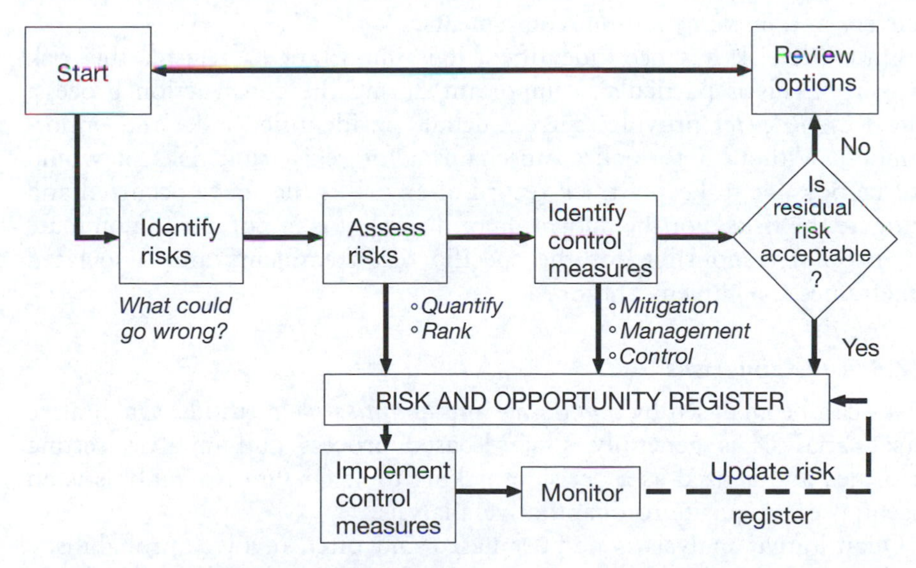

Figure 6.2 Risk and opportunity management flowchart (after Caiden 2008, used with permission)

mitigated or transferred. Further, it is critical that the risks are monitored during all stages of the project including for example cost estimates, labour issues, site and weather. Ultimately, it is important to react to any risks that may occur.

6.2.2 *Risk identification*

The process of risk identification may rely upon: (a) a review of worldwide operational experience of similar projects with written submissions from partner companies, (b) the study of generic guidance on hazards associated with the type of work being undertaken, and (c) discussions with qualified and experienced staff from the project team and other organizations around the world. It is important to identify the potential hazards in a structured process (Eskesen *et al.* 2004).

The identification and classification of the risks is best carried out through brainstorming sessions with risk screening teams consisting of multidisciplinary, technically and practically experienced experts guided by experienced risk analysts. The aim should be to identify all conceivable hazardous events threatening the project including those risks of low frequency but high possible consequence.

In section 6.2.1 it was mentioned that uncertainty in a project is one significant contributor to risk (although not the only one), with the uncertainty related to the ground characteristics being one of the key contributing factors. However, examples of other common areas of uncertainty affected by any civil engineering project include political and economic environment; planning, regulatory and approvals procedures; environmental and sustainability requirements; construction and buildability issues; safety; project delivery and implementation requirements.

Once the risk has been identified, it is important to register this risk properly. This is particularly important during the construction process. The *Risk Register* provides current details on identified risks and opportunities. Without a formalized mechanism for registering risks, it would not be possible to keep a track record of any risks that have occurred and mitigate the risks. For the mitigation of the risk it is important to nominate a person(s) responsible for the specific risk treatments and associated timeframes for implementation.

6.2.3 *Analyzing risks*

Risks can be analysed by *qualitative* and *quantitative* methods. Qualitative based analysis is generally a word-based process and involves setting priorities and is used as a decision-making tool. Qualitative analysis is an essential pre-requisite to quantitative analysis.

Quantitative analysis is number based and often involves probabilistic analysis techniques such as those developed in the 1990s (for example for the Adler tunnel, Einstein *et al.* 1994). It can provide an aggregate view of risks and is focussed on targets and contingencies. This can be conducted using Monte Carlo type simulations, which involve fitting probability distributions to the risks according to surrounding conditions of the variable and then running a large number of analyses taking random variables from each of these probability distributions to assess the overall risks (Rubinstein and Kroese 2007). There is also scheduling and estimating software available based on methods such as fault, event or decision tree analysis or multirisk analysis. However, all these quantitative analysis methods rely on input parameters and hence the expertise of the person doing the analysis.

6.2.4 Evaluating risks

When evaluating the risk, it is critical to look at the identified risks and determine the frequency that these may occur. Table 6.1 shows an example of what type of risk frequency can be utilized.

Once the frequency of a risk has been identified, the consequence of such a risk needs to be determined, which could range from insignificant to catastrophic (Table 6.2). It is now important to develop a matrix to determine a risk rating (for example from low to extreme). This can be achieved qualitatively by combining the likelihood of an event occurring with the resulting consequences. An example of such a matrix is shown in Table 6.2. It should be noted that Table 6.2 is only an example and the likelihood-consequence matrix has to be developed for each project.

6.2.5 Risk monitoring and reviewing

After identification of the risks, their likelihood of occurring and the consequence of them occurring, it is vital to determine various ways of treating these risks. It is always desirable to avoid the risk. Options here include changing the project plan/scope of the works to eliminate the risk. However, in the event that this is not possible, ways have to be considered to mitigate

Table 6.1 Example of risk evaluation (after Caiden 2008, used with permission)

Descriptor	*Description of frequency*
Rare	May occur only in exceptional circumstances – can be assumed not to occur during the period of the project (or life of the facility)
Unlikely	Event is unlikely to occur, but it is possible during the period of the project (or life of the facility)
Possible	Event could occur during the period of the project (or the life of the facility)
Likely	Event likely to occur once or more during the period of the project (or life of the facility)
Frequent/ almost certain	Event occurs many times during the period of the project (or the life of the facility)

Table 6.2 Example of qualitative risk rating (after Caiden 2008, used with permission)

		Consequence				
		Insignificant	*Minor*	*Moderate*	*Major*	*Catastrophic*
Probability (likelihood)	Rare	Low	Low	Low	Medium	Medium
	Unlikely	Low	Low	Medium	Medium	High
	Possible	Low	Medium	Medium	High	High
	Likely	Medium	Medium	High	High	High
	Frequent	Medium	High	High	Very high	Extreme

the risk with either a reduction in probability of occurrence or reduction in consequence if something happens. If this is not possible, then the option of risk transfer or sharing needs to be considered. Finally, if all else fails, the risk needs to be accepted, but a contingency needs to be created. This option involves agreeing to accept the consequences of a risk. It is important that accepted risks are considered carefully and the consequences of these risks are prepared for in advance and appropriate contingency or fallback plans developed. It is also important that these risks are continually monitored carefully and reviewed (after Caiden 2008).

EXAMPLE

The risk assessment carried out as part of the Piccadilly Extension at T5, London UK included: confined working space, hot works, use of electrical equipment, working at height, plant operations, manual handling and lifting, dust, noise, tunnel construction, COSH survey, shotcreting overhead and others (details of this case history are provided in section 8.2). Each of these were analysed with respect to the hazard they posed, for example the hazards related to tunnel construction in this case were:

- excavation;
- face collapse;
- vault collapse;
- unforeseen ground conditions;
- application of sprayed concrete;
- failure of sprayed concrete due to insufficient strength;
- excessive deformations causing failure of the lining.

Each of these hazards were analysed, identifying the level of risk (high or medium), and the persons at risk, which included employees, the public, subcontractors and others. Control measures were established to address these key issues. Examples of the control measures for the excavation hazards included applying an initial layer of sprayed concrete to the face, using an inclined face where possible and monitoring ground conditions.

Table 6.3 Example headings for risk assessment table

No.	Operation	Hazard	Persons at risk				Risk (H/M)	Key issues addressed (control measures)	Residual risk (H/M/L)
			Employees	The public	Sub-contractors	Others			

Control measures for the application of sprayed concrete included having quality control measures in place, systematic testing to assess strength gain, monitoring of the structure and establishment of trigger levels. Based on this analysis and the control measures, the residual risks could be determined, which were assessed as high (H), medium (M) or low (L). Assuming all the control measures were implemented correctly and all the key issues addressed, the residual risk should be low in all cases. The headings used in the risk assessment table are shown in Table 6.3.

7 Ground movements and monitoring

This chapter introduces the topics of how to estimate ground movements caused by the construction of tunnels in soft ground, the importance of these ground movements with respect to their effects on adjacent structures and how these movements are monitored during the construction process. In addition, the important topic of how assessment of the stability of the tunnel during the construction for open face tunnels, such as NATM, via in-tunnel monitoring is introduced.

7.1 Ground deformation in soft ground

When tunnelling in hard ground (rock), ground movements are not normally a problem, except in squeezing ground conditions, and ground movements propagating up to the ground surface as a result of the excavation are unlikely unless the cover depth of the tunnel is relatively small, i.e. in portal areas, or where the groundwater in the overlying soft ground may be affected. In soft ground, however, displacements can occur due to a number of reasons and these are shown for a shield tunnel on Figure 7.1. These components are (after Mair and Taylor 1997):

1 deformation of the ground towards the face due to stress relief;
2 radial ground movements due to the passage of the shield, possibly due to an overcutting edge (bead) used to help steering, or whilst trying to maintain alignment of the shield (pitching and yawing angles);
3 tail void due to the difference in diameter of the tail of the shield and the installed lining, and hence the tendency for ground to move into this gap;
4 distortion of the tunnel lining as it starts to take the ground loading;
5 time dependent consolidation in fine grained soils.

Component 1 is particularly important with open face tunnelling methods. However, if TBMs with pressurized faces, such as EPB and slurry TBMs, are used this component can be negligible if good face control is achieved. It should be noted that over-pressurization at the face can lead

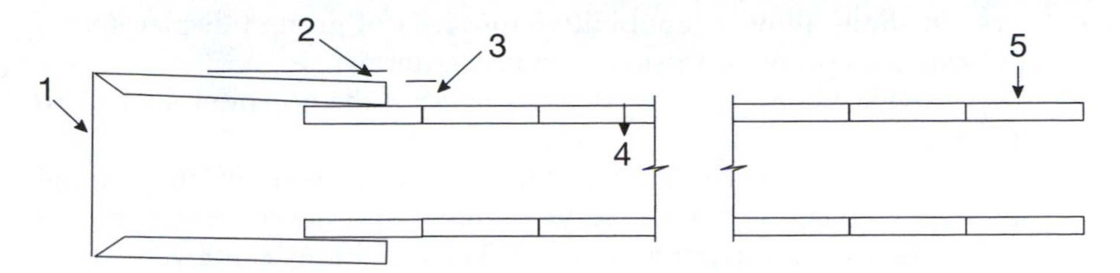

Figure 7.1 Primary components of ground movements associated with shield tunnelling (after Mair and Taylor 1997, from Cording 1991, used with permission from Professor E.J. Cording)

to outward movements and heave at the ground surface. 'Ravelling' or re-compaction due to local loss of ground at the face can also contribute to this component. *Component 2* can result if there is difficulty keeping the tunnelling shield on the correct alignment, or if there is a need to tilt the shield up slightly to prevent it from diving into the ground. *Component 3* can be minimized by immediate grouting of the void. *Component 4* is usually small compared to the other components once the lining ring is completed. *Component 5* can be important for soft clays, and results from the fact that the construction process changes the stress regime locally around the tunnel. This causes changes in the water pressure within the pores between the soil particles. As these excess pore water pressures equilibriate over time the ground will change volume and consolidate (see section 3.2 for more information on stresses around tunnels). It should be noted that when tunnels are constructed with no shield, for example NATM, components 1, 4 and 5 are still applicable.

These components can result in displacements reaching the ground surface, which can be particularly significant in urban areas, where they can influence overlying or adjacent structures such as buildings, other tunnels and services. In contrast, if there are no ground-structure-interaction effects, these ground movements are termed 'greenfield' movements.

It is important to estimate these ground movements so that tunnelling techniques can be optimized in order to control the movements of overlying or adjacent structures. In addition, other measures can be implemented to control these movements, for example the use of compensation grouting. These ground movements can be estimated using numerical methods, as described in section 3.6, or semi-empirical methods as described below.

7.1.1 Surface settlement profiles

Although these days enormous advances in computer based numerical methods for calculating ground displacements are being made, there are still some advantages of using simple empirical based methods in soft ground (Devriendt 2006):

- these methods allow a rapid initial appraisal of ground displacements and can use established risk assessment criteria;
- they provide a conservative risk assessment of the potential damage to structures;
- for 'flexible' structures such as long masonry walls at the ground surface, interaction effects may be minimal and hence semi-empirical approaches based on assuming greenfield conditions can give realistic results.

A number of reviews have been conducted of this subject, for example Mair and Taylor (1997), BTS/ICE (2004) and ITA/AITES (2007). However, a brief overview of the subject is provided in this section.

Schmidt (1969) and Peck (1969b) established, via case history data, that the ground surface settlement 'trough' above tunnels, i.e. normal (or '*transverse*') to the direction of the tunnel, can be described by an inverted normal probability (or 'Gaussian') curve (equation 7.1 and Figure 7.2).

$$S(y) = S_{max} \exp\left(-\frac{y^2}{2\,i^2}\right) \quad (7.1)$$

where S(y) is the vertical settlement at point y, S_{max} is the maximum settlement directly above the tunnel centreline, y is the transverse horizontal distance from the tunnel centreline of the trough, and i is the trough width parameter, which represents the point of inflection on the transverse profile, equivalent to one standard deviation in a normal probability distribution. This has subsequently been confirmed by numerous authors from other case history data, for example O'Reilly and New (1982) and Attewell *et al.* (1986).

By integrating equation 7.1, the volume of the surface settlement trough (per metre length of tunnel), V_S, can be approximated by equation 7.2.

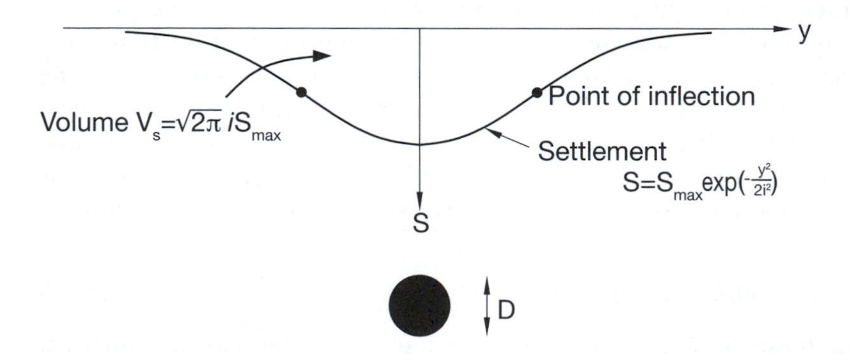

Figure 7.2 Gaussian curve for representing the transverse settlements above a tunnel in soft ground (after Dimmock and Mair 2007a, used with permission from Thomas Telford Ltd and Professor R.J. Mair)

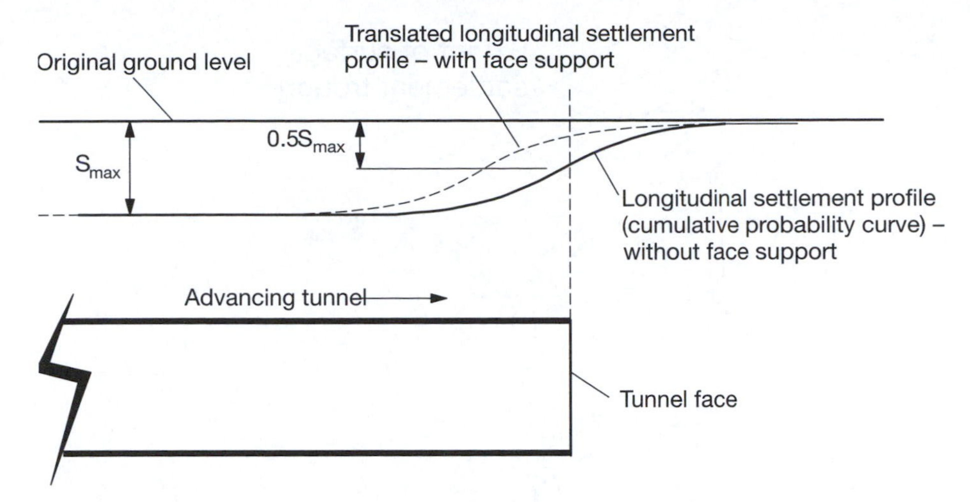

Figure 7.3 Longitudinal settlement profile above tunnels in soft ground, showing the difference in distribution for open face tunnelling and where there is significant face support (after Mair and Taylor 1997)

Equation 7.2 can be rearranged to calculate the maximum vertical settlement, S_{max}, directly above the tunnel.

$$V_S = \sqrt{2\pi}\, i\, S_{max} \tag{7.2}$$

The geometry of the settlement trough is uniquely defined by selecting values for the volume, V_S, and the trough width parameter, i. The choice of these values is discussed later in sections 7.1.1.1 and 7.1.1.2.

In the *longitudinal* direction to the tunnel construction, it has been found that the vertical displacements can be estimated, following examination of a number of tunnel construction case histories in clays (Attewell *et al.* 1986, Attewell and Woodman 1982), by a 'cumulative probability curve' as illustrated in Figure 7.3. For tunnels constructed in stiff clays without face support, the surface settlement directly above the tunnel face corresponds to $0.5S_{max}$. For tunnels in soft clays with face support, for example in EPB or slurry shield machines, the surface settlement directly above the tunnel face is much less than $0.5S_{max}$. In these cases, the major source of the ground movement is further back from the face and this leads effectively to a translation of the cumulative curve (Ng *et al.* 2004).

The *transverse* and *longitudinal* ground displacement profiles can be combined to represent the full three-dimensional surface displacement as shown in Figure 7.4.

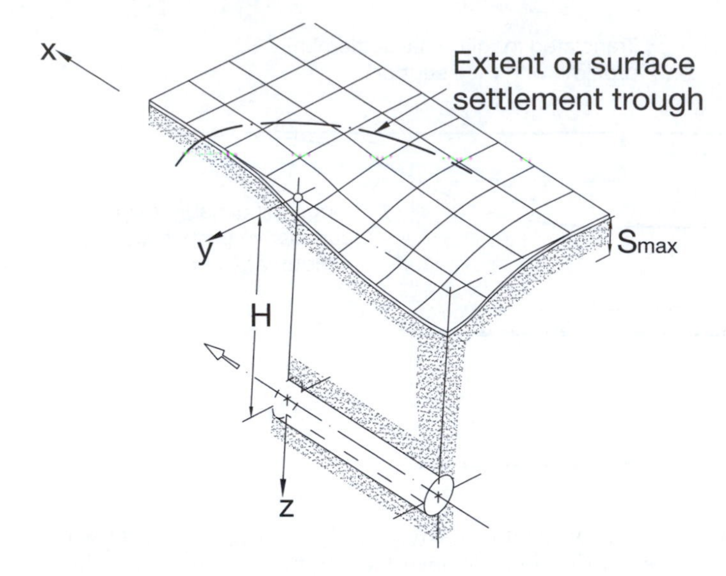

Figure 7.4 Three-dimensional representation of the surface settlement as a tunnel is constructed in soft ground (Attewell 1995, after Yeates 1985)

7.1.1.1 Estimating the trough width parameter, i

There are a number of empirically derived methods for estimating the trough width parameter, i. It has been shown by various researchers on the basis of case history data, for example O'Reilly and New (1982), that the trough width parameter at the ground surface is an approximately linear function of the depth of the tunnel, H, and is largely independent of the tunnel construction method and tunnel diameter (except for very shallow tunnels where the tunnel depth to diameter ratio is less than one). Therefore, the relationship shown in equation 7.3 can be used.

$$i = KH \tag{7.3}$$

where H is the depth from the ground surface to the tunnel axis level, and K can be estimated as shown in Table 7.1

Table 7.1 Typical K values

Soil type	*K*
Stiff fissured clay	0.4–0.5
Glacial deposits	0.5–0.6
Soft silty clay	0.6–0.7
Granular soils above the water table	0.2–0.3

In the urban environment there are often situations where tunnels are constructed close to existing subsurface structures and hence there is a need to estimate the settlements below the ground surface. Mair *et al.* (1993) analysed subsurface data from various tunnel projects in stiff and soft clay, together with centrifuge model test data in soft ground. They showed that subsurface settlement profiles can also be reasonably approximated in the form of a Gaussian curve in the same way as surface settlement profiles. For subsurface regions up to one diameter away from the tunnel, at depth z below the ground surface and above a tunnel at depth H, the trough width parameter can be expressed as shown in equation 7.4.

$$i = K(H - z) \tag{7.4}$$

where z is the depth from the ground surface to the level being considered and K is given by equation 7.5. Equation 7.5 yields shallower, wider and more realistic subsurface settlement troughs at depth compared to those obtained when K is kept constant.

$$K = \frac{0.175 + 0.325\,(1 - z/H)}{(1 - z/H)} \tag{7.5}$$

It should be noted that only one method of estimating the subsurface trough width parameter is presented here and other methods exist.

In the previous equations presented in this section, the assumption is that the ground is homogeneous. However, tunnels are often constructed in layered ground comprising fine and coarse grained soils. Selby (1988) and New and O'Reilly (1991) suggest that the trough width parameter for the surface settlement trough could be estimated from the trough width factor K for each layer and the relative thicknesses of each layer. Hence, for a two layered ground, i would be calculated as shown in equation 7.6.

$$i = K_1 z_1 + K_2 z_2 \ (+ \ldots) \tag{7.6}$$

where K_1 is the trough width factor for soil type 1 of thickness z_1 and K_2 is the trough width factor for soil type 2 for thickness z_2. Field evidence suggests that for sands overlain by clays wider surface settlement profiles are obtained than if tunnels were in sand alone (Ata 1996 and Atahan *et al.* 1996). However, there is less evidence for coarse grained soils overlying fine grained soils. For example, Grant and Taylor (1996) used centrifuge physical model tests to investigate the ground displacements when soft clay is overlain by a sand. In this case they found that the trough was wider than would be expected if there was only soft clay and does not reflect the narrowing predicted by equation 7.6. This is probably due to the overlying sand layer being significantly stiffer than the soft clay (Mair and Taylor 1997).

7.1.1.2 Volume loss

The volume of the surface settlement trough, V_S, must be estimated, together with the trough width parameter, in order to determine the magnitude of the settlements. This trough volume derives from the various short-term components for why the ground movements develop around tunnels during construction, as listed in Figure 7.1, components 1 to 4. These occur mainly in the region close to the tunnel and the term 'volume loss', V_t, (sometimes referred to as ground loss), is used to describe the accumulation of these components, i.e. the volume of 'lost' ground that can propagate up to the ground surface causing the surface settlement trough.

The choice of the volume loss value is fundamental to all the methods of estimating tunnelling displacements. Volume loss can be defined as the ratio of the estimated volume 'losses' (V_t) over the excavated volume of the tunnel (V_o). It is usually defined in the two-dimensional sense as a percentage of the excavated face area, i.e. volume per metre length of tunnel (equation 7.7). If the tunnel is circular then $V_o = (\pi D^2/4)$, where D is the tunnel diameter.

$$V_l(\%) = V_t/V_o \cdot 100\% \tag{7.7}$$

When tunnelling in drained conditions, for example in coarse grained soils such as dense sands, the volume of the surface settlement trough V_S is less than V_t because of volume changes that occur within this type of ground as it moves (Cording and Hansmire 1975). However, when tunnelling in a fine grained soil, such as clay, the ground movements usually occur undrained (constant volume) and hence $V_S = V_t$.

The selection of the volume loss value (V_l) is based on engineering judgement and experience from previous projects in similar ground, or where similar tunnelling techniques were used. Various authors suggest possible values in different ground conditions and/or for different tunnelling methods (O'Reilly and New 1982, Ng *et al.* 2004). Mair and Taylor (1997) concluded the following based on projects conducted at that time:

- volume losses in stiff clays such as London clay using open-face tunnelling are generally between 1% and 2%;
- recent project in London clay using sprayed concrete linings can produce volume losses between 0.5% and 1.5%;
- EPB and slurry machines can achieve a high degree of settlement control, particularly in sands with volume losses as low as 0.5%. In soft clays, short term volume losses of only 1–2% have been reported;
- in mixed face conditions volume losses may be higher for EPB and slurry machines.

As shown on many recent tunnelling projects involving TBMs, if good control of the face pressures can be achieved, volumes losses of less than 0.5% are achievable. On the Channel Tunnel Rail Link (CTRL) project through London in the UK, volume losses of less than 0.5% were commonly recorded using EPBMs (ITA/AITES 2007). Keeping the volume loss small means smaller ground displacements (refer back to equation 7.2) and hence less impact on overlying and adjacent structures (see section 7.2).

Dimmock and Mair (2007b), after work by Macklin (1999), proposed a relationship between volume loss and load factor (LF) for tunnels constructed in overconsolidated clay. This relationship, shown in equation 7.8, was derived empirically from field monitoring data. It is recommended that for design purposes a range of values should be considered for these parameters.

$$V_l(\%) = 0.23\ e^{4.8(LF)}\ (\text{for LF} \geq 0.2) \qquad (7.8)$$

where LF is defined as the ratio of the stability ratio (N) over the stability ratio at collapse (N_c) (stability numbers are described in section 3.3).

7.1.2 Horizontal displacements

From the point of view of damage to structures and services it is not only important to determine vertical displacements within the ground, but also the horizontal movements (Burland *et al.* 2001b). In the transverse direction to the tunnel construction, the surface (and subsurface) horizontal displacements can be estimated by various assumptions. The simplest is to assume that the ground movements are radial, i.e. directed to the tunnel axis (equation 7.9 and illustrated in Figure 7.5a).

$$S_h = S_v y/H \qquad (7.9)$$

where S_v is the vertical ground displacement, S_h is the horizontal ground displacement and y is the transverse horizontal distance from the tunnel centreline.

Alternative methods have been proposed. For example, Mair *et al.* (1993) demonstrated that in London Clay the radial movement assumption can over-estimate the horizontal movements, particularly those close to the tunnel, and proposed the relationship shown in equation 7.10. This assumes that the displacements are directed to a point 0.175H/0.325 below the tunnel axis (Figure 7.5b).

$$S_h = S_v y/(1 + 0.175/0.325)H \qquad (7.10)$$

New and Bowers (1994) developed the idea of the cumulative probability distribution model to provide a full array of equations for the prediction

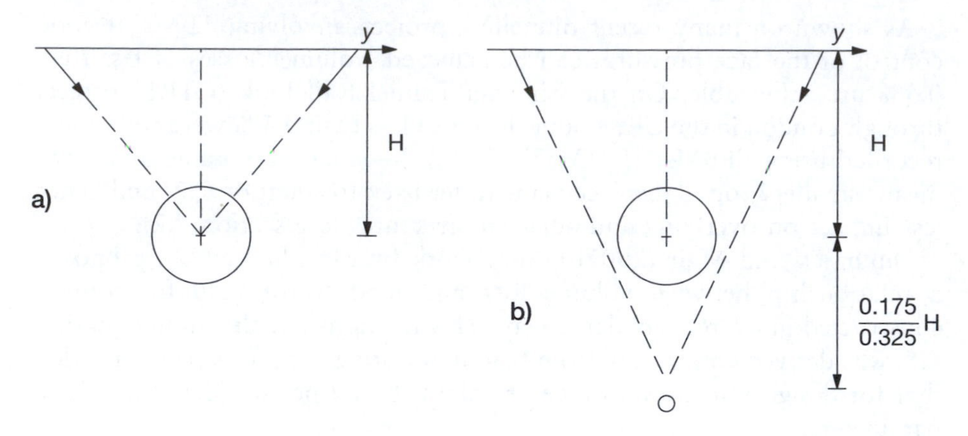

Figure 7.5 Direction of the ground displacement vectors above tunnels in clay (after Mair and Taylor 1997), a) vectors directed towards axis (Attewell 1978, O'Reilly and New 1982), b) vectors directed towards point O (Taylor 1995b)

of ground movements in three dimensions. This approach has been shown to give significantly improved predictions in the vicinity of the tunnel. (BTS/ICE 2004)

7.1.3 Long-term settlements

Long-term settlement is a phenomenon predominantly associated with fine grained soils and is associated with component 5 in Figure 7.1. As described previously, it is the result of the equilibrium of excess pore water pressures within the soil over time and the associated volume changes that occur. Other contributory factors to the total long-term settlements could be (after Devriendt 2006):

- the tunnel acting as a drain (depends on the permeability of the tunnel lining relative to the ground);
- time dependent distortion of the tunnel lining;
- time dependent dissipation of excess pore water pressures due to grouting behind the lining or due to mitigation measures such as compensation grouting;
- creep and secondary consolidation processes in soils;
- time dependent closure of the grouted annular gap due to: bleeding and curing (hardening and shrinkage) of the grout, insufficient grout or loss of grout.

As well as the maximum settlement increasing in the long-term, these effects cause the settlement trough to widen (Burland *et al.* 2001a). For soils such as soft clays and silts the component of volume loss due to initial 'undrained' movements may be small compared to the time dependent processes. The maximum settlement may reach between two to four times the short-term value and the trough width parameter at the surface from 1 to 2.5 times the short-term value. Fang *et al.* (1993) proposed a hyperbolic time settlement relationship to describe the increase in maximum settlement with time for earth pressure balance tunnelling machines commonly used in these soils. Alternatively, when tunnelling with traditional compressed air methods, a number of authors have suggested that long-term settlements in soft soils increase linearly per logarithm of time (Devriendt 2006). Work in this area is continuing, for example Wongsaroj *et al.* (2007) investigated the long-term ground movements in London resulting from the construction of the Jubilee Line Extension, which opened in 1999.

7.1.4 Multiple tunnels

The previous work described in this chapter has related to single tunnels. However, in many situations rather than constructing a single large tunnel, twin tunnels are constructed. This has to be taken into account when determining the ground movements generated by the tunnel constructions. New and O'Reilly (1991) proposed equations for the prediction of cumulative displacements for parallel tunnels with a given separation based on the principle of superposition. However, Attewell *et al.* (1986) discuss the 'interference volume' effect where the volume loss is commonly greater when a second tunnel is excavated adjacent to the completed tunnel (asymmetric effect in the final settlement trough will occur for two side-by-side tunnels of the same cross section and depth) (Devriendt 2006).

Addenbrooke and Potts (2001) considered this phenomenon numerically and found it to be due to the accumulation of shear strain adjacent to the first tunnel. This results in a lower stiffness and hence greater displacements where subsequent tunnels are constructed close to the first tunnel (Devriendt 2006). This has also been investigated by Cooper *et al.* (2002), Hunt (2005) and Chapman *et al.* (2007), with observations from case history data, numerical analyses and small-scale physical modelling, respectively. These authors propose a method for estimating the ground movements (both surface and subsurface) above closely spaced twin side-by-side tunnel constructions in soft ground. This method should be considered if the clear separation between the tunnels is approximately three diameters or less.

7.2 Effects of tunnelling on surface and subsurface structures

One of the most important aspects of tunnelling in soft ground is assessing the effect of any ground disturbance caused by the tunnelling operations on surface or subsurface structures. This is particularly important in urban areas. It is important to recognize that the ground movements, caused by the tunnelling, and the structure interact. The impact of these movements on the structure depends on the size, shape and material of the structure, as well as its position relative to the tunnel. It is the stiffness of the structure that is crucial when assessing the effect of these ground displacements, as a stiffer structure can reduce the effects. Old masonry structures tend to follow the ground displacements closely, as do structures on pad footings. Conversely, structures constructed of reinforced concrete will experience smaller horizontal displacements than the ground, due to their higher longitudinal stiffness, i.e. they do not stretch as much. These structures also experience reduced distortions due to their higher flexural stiffness, i.e. they do not bend as much. Stiff structures exhibit a high level of shear resistance, i.e. relative movements within the structure are small and tend to be subject to tilt rather than distortion. The response depends on such factors as the height of the structure, the number of openings and the design of the structure, for example concrete walls or beam and column construction (ITA/AITES 2007).

In addition, the location of the structure in relation to the ground deformations influences the movements it experiences. For a Gaussian shaped settlement trough, the structure will experience extension and hogging over the convex parts of the settlement trough and compression and sagging over the concave parts. Figures 7.6a and b show the typical response of short buildings. As the tunnel approaches, the short building rides the forward settlement wave with little significant sagging or hogging deformation (Figure 7.6a). Furthermore, short buildings experience tilt as a rigid body, but little significant sagging or hogging deformation across a transverse settlement profile (Figure 7.6b). Figures 7.6c to e show the response of a stiff long building as the tunnel advances, i.e. it experiences progressive deformation and differential settlements. Figures 7.6f and g show the potential sagging and hogging of a long building across a transverse settlement trough when it is directly above the centreline of the tunnel and also offset from the centreline. It should be noted that structures are more sensitive to differential movements, and hence it is important to assess the position of the structure in relation to the ground deformations caused by the tunnel.

Cracking is often used as an indication of distress in structures. Several researchers have investigated cracks and developed classification methods to assess structural damage, for example the reader is referred to ITA/AITES (2007), Burland and Wroth (1975) and Boscardin and Cording (1989). Further information is provided in section 7.2.2.

7.2.1 Effect of tunnelling on existing tunnels, buried utilities and piled foundations

In addition to buildings, it is important not to forget about structures that lie beneath the ground surface, for example existing tunnels and buried

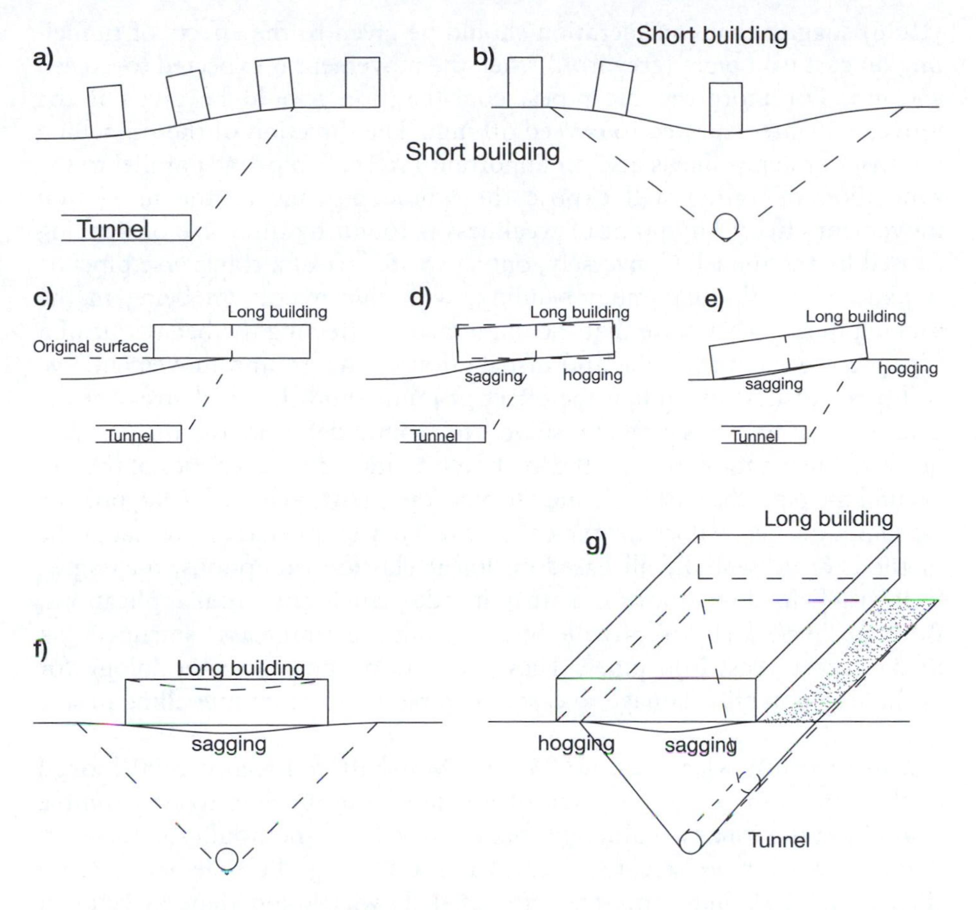

Figure 7.6 Some idealized modes of behaviour for short buildings and long buildings due to a tunnel construction below (Attewell 1995, after Attewell *et al.* 1986)

utilities. Pipelines in particular can be vulnerable to ground displacements caused by tunnelling activities, particularly the older, more brittle, cast iron gas and water pipes.

Attewell *et al.* (1986) provide a comprehensive investigation of pipeline response to tunnelling displacements. They state that in the vicinity of a buried cast iron pipeline the greenfield ground movement is modified, since the pipe stiffness is typically 1000 to 3000 times the soil stiffness. The resistance of a pipeline to system disturbance depends on the longitudinal flexural rigidity (including the flexural rigidity of the pipeline joints), longitudinal bending strength (including reduction in strength caused by corrosion or service holes) and the pipe diameter. The true risk in any particular case will largely depend on existing local stress levels due to previous disturbance, stress due to internal pressure, stress due to external loading such as traffic, and seasonal effects associated with ground temperature and moisture changes. Attewell *et al.*

(1986) suggest that consideration should be given to the effects of tunnelling on cast iron pipes (grey iron) when the movement is expected to exceed 10 mm. For more flexible pipes, consideration should be given if the movements are expected to exceed 50 mm. The direction of the tunnelling relative to the pipeline is also an important factor. A pipeline parallel to the tunnelling operation will expose the whole pipeline to the maximum movement effect. Any point of weakness is found by the wave of bending caused by the tunnel. Conversely, only a small part of a transverse pipeline is exposed to the maximum bending, with this maximum being in the sagging mode. They state that the main factors affecting the behaviour of a pipeline are the magnitude and distribution of the ground movement, the soil-pipe stiffness (including the effect of joints) and the yield stress of the soil. In order to assess pipelines subject to ground deformation, the designer needs to have information on the load-deformation characteristics of the soil around the pipe, the pipe itself and the pipe joints (Attewell *et al.* 1986 provide estimates of these values for typical cases). They show that even though the methods are essentially all based on linear elastic assumptions, they agree well with field observations and are quite adequate for practical applications. Bracegirdle *et al.* (1996) provide further guidance on the assessment of risk of damage to cast iron pipes. These authors propose a methodology for evaluating potential damage to cast iron pipes induced by tunnelling in soft ground.

More recently Klar *et al.* (2005) and Marshall and Klar (2008) looked at the commonly used approach of assuming the pipeline to be a simple beam by comparing two different theories; the Euler–Bernoulli simple beam theory, and a more accurate representation using shell element theory (FLAC3D® 3-D finite difference analysis). It was found that, in general, steel and concrete pipes are well represented using beam theory (due to the high relative pipe-soil material stiffness for steel pipes, and large wall thickness to diameter ratio for concrete pipes). For polyethylene pipes (which have a low value of relative pipe-soil material stiffness and can have both large and small values of wall thickness to diameter ratios) predictions using the beam theory deviate significantly from the shell element predictions. Therefore, it appears that the shell element formulation is better suited for the analysis of polyethylene (or similar) pipelines. Although it should be noted that these analyses were linear elastic and did not take into account possible 'sliding' between the ground and pipe (possibly important for the smooth wall polyethylene pipes).

In terms of the effects of tunnelling on existing tunnels, an extensive investigation was conducted by Cooper (2002) based on a number of case histories in London, UK, but particularly focusing on one associated with the Heathrow Express Tunnel to Terminal 4 in 1995 (Cooper *et al.* 2002). The research looked at the extensive monitoring data taken inside the London Underground Piccadilly Line tunnels (unbolted segmental concrete lined tunnel) as the new tunnels were constructed underneath. It showed how the tunnels behaved in terms of potential disturbance as the tunnelling passed underneath. It also showed that in this case, the empirical settlement prediction method, described in section 7.1, could be used to estimate the settlements, rotations and deformations of the existing tunnel lining.

There are other examples in the literature of monitoring existing tunnels during new tunnelling works. For example, Moss and Bowers (2006) describe the approach adopted on the Channel Tunnel Rail Link (CTRL) in the UK with respect to monitoring the effects on London Underground lines in London. This staged approach is similar to that described in section 7.2.2. The overall philosophy, though, was to minimize the ground movements at source, i.e. using high specification EPB tunnelling machines, and to use a risk-based engineering assessment of the effect of the tunnelling works on the existing tunnels. There have also been studies to investigate the behaviour of existing tunnels due to adjacent excavations, for example, Chang *et al.* (2001).

EFFECTS OF TUNNELLING ON PILED FOUNDATIONS

There have been a number of investigations into the effects of tunnelling on piled foundations. These include studies using small-scale physical modelling (for example Lee and Bassett 2007), centrifuge modelling (for example Jacobsz 2002, Jacobsz *et al.* 2004), three-dimensional numerical modelling (for example Loganathan *et al.* 2001, Mroueh and Shahrour 2002, Lee and Ng 2005), and also full-scale field monitoring (for example Selemetas *et al.* 2006, Pang *et al.* 2006). There has also been work conducted into analytical analysis methods for pile groups affected by tunnel construction, for example Huang *et al.* (2009).

The findings from this research have indicated that there are 'zones' of influence that affect the piles in different ways depending on their relative position to the tunnel centreline. Figure 7.7 shows the findings of Selemetas

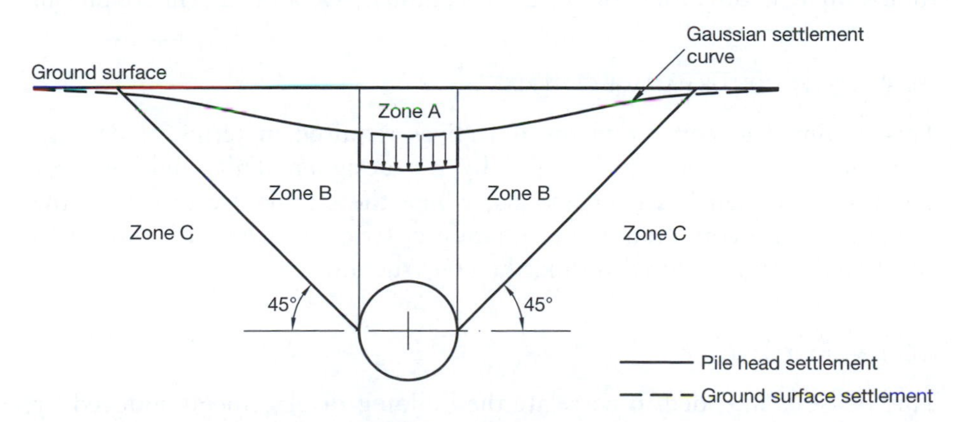

Figure 7.7 Zones of influence of pile settlement due to earth pressure balance shield tunnelling in London Clay (after Selemetas *et al.* 2006)

et al. (2006), although they stress that the boundaries of Zone B are simplified and probably are a function of the shearing resistance of the soil and the volume loss during tunnelling and are therefore not constant (see also Jacobsz *et al.* 2004, and Lee and Bassett 2007). Piles in Zone A were found to settle more, those in Zone B settled by the same amount and those in Zone C settled less than the ground surface. In addition, piles in Zone A experienced a considerable reduction in their base loads during the tunnelling and this was accompanied by differential pile settlement. Piles in Zone B and C only experience small changes in their base loads and actually showed a net gradual increase with time due to the ground-induced negative shaft friction (Selemetas *et al.* 2006).

7.2.2 Design methodology

This section describes a methodology recommended by ITA/AITES (2007) for studying the effect of underground works on existing structures. This methodology is broken down into six stages and a brief summary is provided below.

PHASE 1: INVESTIGATION OF EXISTING STRUCTURES

This phase involves surveying and data collection on the nature, configuration and condition of buildings and utilities together with topographic measurements and technical expert reviews. This is essential to assess the baseline or zero condition of each structure prior to the start of construction works.

PHASE 2: INFORMATION SUMMARY

This involves a 'typological' classification of the structures according to, for example, nature, function, value, size, design, age and current condition.

PHASE 3: SELECTION OF DAMAGE CRITERIA

This is aimed at converting the objectives required in terms of damage limitation into criteria to be used by the designer. This could be, for example, a system based on strains, where the strains are based on the initial reference condition and evaluation criteria are used with respect to additional strains induced during the construction.

PHASE 4: MODELLING

This phase is intended to correlate the building displacements induced by ground movements to its structural deformations. Mair *et al.* (1996) suggested a three stage assessment process, as follows.

Preliminary assessment Based on the empirical predictive methods described in section 7.1.2 (i.e. assuming greenfield settlements), an assessment is made on the level of ground movement and hence the effect on the surface structures. This is often done using contours of settlement and it is considered that any settlements of less than 10 mm cause negligible risk of damage to the structure. In addition, it should also be checked that no structure experiences a slope, due to differential settlements, in excess of 1 in 500. Table 7.2 shows the damage risk assessment and suggested actions.

Second stage assessment The ground deformations will comprise both sagging and hogging zones and so, depending on the relative position of the structure to the point of inflection of the settlement trough, there will be zones of tensile and compressive strains. The second stage assumes that the structure is weightless and fully flexible, i.e. follows the greenfield displacements exactly (refer to Burland 1995, Franzius *et al.* 2006, Burland *et al.* 1977 and Boscardin and Cording 1989). This is usually conservative as the stiffness of the structure will reduce the actual movements. An example of how the potential damage can be classified according to the maximum tensile strain as shown in Table 7.3.

Detailed evaluation This stage would normally be undertaken for structures where a 'moderate' or greater level of damage has been predicted in stage 2. It considers the existing condition of the structure, the tunnelling sequence, three-dimensional aspects, characteristics of the structure and the soil-structure interaction effects. Protective measures (for example compensation grouting, section 4.2.8) would then be considered for structures remaining in the 'moderate' or higher damage categories. It is possible at this stage to use numerical methods and the method proposed by Potts and Addenbrooke (1997), updated by Franzius *et al.* (2006).

PHASE 5: DETERMINATION OF THE ALLOWABLE DISPLACEMENT THRESHOLDS

ITA/AITES (2007) states that the purpose of this phase is to determine the contractual threshold requirements that will have to be met during construction. The summary, prepared as part of Phase 2, is essential to allow contractual criteria to be developed that meet the needs, in terms of protection, of the structures.

Threshold values must never be taken as constant. They should be essentially treated as alarm indicators and continuously reviewed based on the actual building response due to the tunnelling works. However, it is important that 'alarm' thresholds and 'stopping' thresholds exist for each project. A system based on amber and red trigger values is often used (see section 7.3.2 on trigger values).

Table 7.2 Typical values of maximum building slope and settlement for damage risk assessment, and suggested action for various risk categories (after Rankin 1988, used with permission of the British Geological Society)

Risk category	*Maximum slope of building*	*Maximum settlement of building (mm)*	*Description of risk*	*Description of action required*
1	Less than 1/500	Less than 10	*Negligible*: superficial damage unlikely.	No action. Except for buildings identified as particularly sensitive for which an individual assessment should be made.
2	1/500–1/200	10 to 50	*Slight*: possible superficial damage which is unlikely to have structural significance.	Crack survey and schedule of defects, so that any resulting damage can be fairly assessed and compensated. Identify any buildings and pipelines that may be particularly vulnerable to structural damage and assess separately.
3	1/200 to 1/50	50 to 75	*Moderate*: expected superficial damage and possible structural damage to buildings. Possible damage to relatively rigid pipelines.	Crack survey: a schedule of defects and a structural assessment. Predict extent of structural damage. Assess safety risk. Choose whether to accept damage and repair, take precautions to control damage or, in extreme cases, demolish. Buried pipelines at risk: identify vulnerable services and decide whether to repair, replace with a type less likely to suffer damage, or divert.
4	Greater than 1/50	Greater than 75	*High*: expected structural damage to buildings. Expected damage to rigid pipelines. Possible damage to other pipelines.	

Note: The above criteria relate to *near surface* foundations or pipelines.

Table 7.3 Classification of building damage (from Devriendt 2006, after Burland *et al.* 1977 and Boscardin and Cording 1989)

Damage category	*Description of typical damage (ease of repair in italics)*[a]	*Approx. crack width (mm)*[b]	*Limiting tensile strain (%)*
0 Negligible	Hairline cracks of less than about 0.1 mm are classed as negligible	< 0.1 mm	0–0.05
1 Very slight	*Fine cracks which can easily be treated during normal decoration.* Perhaps isolated slight fracture in building. Cracks in external brickwork visible on inspection.	1 mm	0.05–0.0075
2 Slight	*Cracks easily filled. Redecoration probably required.* Several slight fractures showing inside of the building. Cracks are visible externally and some re-pointing may be required externally to ensure weather tightness. Doors and windows may stick slightly.	5 mm	0.075–0.15
3 Moderate	*The cracks require some opening up and can be patched by a mason. Recurrent cracks can be masked by suitable linings. Re-pointing of external brickwork and possibly a small amount of brickwork to be replaced.* Doors and windows sticking. Service pipes may fracture. Weather tightness often impaired.	5 to 15 mm, or a number of cracks > 3 mm	0.15–0.3
4 Severe	*Extensive repair work involving breaking-out and replacing sections of walls, especially over doors and windows.* Windows and door frames distorted, floors sloping noticeably. Walls leaning or bulging noticeably, some loss of bearing on beams. Service pipes disrupted.	15–25 mm	> 0.3
5 Very severe	*This requires a major repair job involving partial or complete rebuilding.* Beams lose bearings, walls lean badly and require shoring. Windows broken with distortion. Danger of instability.	usually > 25 mm, but depends on number of cracks	> 0.3

Notes:

(a) In assessing the degree of damage account must be taken of its location in the building or structure.

(b) Crack width is only one aspect of damage and should not be used on its own as a direct measure.

PHASE 6: BACK ANALYSIS AND CALIBRATION OF MODELS WITH OBSERVED DATA

It is essential to check the displacement estimates (Phase 4) by monitoring the construction works and the effects on the surrounding ground and structures. Validating the design assumptions should be a routine part of all construction management planning.

A practical example of how the buildings along the route of a new metro in Amsterdam (North/South Metroline) were assessed for potential damage prior to the tunnelling works, and also how the monitoring during construction was linked to the tunnelling operations, i.e. the TBM operations, is described in van Hasselt *et al.* (1999). The so called 'Interactive Boring Control System' (IBCS) meant that the TBMs were not controlled solely on the basis of tunnel and machine data, but also by using the settlement monitoring data as an additional criterion. The machine parameters were used to make virtual predictions of the subsequent excavation section (based on, amongst other factors, the face control pressures and tail void grouting) and if this prediction showed the likelihood of unacceptable damage to the overlying structures, the machine parameters were adjusted.

7.3 Monitoring

7.3.1 Challenges and purpose

The challenge for tunnel engineers is to achieve the completion of a tunnelling project without the general public realizing that the tunnelling operations are taking place beneath/around them. In order to achieve this, there must be no inconvenience to people going about their daily lives in terms of disruption, noise and dust (this is not always possible when using cut-and-cover construction) and no noticeable effect on structures, e.g. buildings, other tunnels or utilities in the vicinity of the tunnel construction (see section 7.2). At the current time no tunnelling operation can completely eliminate disruption or ground movements (although with modern tunnelling machines and good quality control very small movements can be achieved) and so monitoring of affected structures is usually required.

In tunnelling using NATM methods monitoring inside the tunnel as construction proceeds is an integral part of the construction process, and this is described further in section 7.3.4.

Monitoring can be done on a number of levels depending on the purpose. It is therefore important to ask the questions 'what is the purpose of the monitoring?', and hence 'what information is required?'. This needs to be clearly understood prior to designing the monitoring regime. It is also beneficial to keep the instrumentation regime as simple as possible for the required purpose. This does not mean that the instruments should not be state-of-the-art, but their arrangement should not be overcomplicated. It should be borne in mind that the data from the instruments must be analysed regularly and it is important that the data inform this process in the most effective way. It is also important that there is suitable redundancy of instrumentation in critical or inaccessible areas to insure against failures.

BTS/ICE (2004) states that instrumentation is typically installed to:

- obtain 'baseline' ground characteristics;
- provide construction control;
- verify design parameters;
- measure performance of the lining during, and after, construction;
- monitor environmental conditions (for example settlement, air quality and effects on the groundwater regime);
- carry out research to enhance future designs;
- monitor mitigation measures, for example compensation grouting and ground freezing.

It is important to understand the quality of the data received from the instrumentation. For this reason some important definitions related to monitoring and instrumentation are given below (after Dunnicliff and Green 1993):

- *Conformance*: the presence of the measuring instrument should not alter the value of the parameter being measured. The degree by which the parameter is altered by the instrument is known as its *conformance*.
- *Accuracy*: this is the closeness of a measurement to the true value of the quantity measured. Accuracy is synonymous with *degree of correctness*. The accuracy of an instrument is evaluated during calibration to a known standard value. It is customary to express accuracy as a ± *number*.
- *Precision*: this is the closeness of each of a number of similar measurements to the arithmetic mean. Precision is synonymous with *reproducibility and repeatability*. The number of significant figures associated with the measurement indicates precision. For example, ±1.00 indicates a higher precision than ±1.0.

The difference between accuracy and precision is illustrated in Figure 7.8.

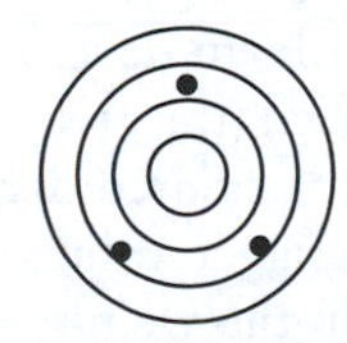

Figure 7.8 Accuracy and precision (after Dunnicliff and Green 1993)

- *Resolution*: this is the smallest division on the instrument readout scale.
- *Error*: this is the deviation between the measured value and the true value, i.e. it is mathematically equivalent to accuracy. Errors can occur due to human carelessness, fatigue or inexperience, or can be due to improper calibration, poor installation procedures or environmental conditions such as heat, humidity or vibration.

The data from the instrumentation and monitoring programme need to be appropriately managed to ensure that it is in a suitable form to be clearly understood. It should be regularly reviewed by qualified people, experts in monitoring so that unexpected trends can be identified easily and appropriate actions taken. It should be routine on any project that comparison is made between the predicted and the observed values in order to understand the behaviour of the structure and the ground being monitored.

7.3.2 Trigger values

It is common practice to establish 'trigger values' for key measurement parameters associated with a project, for example displacement. If these values are exceeded during the tunnelling project, then certain actions need to be clearly defined. Two trigger values are normally established (BTS/ICE 2004):

- *amber or warning value*: this could be a pre-determined value or a rate of change in a parameter that is considered to indicate a problem;
- *red or action value*: this could be where threshold values for safe operation are exceeded. This should initiate an immediate check of the instrument function and a visual inspection, and initiation of pre-determined action, for example temporary cessation of the tunnel work.

The determination of trigger levels is often specific to the project. A widely used approach is, for example amber trigger value = the calculated displacements exceeded by 50%: red trigger value = the calculated displacements exceeded by 100%.

Even if the trigger values are not exceeded, the monitoring data should be carefully checked to highlight any unexpected trends so that these can be acted on appropriately before a problem develops. For this reason there is often the green trigger value established (the green or early warning value), for example green trigger value = the calculated displacements are reached.

It should be noted, however, that trigger values are more easily applied when monitoring existing structures in the vicinity of the tunnelling operation, where specific limits must not be exceeded in order to avoid damage to the structure. It is more difficult to apply trigger values to in-tunnel monitoring data because significant estimations have to be made regarding the behaviour of the ground and the tunnel lining. Therefore, the calculated displacements are only estimates based on engineering judgement. This is discussed in more detail in the next section on the observational method as well as in the section on in-tunnel monitoring for NATM (section 7.3.4).

7.3.3 Observational method

Instrumentation and monitoring forms an integral part of the observational method. This method is very important in civil engineering and particularly ground engineering projects, such as tunnelling, and it is carried out, whether formally or informally, on all projects. The ground cannot be characterized exactly and even if a good site investigation has been conducted, it is possible that the predicted behaviour is not observed in practice. The observational method allows deviations to be identified early (these can be both positive and negative) and appropriate actions taken.

Nicholson *et al.* (1999) carried out a comprehensive review of this approach and they define the method as 'a continuous, managed, integrated process of design, construction control, monitoring and review that enables previously defined modifications to be incorporated during or after construction'.

Peck (1969a) considered that the complete application of the observational method embodies the following aspects:

- sufficient exploration to establish at least the general nature, pattern and properties of the ground, but not necessarily in detail;
- assessment of the most probable conditions and the most unfavourable conceivable deviations from these conditions. In this assessment geology often plays a major role;
- establishment of the design based on a working hypothesis of behaviour anticipated under the most probable conditions;
- selection of quantities to be observed as construction proceeds and calculation of their anticipated values on the basis of the working hypothesis;
- calculation of values of the same quantities under the most unfavourable conditions compatible with the available data concerning the subsurface conditions;
- selection in advance of a course of action or modification of design for every foreseeable significant deviation of the observational findings from those predicted on the basis of the working hypothesis;
- measurement of quantities to be observed and evaluation of actual conditions;
- modification of design to suit actual conditions.

In terms of applying the observational method to uncertainties in the ground, the principal objective is to apply sufficient resources to prevent uncertainty to unacceptable levels of risk. In this respect there are three types of uncertainties (after Nicholson *et al.* 1999), as follows.

GEOLOGICAL UNCERTAINTY

On projects where there are complex geological and hydrological conditions, there may be unexpected variations in the ground conditions between boreholes. A conceptual model of the geological conditions will have been developed at the design stage. Based on this model, modifiable design solutions are developed for a range of conditions. The actual ground conditions are determined during the works and the appropriate design solution selected to suit these actual conditions.

PARAMETER UNCERTAINTY

Uncertainties exist in the knowledge of the ground characteristics and the modelling of its behaviour, and hence it is not possible to accurately determine ground parameters. The observational method involves developing flexible and robust designs for a range of parameters. Monitoring results are reviewed during construction and the design modified as appropriate.

GROUND TREATMENT UNCERTAINTY

There are a number of ground treatments available to improve specific properties of the ground. The use of these techniques is often based on a performance specification identified by the designer. The effectiveness of the technique is monitored and reviewed during the treatment and modifications implemented where necessary to meet the specification.

There is also a fourth area of potential uncertainty.

SUPPORT UNCERTAINTY

The time dependent behaviour of sprayed concrete (i.e. the development of strength, ultimate strain, creep) is difficult to simulate accurately, and hence estimate, during the design process. In addition, the load transmission in the joints of segmental linings is difficult to estimate.

For tunnelling in soft ground using a sprayed concrete lining, the ground treatment, for example compensation grouting, can also be an uncertainty with respect to influencing the stresses on the tunnel support. The geological uncertainty could apply to the heading excavation and result in the need for face logging and possibly horizontal boreholes to log ahead of the face. It is important to have sufficient time to support the excavation, but there is potential uncertainty in the associated ground parameters. Convergence monitoring within the tunnel can help to provide data to review parameter uncertainty.

NATM tunnelling is a prime example of where the observational method is an integral part of the philosophy for this method and hence the construction process. The monitoring associated with NATM is described in the next section. Further information on the application of the observational method to tunnelling can be found in Powderham (1994 and 2002).

7.3.4 In-tunnel monitoring during New Austrian Tunnelling Method tunnelling operations

7.3.4.1 Measurements

Monitoring cross sections or measuring profiles are installed in the tunnel while it is being constructed. The distance between monitoring cross sections can vary from approximately 3 m to more than 50 m, and is dependent on the ground conditions, the ground-lining interaction, and the sensitivity of service structures around the tunnel and buildings at the surface. A common distance in inner city tunnels is 10 m. Each monitoring cross section contains a number of monitoring points at which targets are positioned for the laser theodolite. Two commonly used targets are shown in Figure 7.9a. The left-hand side of Figure 7.9a shows a Bireflex target. The name derives from the reflective mirror foil on both sides ('bi'-reflex) of the target and it has a diameter of 50 mm or 60 mm. The actual target is a small hole in the middle. The reflective foil helps the laser of the theodolite to find this hole. The right-hand side of Figure 7.9a shows a prism target. In this target, three prisms form a circle of 25 mm in diameter. The prisms guide the laser by the intensity of the reflexion to the centre. The targets are screwed on top of short bolts (15 to 30 cm), which have previously been sprayed firmly into the sprayed concrete lining (Figure 7.9b). Prism targets are five times more expensive than Bireflex targets. They are, however, more accurate and they can be used for automated monitoring.

Typical monitoring arrays are shown in Figure 7.10. For a typical tunnel construction, in which the excavation is divided into crown, bench and

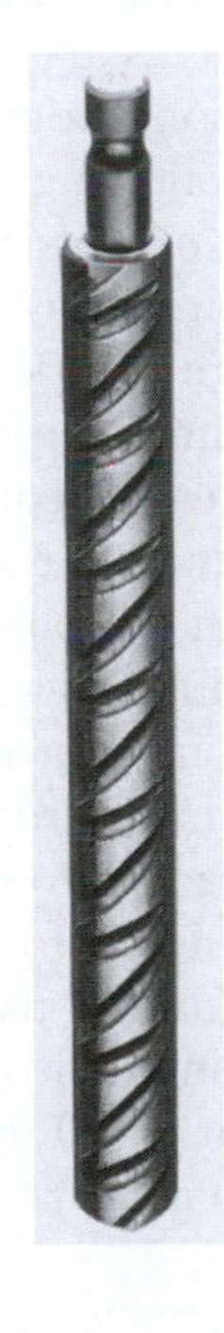

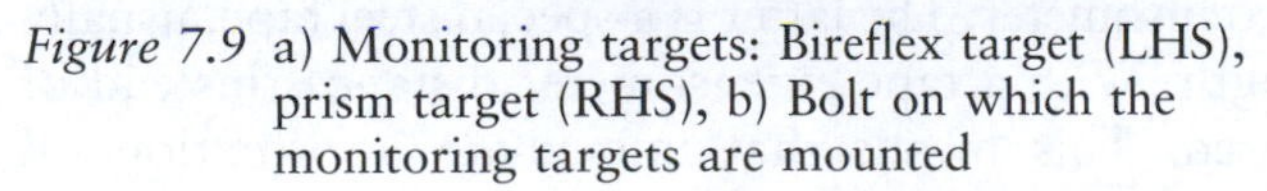
Figure 7.9 a) Monitoring targets: Bireflex target (LHS), prism target (RHS), b) Bolt on which the monitoring targets are mounted

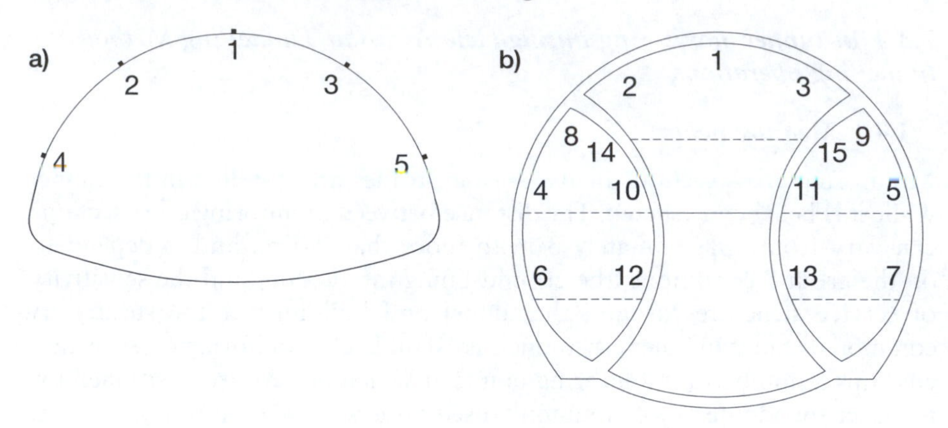

Figure 7.10 Examples of a measuring profile a) for a crown and, b) a double track tunnel excavated with side wall drifts

invert, five monitoring points are installed in the crown (Figure 7.10a), and one or two points on each side in the bench (not shown). The monitoring points are numbered so as to distinguish them from each other. It is common to number the highest monitoring point in the middle of the crown as number '1'. Usually numbering continues on the left-hand side with even numbers, and on the right-hand side with odd numbers (line of vision in the direction of excavation). The invert is not usually monitored because it is covered with backfill that provides a track for the construction equipment. For information on monitoring the invert see section 8.1 on the case history of the Eggetunnel.

Readings are taken by a laser theodolite, which digitally stores the three-dimensional coordinates of each target: vertical, horizontal and longitudinal. After collecting the readings, these data are transferred to a computer for further processing. From the change in coordinates, displacements in all three directions can be calculated and the deformation behaviour of the structure can be derived (this is discussed later in this section).

It is important to have a few stable reference points at the beginning of the measurements, which are not affected by the tunnel construction. The accuracy of the laser theodolite measurement should be ±1 mm or better for ideal measuring conditions. The accuracy can go down to ±3 mm or worse in difficult conditions, e.g. dust or large temperature differences inside the tunnel. This can happen for example, if the monitoring is done from a large cross section into a small one with a significantly higher temperature due to the hydrating of the sprayed concrete or if the ventilation is shut down for maintenance.

The digital technique of using a laser theodolite has widely replaced measuring with a tape extensometer. The latter is a special steel tape, usually of 20 m or 30 m in length. With a tape extensometer distances instead of coordinates are measured. This means that immediate information on convergence (distances becoming shorter) or divergence (distances becoming longer) is available. Although the resolution can be as good as 0.5 mm, in practice reasonable accuracy can be as low as approximately ±10 mm. This is especially true when measuring longer distances, as the tape extensometer sags due to its own weight. It has to be tensioned to mitigate this effect. Although the force necessary to tighten the tape must stay in a defined range (lots of tapes have been ripped apart), the sag and the tightening reduces the accuracy of a tape extensometer significantly. Since digital processing of data is not common with tape extensometers and the readings are time consuming, tape extensometers are only used in situations where targets are difficult or impossible to focus in on with a theodolite, e.g. in the bottom of a shaft.

7.3.4.2 General development of displacements

The ground usually reacts to the approaching tunnel excavation before the heading actually reaches a particular point (see section 7.1.1). In addition, the influence of the tunnel excavation does not stop immediately, but slowly fades out as the face moves forward. In general, the weaker and softer the ground is, the earlier the displacements will start and the longer they will last. This is particularly true when the ground has time dependent behaviour such as creeping (this is often observed in clay). As a rule of thumb the influence of the tunnelling excavation is about ±2D before and behind the monitoring cross section where D is the tunnel diameter. Figure 7.11 shows the idealized displacement curve of one monitoring point reacting to an approaching and disappearing tunnel heading.

Figure 7.11 indicates the tunnel face moving along a number of chainages and the related displacements (Stärk 2009).

- Tunnel face is at Chainage 1: the face is far away from the monitoring cross section so no displacements can be determined at the monitoring cross section. The primary stress condition is undisturbed.
- Tunnel face is at Chainage 2: the face is closer to the monitoring cross section and the first small displacements occur.
- Tunnel face is at Chainage 3 and 4: the face is even closer to the monitoring cross section and the displacements have increased significantly.
- Tunnel face is between Chainage 4 and 5: the tunnel excavation passes the monitoring cross section. The displacements increase further, but are not yet measurable from inside the tunnel.
- Tunnel face is at Chainage 5: the measuring bolts in the monitoring cross section are installed. Thereafter the first measurement ('base reading') is taken.
- Tunnel face is between Chainage 6 and 7: further readings are taken. The increase in the displacements between the measurements is slowly reducing, which means that the influence of the excavation is fading out.

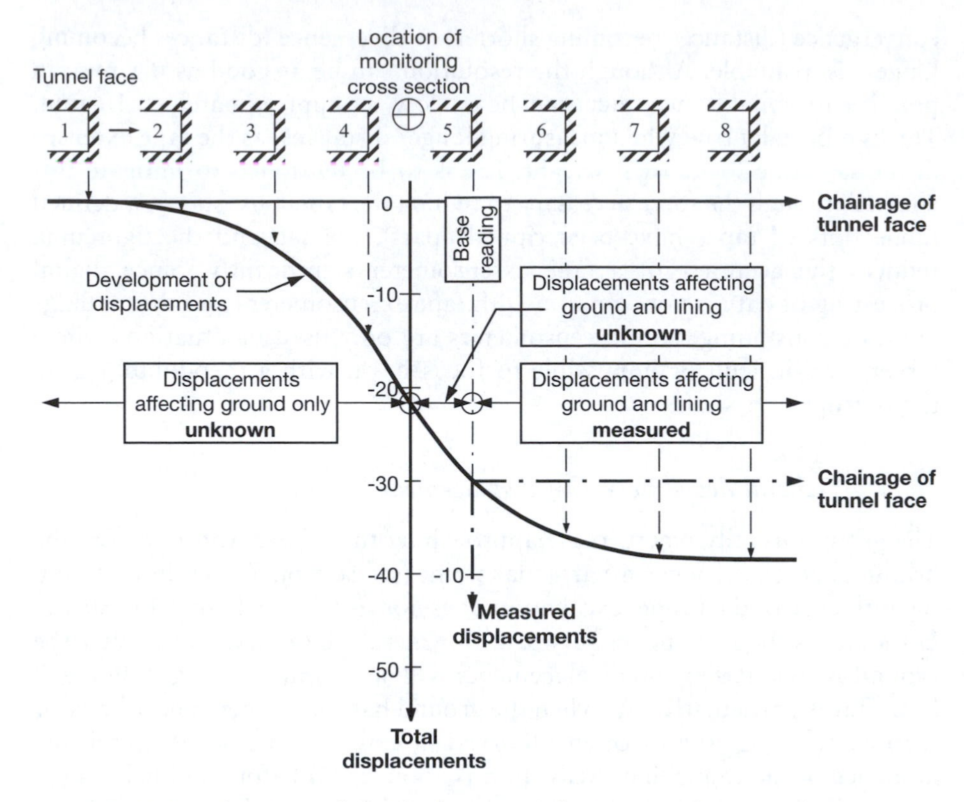

Figure 7.11 Development of the displacements for one monitoring point in relation to the position of the tunnel face. The scaling and magnitude of the displacements are arbitrary and only by way of example

- Tunnel face is at Chainage 8: the increase in the deformations has stopped. At this distance the excavation has no more influence on the displacements.

It should be noted that the monitoring cross section cannot be installed until the tunnel construction has reached Chainage 5. The displacement data are therefore only available from inside the tunnel after the base reading at Chainage 5 has been taken. The curve prior to the base reading, from Chainage 1 to Chainage 5, can be measured using external instruments, such as vertical extensometers, which can be installed from the ground surface (see section 7.3.5 for a description of vertical extensometers). However, in general this section of the displacement curve remains unknown. This means that only a fraction of the total displacement can actually be measured. Depending on the geological conditions, the displacements affecting only the ground range typically from approximately 30% to 70% of the total displacements (Chainage 1 to the monitoring cross section). The better the ground, the better its ability for stress redistribution and the larger the displacements affecting only the ground (maximum 100% if no support is required). The displacements from the monitoring cross section to Chainage 8 affect both the ground and the tunnel lining. These displacements load the lining and it is therefore important to determine these with respect to the stability of the tunnel.

There is a period when there is a lack of monitoring information. This period is between the face passing the monitoring cross section and the actual base reading. Therefore the base reading must be taken as quickly as possible and should be positioned as close as possible to the tunnel face. The monitoring cross section is usually installed in the face advance currently under construction and the base reading is taken before the excavation of the next advance commences (the base reading is taken within six hours or less). If the base reading is done in this way, the previously mentioned lack of information is negligible with respect to the loading of the sprayed concrete; because the 'young' sprayed concrete cannot take much load. It is therefore very important for the interpretation of the measuring results that there is an indication of the place and time of the base reading in relation to the location of the face.

7.3.4.3 Interpretation of the measurements: displacements

The basic graph derived from the measurements is displacement versus distance to 'face – monitoring cross section', or displacement versus time. By default, a distance-dependent graph is used to control the effect of the tunnelling progress on the already excavated section. However, if the advance rate is constant and the excavation process is uniform, a time-dependent graph can be used. In the case of a hiatus in the excavation, for example the Christmas shutdown, one also has to switch to a time-dependent graph.

The displacements increase quickly immediately after the base reading. The influence of the tunnel construction then decreases, and finally the displacements do not increase any further. A new stable equilibrium between the support and the ground has then been established and the displacements must remain constant. This statement is very important with respect to settlement control and stability. Figure 7.12 shows, by way of example, the vertical displacements of the crown.

CRITICAL TRENDS OF THE MEASURING CURVE

The displacement graph is also important to identify adverse situations, for example when the displacements do not remain constant or increase again after a period of stability. It is important that the measurements are not stopped too early as settlements of the crown can occur after a hiatus. Possible adverse causes for increasing displacements are listed below:

- modification of the ground behaviour due to ingress of water (in the joints, fissures etc.). This can result in a reduction of the internal friction;

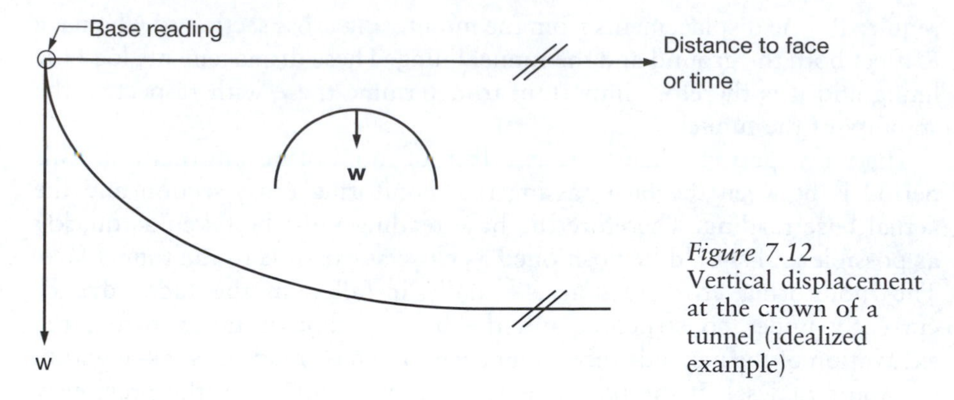

Figure 7.12 Vertical displacement at the crown of a tunnel (idealized example)

- deformation behaviour of the ground is heavily time dependent;
- sudden failure of the sprayed concrete lining;
- the horizontal shape of the measuring curve only 'pretends' that the settlements have stopped due to the intervals between measurements.

The design and scheduling of the excavation sequence can also create additional new displacements after an apparent halt, for example:

- restart of an excavation after a longer pause (e.g. the Christmas shut-down);
- the time delayed excavation of successive headings (e.g. bench excavation follows crown) creates a new stress redistribution and thus leads to further displacements of the leading heading.

The bedding of the footing of the crown is being removed during the excavation of the bench so that the sprayed concrete has to arch over the excavated area. The area of this longitudinal support is thus likely to be affected by further displacements in addition to the stress redistribution due to the excavation of the ground (Figure 7.13).

Figure 7.14 shows by way of example the vertical displacements of the monitoring cross section at Chainage 650 of the leading side wall drift in Section W of the Lainzer Tunnel LT31, Vienna (for detailed information

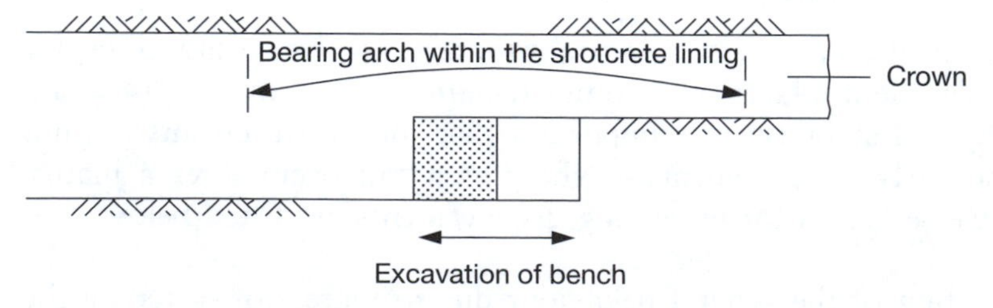

Figure 7.13 Longitudinal support within the lining during bench excavation

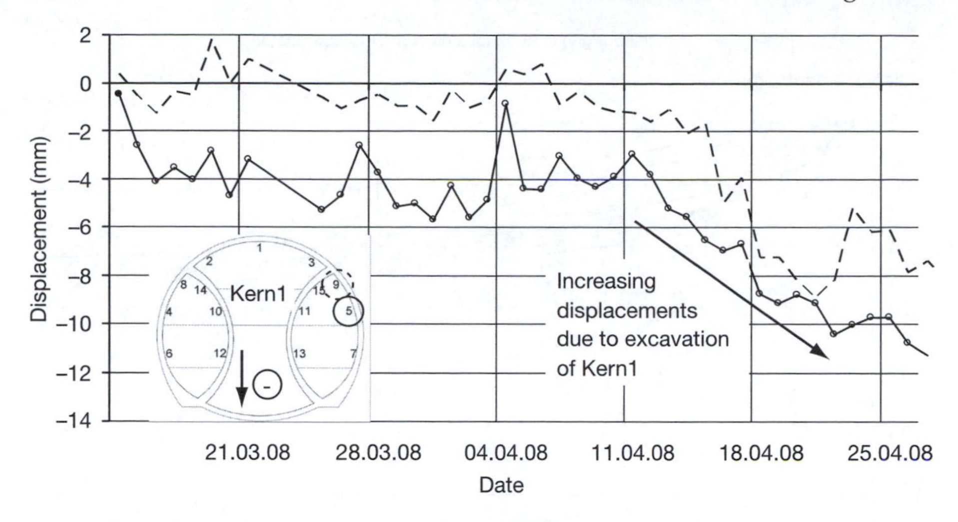

Figure 7.14 Displacement versus time graph showing the development of new displacements in the right-hand side wall drift following excavation of Kern1 (LT31, Vienna, Section W, Chainage 650)

on this project see section 8.3 of the case histories). For reasons of clarity only two points (5, 9) are displayed in the figure. The displacements remained constant for approximately four weeks, until mid April, when new displacements appeared, nearly 2.5 times as large as the old ones. The reason for this was the excavation of the remaining crown (Kern1) following the leading side wall drift after a scheduled delay of four weeks.

7.3.4.4 Interpretation of the measurements: comparative observation

The task of the interpretation of the measurements is not only to see whether a new stress state in the combined system ground-lining has been established, but also to derive a statement with respect to the stability and to check the measured displacements against triggers. At first, one would compare the measured displacements with those calculated in the design. In section 8.2, the London Heathrow T5 PiccEx Junction case history, an example is given in which the comparison of calculation and measurement successfully led to an improvement in the safety of a neighbouring tunnel. However, in general some difficulties do arise. During the design a tunnel is divided into sections, in which similar geological conditions are anticipated. Calculations are prepared for each section, the results of which include expected displacement in the vertical and horizontal directions. During excavation a number of monitoring cross sections are installed within the respective sections. However, each monitoring cross section will record different displacement values. The reason for this is obvious: the *in situ* displacements depend on the geological conditions, on the type and quantity of the support, the time the support is installed and on the

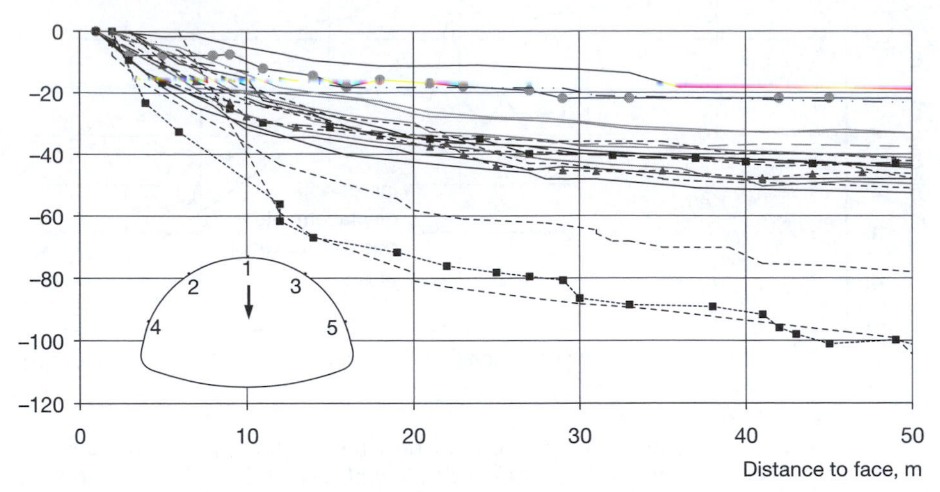

Figure 7.15 Vertical displacements for point 1 at different cross sections in the Eggetunnel for Chainage 482–684 in similar geological conditions

excavation sequence. The combination of these factors is never the same, and as a consequence there will be different displacements in each monitoring cross section. Due to the time and expense involved in computing these displacements, and the limited accuracy of the input data, the technique of parameter variation can only cover a small spectrum of the possible permutations. Quick *et al.* (2001) give an example of such a parametric study using three-dimensional finite element method calculations to simulate actual displacements measured. In their presentation they come to the conclusion that, because of the wide scatter of measured displacements, a comparison of the measured and the theoretical displacements makes little sense, as it is not known which of the monitoring cross sections is actually represented by the calculations.

This result reflects the authors' own experience. Figure 7.15 shows, by way of example, some monitoring results of the Eggetunnel (for further details on this project see section 8.1). This shows the vertical displacements of monitoring point No. 1 (crown) at different cross sections between Chainage 482 and 684. In this section similar geological conditions were predicted and encountered. Nevertheless the displacements for the different cross sections range from 20 mm to more than 100 mm; in other words, the displacements scatter by more than 500% for just one monitoring point. Monitoring results always come with a range and never with a single value. This makes it impossible to derive a statement with respect to stability. In addition, all the other monitoring points have to be taken into account as well.

In conclusion, therefore, a comparison of calculated and measured displacements is often not possible because tunnelling conditions are too variable to be modelled adequately in the calculations. In addition, it must not be forgotten that the measured displacements are the true displacements, and not the calculated values. If the measured displacements cannot be checked against calculated triggers, the monitoring cross sections have to be checked against each other (comparative observation). The aim is to filter a 'normal' range of displacements and to identify adverse trends well in advance, i.e. before any critical situation arises. This can be done by adopting a monitoring cross section as a reference section. A precondition for this is of course that no complications arise in the reference section. Alternatively it could be done by looking at average displacements over a couple of monitoring cross sections.

Figure 7.15 indicates that there is obviously an accumulation of curves around 40 mm. One could possibly refer to this as the 'normal' behaviour of the displacements. This would cut the scatter to a mere 40 mm ± 10 mm. However, the deviation outside 'normal' behaviour must not be ignored.

The average displacement can be set as a trigger. But one must be aware that it takes some time and a couple of monitoring cross sections to establish average values. In consequence, triggers must be adjusted during excavation (further information on trigger values can be found in section 7.3.2).

7.3.4.5 Interpretation of the measurements: deformation

Another important reason for monitoring is to aid the estimation of the residual bearing capacity of the lining. When new displacements occur, it is necessary to know if these additional displacements are likely to overstress the lining. If the calculated and measured displacements differ considerably, or if the displacements scatter widely as shown in Figure 7.15, it is also necessary to check the bearing capacity of the lining against collapse. Concrete often has cracks, and so does sprayed concrete. However, it is essential to investigate the cracks in the sprayed concrete because of the potentially severe consequences in this case (an example is given in section 8.3.6). There are many reasons for determining the bearing capacity of the lining, especially in a shallow tunnel in soft ground. To do this, deformations are needed. Deformations result from different displacements within a monitoring cross section (vertical, horizontal and longitudinal) and they are an indication of the stress state of the lining. It is possible to differentiate four distinct and different stress states depending on the observed displacements (Figures 7.16 to 7.20) (Rokahr *et al.* 2002). The dashed line in these figures represents the deformed lining. The displacements shown are not to scale and are exaggerated to make the deformation visible. In these figures, 'w' is the vertical displacement of the crown, 'v' is the vertical displacement of the footing, and 'h' is the horizontal displacement.

Figure 7.16 Rigid body displacement If the vertical displacement of the roof is equal to the vertical displacement of the footing, then the lining is just subject to a rigid body displacement; it is not deformed and therefore not stressed.

Figure 7.17 Roof settlement Roof settlement with none or negligible settlement of the footing is a deformation which is often encountered. The deformation generates normal forces and bending moments in the lining.

Figure 7.18 Divergence Roof settlement accompanied by an outwards movement of the footing is called divergence (the distance lengthens). This deformation is common in a ground with a low horizontal earth pressure. The deformation generates similar stress as the roof settlement shown in Figure 7.17, but with higher bending moments developing.

Figure 7.19 Convergence A combination of roof settlement and an inward movement of the footing is called convergence (the distance shortens). In the schematic shown, the vertical displacement is equal to the horizontal displacement. The original circular lining is not deformed but shortened. The radius of the circle becomes smaller. The lining is stressed by normal forces only, and no bending moments develop.

It should be noted that the deformations do not necessarily remain circular; theoretically an unlimited number of possible deformation shapes

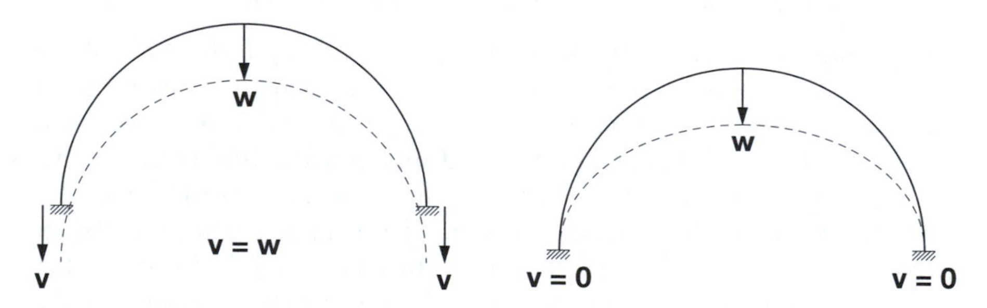

Figure 7.16 Rigid body displacement

Figure 7.17 Roof settlement

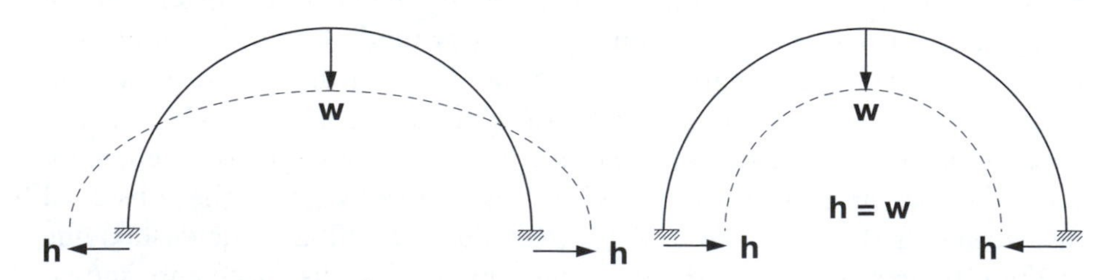

Figure 7.18 Divergence in combination with roof settlement

Figure 7.19 Convergence in combination with roof settlement

exist. In Figure 7.20 two possible shapes are shown, with the displacements of the roof and the footings being the same as shown in Figure 7.19. Both generate totally different stress states in the lining. To be sure of the actual deflected shape, the number of monitoring points in each monitoring cross section must be sufficient to derive an unambiguous shape of the tunnel lining from the displacements. In the example shown in Figure 7.20, representing a crown, monitoring is also required at the shoulders of the tunnel to confirm the deformed shape. Hence, at least five measuring points per monitoring cross section are necessary (see also Figure 7.10). The displacements in all three directions must be measured, i.e. horizontally, vertically, and in the longitudinal direction (x, y, z).

In order to get a better idea of the deflected shape, it is often useful to plot the displacements as vectors on the cross section, as shown in Figure 7.22a. It should be noted that depending on the scale used, it is possible to get a completely different impression from the displacements or deformations (as with all graphs!). Figure 7.22b shows the same displacements plotted with a different scale. Whereas the displacements in Figure 7.22a seem to be negligible, the ones in Figure 7.22b look worrying. Therefore, it is always a good idea to have a look at the actual values too.

It has to be emphasized, however, that it is generally difficult to make exact statements regarding the loading of a lining by looking at measurement curves

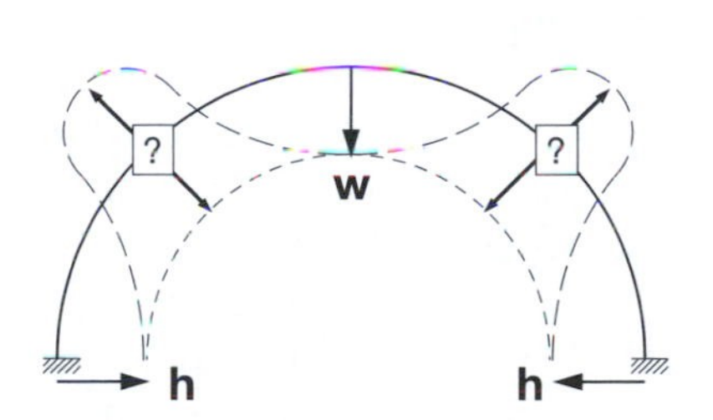

Figure 7.20 Shoulder displacements

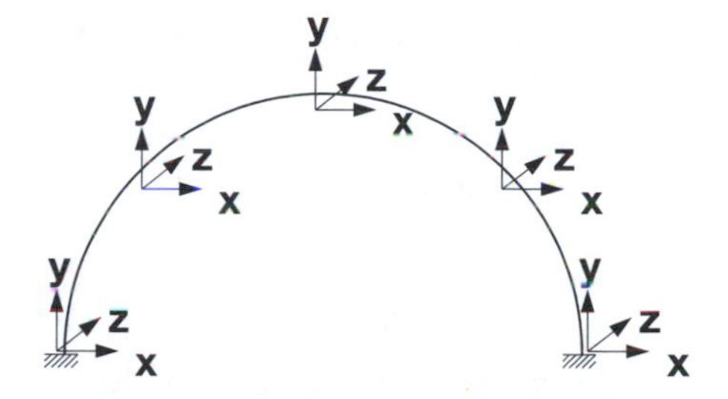

Figure 7.21 Minimum required monitoring points to create a deflected shape

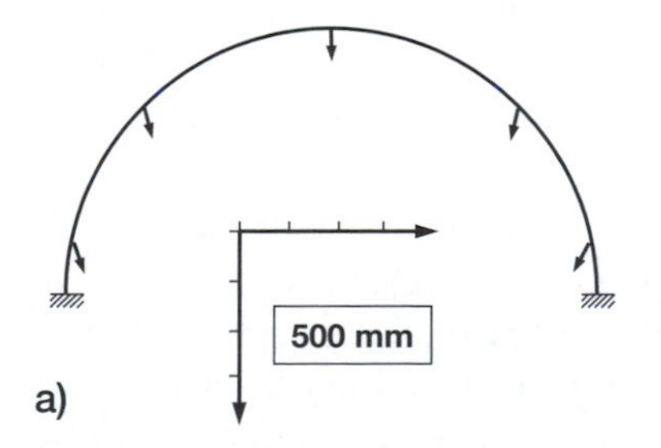

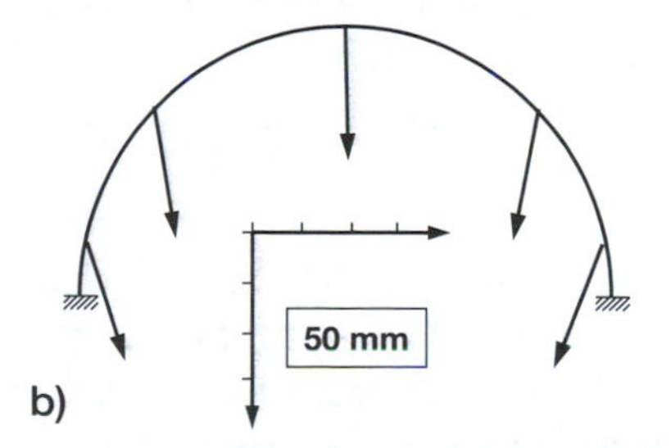

Figure 7.22 a) and b) Vertical and horizontal displacements plotted as x-y-vectors using different scales (Rokahr *et al.* 2002)

or deflected shapes, especially with regard to the level of the lining capacity used, and conversely how much capacity is still left. It is therefore necessary to determine a stress-intensity-index.

7.3.4.6 Interpretation of the measurements: stress-intensity-index

The stress-intensity-index, η, is based on the measured displacements and is defined as the ratio of the existing stress and the permissible stress in the tunnel lining at a certain point in time. If the stress-intensity-index is 100%, the ultimate strength of the sprayed concrete has been reached (Zachow and Vavrovsky 1995, Stärk *et al.* 2001, Rokahr and Zachow 1997).

Unfortunately it is not possible to transfer the deformation directly into a stress state of the sprayed concrete lining. The sprayed concrete is applied to the tunnel wall when it is still soft, and in this state the sprayed concrete is already loaded by the stress redistribution in the ground and by further excavation. The young sprayed concrete is still a long way from the standardized 28-day-values for compressive stress, compressive strain, and Young's modulus. However, the sprayed concrete gains strength, i.e. its material parameters and stress-strain behaviour change rapidly during the first few days as a function of time. Furthermore, the stress will generate a significant time- and stress-dependent creep strain (or relaxation) in the sprayed concrete lining. A determination of the actual stress state must take into account the time, importantly the time of the actual reading, the age of the sprayed concrete at the time of this reading, and the deformation history before this reading.

Figure 7.23a gives a simple example to help clarify this. It is assumed in this case that there is no displacement other than the roof settlement. Therefore the vertical axis displays the relative settlement between the roof and the footing or the deformation, respectively. After comparing both of the measuring curves one is tempted to declare the sprayed concrete lining belonging to curve 2 as less loaded than the sprayed concrete lining belonging to curve 1. However, exactly the opposite is the case. Despite the larger total settlement, the sprayed concrete lining 1 is less loaded than lining 2 and although both deformation curves are geometrically similar, the stress-intensity is quite different (Figure 7.23b). The reason for this surprising statement is the creep ability of the young sprayed concrete. Creep and relaxation in the first 8 to 12 days avoids a large proportion of the load.

Sprayed concrete lining 1 starts at the first day with, in this case, a high deformation of approximately 15 mm. This results in a high stress-intensity-index of more than 50%. Since creeping of the sprayed concrete depends on several factors, such as the sprayed concrete mix, its age and also on the stress-intensity, the high stress generates significant creeping. In addition, the increase in strength is the highest during the first day. Hence, the stress-intensity reduces on the second day. During subsequent days the

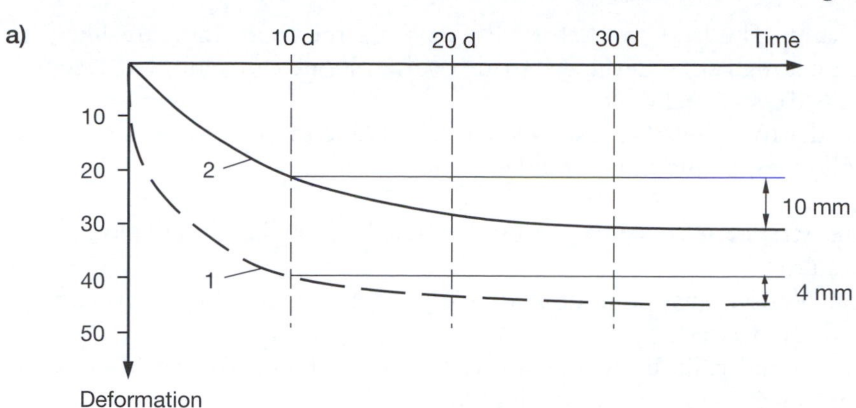

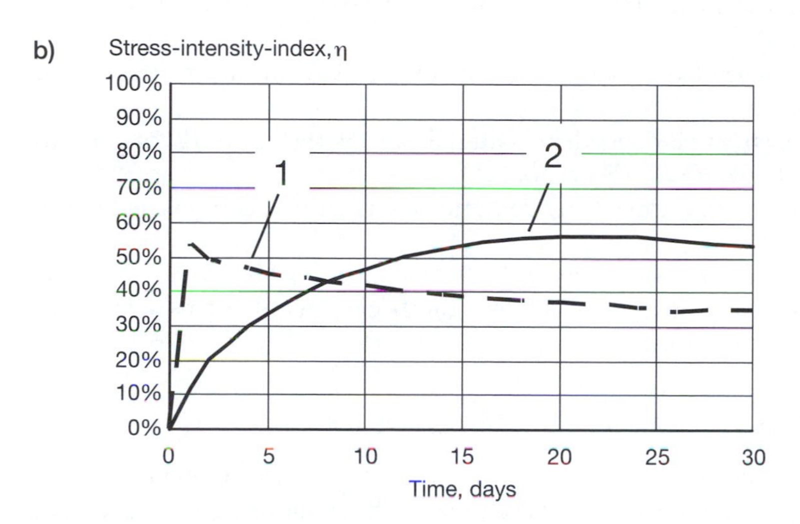

Figure 7.23 a) Example deformations of the crown depending on time (d = time in days), b) corresponding stress-intensity-index

deformation still increases, but with ongoing creeping and increasing strength the stress-intensity reduces steadily down to approximately 35%.

Sprayed concrete lining 2 starts with a small value of approximately 2 mm deformation, which generates neither a high stress-intensity nor a high creep rate. During subsequent days the deformation increases in a more regular way (compared to curve 1). As a result the stress-intensity grows slowly. The creep ability reduces with time and the gain of strength also slows down. A large proportion of the deformation affects sprayed concrete lining 2 when the creep ability is already reduced. To highlight this, the deformation after 10 days of both sprayed concrete linings is marked on the figure. In this example, lining 1 shows only 4 mm of settlement between the 10th and 30th day, but lining 2 shows 10 mm.

Thus, in order to determine the level of loading of a sprayed concrete lining in tunnelling, the age at which the sprayed concrete is deformed is significant. The later the deformation occurs the more they are likely to generate a high stress level. As a rule of thumb one can state that creeping after 28 days is negligible.

In order to get reliable results when calculating the stress-intensity-index, the following points must be taken into account:

- the stress-intensity-index must be based upon the actual monitoring results;
- calculation must be done with every displacement reading, usually on a daily basis;
- the actual contour of the sprayed concrete lining (which is given by the position of the monitoring points);
- the actual absolute location of the monitoring points relative to the tunnel axis;
- the time-dependent development of the ultimate strain of the sprayed concrete:
- a non-linear, time-dependent material law of the sprayed concrete to cope with creep and relaxation;
- since everything is time dependent, the deformation history is necessary.

The use of the stress-intensity-index is common on NATM/SCL tunnels in soft ground. Rokahr and Zachow (2009) give details of two different methods currently in use.

7.3.4.7 Measuring frequency and duration

As long as the displacements change significantly, a measurement is taken at least every day. After they have stabilized, the monitoring frequency can be reduced stepwise to every other day, twice a week, weekly and finally monthly. If the total cross section is excavated by a staggered multi-face heading, the monitoring frequency should be increased again to daily measurements well in advance of the following heading. If adverse trends are detected or displacements begin to increase again, readings should be taken at least once a day, or even potentially twice a day.

Generally, measurements must be taken until the displacements have become stable. The measurements should be extended until the face is at least a distance of 4D from the monitoring cross section. An exactly defined period cannot be given as the time over which one has to measure is mainly dependent on the ground type and the advance rate. Possible stoppages during excavation can increase the period of the measurements.

7.3.4.8 Contingency measures

When constructing a tunnel, it is in the nature of the project that there are gaps in the knowledge of the main construction material, i.e. the ground (see Chapter 2). The geological model for the ground is only an estimation with the assessment of its bearing capacity resting largely on experience, and 'even the most complex calculation is still only an approximation of reality' (Thomas 2009b). Therefore, during the excavation, it is possible that the interaction between the ground and the support does not behave according to the estimations and calculations, but potentially diverges from this in a negative way. This presents an exceptional circumstance, for which one has to be prepared with contingency measures.

Contingency measures can become necessary if the support is not sufficient to ensure quality, usability or stability. This can manifest itself by larger than expected surface settlements, larger than expected in-tunnel displacements, an unstable face, overloading of the ground, overloading of the lining or other performance indicators, which are not in accordance with the design or expectation. Special attention should be paid to the overloading of the lining because of the potentially severe consequences for the overall stability of the tunnel. There are visual signs (cracks) and acoustic signs (the concrete sounds dull and hollow) that give an indication of an overloading of the lining. However, not every crack in the sprayed concrete lining results in a collapse or is the result of overloading of the system. But every crack should be taken seriously due to its potential origin and the resulting serious consequences.

Prior to taking further measures, it is normal to observe the cracks: are they getting longer, do they open up? The end of the crack should be marked with a colour and the date. It is then easy to check if the crack is getting longer. If it is, then a new mark has to be made. Measuring the width of the crack (to see whether the crack opens) is difficult due to the rough surface of the sprayed concrete. It is possible to place gypsum patches over the crack. Gypsum hardens quickly and is very brittle and hence it will break immediately if the crack opens further. A better option is to install a Tell Tale™ over the crack (Figure 7.24). These consist of two overlying transparent plates which can move relative to each other. A scale on the plates shows exactly if the crack under the Tell Tale™ is changing and by how much.

The sprayed concrete around the crack should be hit with a hammer or similar. If sprayed concrete layers have separated from each other, then the hammer blow sounds dull and hollow. This is a clear indication of a weakened lining. Taking cores from the sprayed concrete lining just over the crack provides good information on whether the crack is due to bending or shear forces, or just due to shrinkage of the sprayed concrete (which is not a matter of concern with respect to stability). Cores will also show if the sprayed concrete is seriously damaged or still intact.

If overstressing of the lining is suspected, then an intensified monitoring regime must be implemented: the measuring intervals should be shortened and the monitoring data from the particular area must be analysed very thoroughly. It is not only overstressing of the lining and a potential collapse

Figure 7.24 Tell Tale™ placed over a crack in a sprayed concrete lining

which make contingency measures necessary. Larger than expected displacements, at the ground surface or in the tunnel, can cause potential harm to third party structures or can reduce the usability of the tunnel.

A list of common contingency measures is given below. This does not claim to include all measures available:

ELEPHANT'S FEET (SETTLEMENT REDUCTION)

Elephant's feet are enlargements, which look similar to the foot of an elephant, of the sprayed concrete lining where it bears onto the ground at the sides of the tunnel, i.e. they enlarge the foundation or footing. They reduce the ground pressure under the footing and thus help to reduce the likely settlement. In order to avoid the lining protruding into the internal tunnel space, the elephant's feet are usually constructed towards the extrados, i.e. the outside of the lining.

TEMPORARY INVERT (SETTLEMENT AND CONVERGENCE REDUCTION)

A temporary invert makes the sprayed concrete lining into a closed ring. A ring closure of the lining stabilizes the whole system. The temporary invert reduces the ground pressure under the side footings and thus helps to reduce the likely settlement. In addition, it reduces convergence since it stiffens the lining horizontally. This is usually used in addition to elephant's feet.

FOOTING PILES (SETTLEMENT REDUCTION)

Footing piles are steel rods located in the footing of the lining. Boreholes are drilled vertically from the elephant's feet down into the ground. Usually two piles are placed on either side with each advance. The piles distribute the load deeper into the ground and thus reduce the bearing pressure under the footing. The steel rods have a diameter of approximately 30 mm to 70 mm, and the borehole is grouted to guarantee friction between the ground and the piles.

FOREPOLING (PROVIDES OVERHEAD PROTECTION OF AN UNSUPPORTED HEADING)

Steel rods are placed around the circumference of the roof at a spacing of 20 cm to 40 cm in the direction of the excavation. Steel rods are rammed or bored into the ground: the girders act as an abutment inside the tunnel, and the ground acts as an abutment in front of the face. Thus they protect the work area from any ground falling from the roof of the tunnel until the excavation of the advance has been completed and the support installed. The common lengths of rod are 3 or 4 m, with a diameter of approximately 22 mm to 32 mm. They are placed with every advance. Forepoling needs to have an overlap of one or two advance lengths. Therefore the steel rods are slightly inclined by a few degrees. (See section 4.2.5 for further details.)

SHEET PILING (PROVIDES OVERHEAD PROTECTION OF AN UNSUPPORTED HEADING)

This method is similar to forepoling, but uses steel sheets instead of rods. The steel sheets are rammed into place and are positioned so that they touch. Sheets are used in coarse soil. (See section 4.2.5 for further details.)

FACE SUPPORT (HELPS TO PREVENT FAILURE OF THE FACE AND REDUCES GROUND LOSS)

The face is sprayed with between 3 cm to 20 cm of sprayed concrete, mesh or steel fibre reinforced if the thickness exceeds 10 cm. The sprayed concrete can be accompanied by face anchors (horizontal anchors in the direction of the excavation). Face anchors usually overlap by several times the advance length. (See section 4.2.4 for further details.)

SUPPORT CORE (HELPS TO PREVENT FAILURE OF THE FACE AND REDUCES GROUND LOSS)

The tunnel advance is only excavated around the circumference; the centre of the face is not excavated and acts as an abutment for the face. A support core can be sealed with sprayed concrete and can be bolted against the face with face anchors.

ADVANCE LENGTH

Shorter advances can help prevent overstressing of the ground and reduce settlement. With shorter advances, the disturbance of the ground is reduced, as is the time necessary for excavation and completion of the support. Thus, the load is transferred from the ground to the support more quickly. The installed support potentially needs to be increased to cope with the additional load.

DIVIDED FACE

Dividing the face into smaller cross sections has the same effect as using shorter advance lengths. In addition, this also increases the face stability.

ADVANCE RATE

Limiting the number of advances per day has the same effect as shorter advance lengths. The stress redistribution due to the excavation acts on an older and therefore stiffer lining, which takes a larger part of the load, and thus reduces the stress in the ground.

ANCHORING

Systematic anchoring avoids or reduces the loosening and the weakening of the ground due to the deformation following the excavation. The anchors 'nail' the ground together and thus help it to maintain its bearing capacity. Anchors are only active if there is a rigid bond between the anchor and the ground, which is achieved for the most common anchor types by using injected mortar. (See section 4.2.4 for further details.)

LINING

A thicker and stiffer lining reduces settlement and helps to reduce overstressing of the lining. The stress redistribution due to the excavation acts on a thicker and, therefore, stiffer lining, which takes a larger part of the load, and thus reduces the stress on the ground. As the lining is much stiffer than the (soft) ground, this reduces the settlement. Furthermore, if the lining itself shows symptoms of overstressing, it needs to be strengthened by thickening. In this case, the excavation must become larger to avoid the lining encroaching into the final tunnel profile.

SEALING THE GROUND

A 'flash' coat of sprayed concrete protects against any ground and rocks falling off the unsupported heading. It also hinders groundwater flowing into the tunnel. Flowing water can wash fine particles out of the ground causing the latter to lose cohesion and destabilize. Water must therefore be controlled.

GROUTING

This increases the bearing capacity of the ground and reduces or stops water inflow. (See section 4.2.3 for further details.)

In the case of overstressing of the lining such that a collapse would be unavoidable if nothing were done, there are still usually some options to strengthen the support long before the tunnel has to be evacuated. Some of the most suitable contingency measures are:

STOP EXCAVATION

Any additional load has to be avoided.

POST ANCHORING

For this measure, additional anchors are installed in the already completed tunnel support. If grouted anchors are used it takes approximately a day until the mortar around the anchor hardens and for the method to become effective.

POST SHOTCRETING

In order to strengthen the sprayed concrete lining, an additional layer of sprayed concrete can be sprayed in areas where the lining is already overstressed. This measure can be carried out quickly because the sprayed concrete is available immediately on site. However, it also takes at least 12 hours for the sprayed concrete lining to become reasonably load bearing. It has to be considered that at a later stage it is essential to remove the additional sprayed concrete lining as it encroaches into the final tunnel profile. In many cases it is often necessary to renew the overloaded sprayed concrete so that the additional sprayed concrete lining is removed anyway. Post shotcreting all of the lining of an overstressed section can be very time consuming. In order to gain time it is possible to spray a couple of sprayed concrete ribs placed around the circumference (see section 4.3.2).

BACKFILL

Placing backfill against the face can re-stabilize it. The additional load from the backfill can stop further cracking of a broken invert.

TREE TRUNKS

As an immediate measure against a threatening collapse, this option is well suited for cross sections of up to 6 m in height (for example, in the crown heading, Figure 7.25). For larger heights, tree trunks are not suitable from a practical and structural point of view (danger of flexural bending increases). Tree trunks can easily be adjusted to the actual geometry (by using a saw). If fixed well with wooden wedges and sprayed concrete on either side, tree trunks can immediately take loads and help to avoid a collapse. It is compulsory to have a supply of tree trunks on a tunnel site in order to be able to react immediately if required. If trees have to be placed, it is often necessary to renew the destroyed sprayed concrete lining

Figure 7.25 Emergency support measures using tree trunks during the tunnel excavation (Eggetunnel, crown heading) (courtesy of Professor Dr-Ing. habil. Reinhard B. Rokahr, photograph by Ulrich Mertens DFA DGPh, Atelier für Kunst und Fotografie, Hamburg)

in sections. The tree trunks themselves can be removed relatively quickly if necessary care is taken. The tree trunks should be positioned in such a way that the excavation plant can still pass. Of course, steel beams or other similar elements can also be used, but they are not as easy to cut to the required length.

7.3.5 Instrumentation for in-tunnel and ground monitoring

In addition to laser theodolites, steel tapes and crack monitoring mentioned in the previous section for in-tunnel monitoring during construction, pressure cells can be used to determine the stresses in the tunnel lining. Pressure (or 'stress') cells can be installed between the lining and the ground (total pressure cells, tangential pressure cells), or cast into the lining (radial pressure cells) and can use either liquid pressure or vibrating wire transducers. They need careful installation and experience to interpret the results due to the complexities of the sprayed concrete behaviour, which causes much debate as to their reliability. However, research by Jones (2007) has suggested various procedures to reduce the potential errors when using pressure cells.

Furthermore, it is important to monitor the ground around the tunnel during construction in order to assess its behaviour. There are a number of common instruments available and these are briefly described below (see section 7.3.6 for references).

BOREHOLE MAGNET EXTENSOMETER (RELATIVE VERTICAL MOVEMENT)

These devices consist of a series of circular magnets fixed at certain levels within the borehole to either a rigid or telescopic access pipe. A probe is inserted to record the level of each magnet. The rigid plastic tube and 'spider' magnets can cope with small vertical compressions of up to 1%.

BOREHOLE ROD OR INVAR TAPE EXTENSOMETERS (RELATIVE VERTICAL MOVEMENT)

These can consist of simple rods of different lengths anchored at different levels within the borehole. More sophisticated methods involving linear variable differential transformers (LVDTs) are also available. Wire based extensometers, although more difficult to install than rod extensometers, are useful over long distances.

SATELLITE GEODESY (RELATIVE VERTICAL MOVEMENT)

This method is useful for monitoring relative movements over large areas of the ground surface.

CONVERGENCE GAUGES – (LATERAL DISPLACEMENT)

Gauges, consisting of tape, wire and rods with a deformation indicator, can be used to measure horizontal displacements between permanent anchor points, for example ground surface settlement points.

BOREHOLE INCLINOMETER PROBES (CHANGE IN INCLINATION)

When horizontal deformation measurements are required within the ground, a permanently installed vertical casing is used. A probe containing a gravity-sensing transducer is inserted into the casing. The guide casing usually has tracking grooves for controlling the orientation of the probe. An alternative system uses borehole electrolevels.

HORIZONTAL BOREHOLE DEFLECTOMETER (CHANGE IN INCLINATION)

These rely on angle transducers instead of tilt transducers (as used in inclinometer systems) and this means that they can be used in inclined or horizontal boreholes (as well as vertical) as the sensors are not reliant on gravity.

'PUSH IN' TOTAL PRESSURE CELLS (CHANGES IN EARTH PRESSURE)

These can be either diaphragm or hydraulic cells. Diaphragm cells consist of a circular membrane with strain gauges attached. The membrane deflects under pressure and the strains measured can be related to the change in pressure. The hydraulic cells consist of two membranes sealed around the edge, with the gap between them filled with liquid. The pressure acting on the cell is measured via the pressure of the liquid in the cell. They can be used in combination with piezometers to obtain effective stress values.

STANDPIPE PIEZOMETER (CHANGE IN GROUND WATER PRESSURE)

Used to monitor groundwater pressure at a particular elevation. The main disadvantage of these piezometers is the problem of assessing 'real time' fluctuations in piezometric head due to manual reading and time lags.

PNEUMATIC PIEZOMETER (PORE PRESSURES ARE BALANCED BY APPLIED PNEUMATIC PRESSURES/CHANGE IN WATER PRESSURE)

Uses the pressure of gas on a flexible diaphragm to measure the external pore water pressure, i.e. the external gas pressure is increased until it balances the water pressure. These instruments have a short time lag.

VIBRATING WIRE PIEZOMETER (CHANGE IN WATER PRESSURE)

Uses a vibrating wire strain gauge attached to a diaphragm. As the pore water pressure changes, the diaphragm deflects and registers a change in strain. This can be related to the magnitude of the pressure. These instruments have a short time lag and are easy to read.

STRAIN GAUGED BOREHOLE EXTENSOMETERS INSTALLED FROM WITHIN A TUNNEL (GROUND DEFORMATIONS)

These can be directly measured and have multiple extensometers in one borehole. Using vibrating wire strain gauges they can be automatically data-logged. The longest/deepest extensometer is assumed to be beyond the disturbed zone, otherwise relative movements are underestimated.

Example of an instrumentation layout An example of an instrumentation layout used to monitor the ground behaviour is shown in Figure 7.26. This monitoring arrangement was used as part of a research project conducted by Imperial College on the London Underground Jubilee Line Extension project in the UK, and illustrates the type of instruments that can be used. Further details of this and the other monitoring conducted on this project can be found in Burland *et al.* (2001a and b).

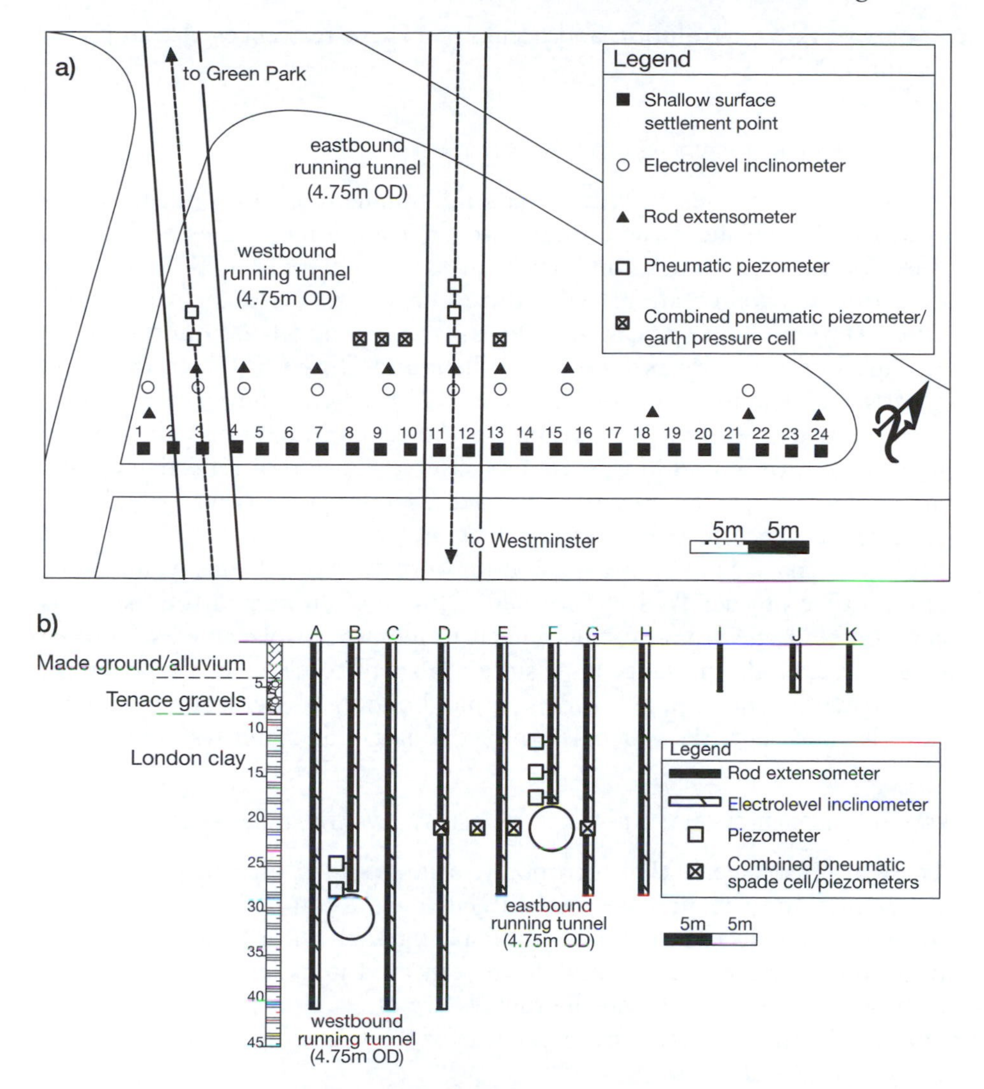

Figure 7.26 An example of the instrumentation used to monitor the ground behaviour resulting from a bored tunnel as part of the Jubilee Line Extension project on the London Underground, UK, a) site plan, and b) instrumented cross section (after Standing *et al.* 1996)

7.3.6 Instrumentation for monitoring of existing structures

A few of the more commonly used instrumentation for monitoring existing structures (both surface and subsurface) are briefly described in this section. There is considerable literature available on this topic, for example Dunnicliff and Green (1993), Clayton *et al.* (2000), BTS/ICE (2004) and Kavvadas (2005). A detailed list of commonly used instrumentation for tunnelling projects is provided by BTS/ICE (2004), together with their respective range, resolution and accuracy. (These references also apply to section 7.3.5.)

AUTOMATED TOTAL STATIONS (RELATIVE MOVEMENT)

These have been used in recent years in conjunction with optical targets attached to existing structures. However, traditional survey techniques (theodolites, total stations and levels) are still commonly used. A network of automated total stations was utilized on the redevelopment of King's Cross Station, London, UK and the associated tunnelled connections to existing infrastructure as reported by Beth and Obre (2005). In this case, a network of automated total stations was used to monitor both above ground structures and also within operating station tunnels during the works. Each total station was used to observe a group of reflective optical prisms located on the structures as well as reference prisms outside the zone of influence of the construction works where possible. This system has also been used in Hong Kong and Amsterdam, Netherlands (van Hasselt *et al.* 1999, van der Poel *et al.* 2006). Although automated total stations can also be used within existing tunnels to monitor displacements, in some metro systems the running tunnels are too small to have such a system in the crown of the tunnel (for example London Underground running tunnels), and hence the system is confined to larger diameter station tunnels.

PRECISE LIQUID LEVEL SETTLEMENT GAUGES (RELATIVE VERTICAL MOVEMENT)

These are instruments that incorporate a liquid-filled tube or pipe for the determination of relative elevation. Relative elevation is determined either from the equivalence of the liquid level in a manometer or from the pressure transmitted by the liquid. These have been used in existing metro tunnels to monitor 'rotations' when the tunnels are affected by new construction works. The tube is passed from one side of the tunnel under the track and up the other side.

PLUMB-LINES (CHANGE IN INCLINATION)

These can be used for monitoring the tilt of structures by measuring the horizontal distance between two points at different elevations. Direct plumb-lines consist of a weight suspended from the highest possible elevation of a structure and measure the horizontal movement of the suspension point relative to a point at the base, about which the weight moves. Inverted plumb-lines have a similar operation, but the plumb-line wire is fixed at both ends and movement is observed at an intermediate elevation. A digitized plumb-line was used as part of the monitoring of the Big Ben Clock Tower in London, UK during the construction of the Jubilee Line Extension project. The plumb-line was suspended from 55 m above the ground, to provide an accurate and precise tilt measurement. The movement of the base of the plumb-line was sensed on a digitizing tablet (placed just below the plumb) and the measurements were processed in real-time. This was important as it provided continuous feedback to the compensation grouting (section 4.2.8) that was being used as a protective measure during these construction works (Kavvadas 2005).

TELL TALE™ AND CALLIPER PINS/MICROMETER (DEMEC™ GAUGES) (CRACK OR JOINT MOVEMENT)

These are manual methods for monitoring structural damage, such as cracks. The distance between measuring studs attached to the structure are measured accurately using a Demec™ gauge (basically a distance measurement device). These devices can also be used to monitor tunnel lining as detailed in section 7.3.4.8, Figure 7.24.

VIBRATING WIRE STRAIN GAUGES (STRAIN IN STRUCTURAL MEMBER OR LINING)

These are the main way of monitoring individual structural members and are accurate, robust and reliable.

FIBRE OPTICS (STRAIN IN STRUCTURAL MEMBER OR LINING)

These are based on the ability of glass fibres to carry light from a source. These fibres can be embedded in concrete or attached directly to a structure, but have to incorporate a light source and optical analyser at one end. Systems can be based on Fibre Bragg Gratings, which use discrete optical strain gauges positioned along the optical fibre, i.e. discrete strain monitoring (Metje *et al.* 2008), or on pulsed light systems which allow axial deflection and bending to be monitored at all points along a fibre, i.e. continuous strain monitoring (Vorster *et al.* 2006 and Mohammad *et al.* 2007).

TAPE EXTENSOMETERS ACROSS FIXED CHORD (TUNNEL LINING DIAMETRICAL DISTORTION)

This is a relatively simple manual method, but obviously needs access to the tunnel to take readings. This limits monitoring within live metro lines to 'engineering hours', i.e. when the trains are not running. It can also disrupt construction processes if used in new tunnels. (see section 7.3.4.1)

BASSETT CONVERGENCE SYSTEM (TUNNEL LINING DIAMETRICAL DISTORTION)

The Bassett Convergence system (Bassett *et al.* 1999) is based on a series of rods and electrolevels, which are attached around the inner circumference of a tunnel. It can be used whilst the tunnel is operational, but does need a suitable clearance between the vehicles and the tunnel wall. Electrolevels are tiltmeters that contain an electrolytic level (a sealed glass vial similar to that used on a conventional builders level, however this contains a conductive liquid and uses contacts within the vial to register changes in resistance as the vial tilts). Electrolevels can be used in 'beam' arrangements for measuring relative movements within tunnels and also on structures affected by tunnelling activities.

Example of monitoring existing tunnels An example of the monitoring arrangement that can be used in existing tunnels to assess the effects of adjacent construction activities, including new tunnels, is shown in Figure 7.27. The tape extensometers are used for monitoring diametrical distortion and levelling either side of the track can be used for relative vertical displacements of the tunnel. In addition, the levelling can be used to obtain a measure of rotation of the tunnel by taking the difference in the levels either side of the tack. Relative movements can also be obtained from electrolevel strings running along the tunnel (Cooper 2002). Automated total stations can also be used to monitor existing tunnels if there is space inside the tunnel, as described earlier in this section.

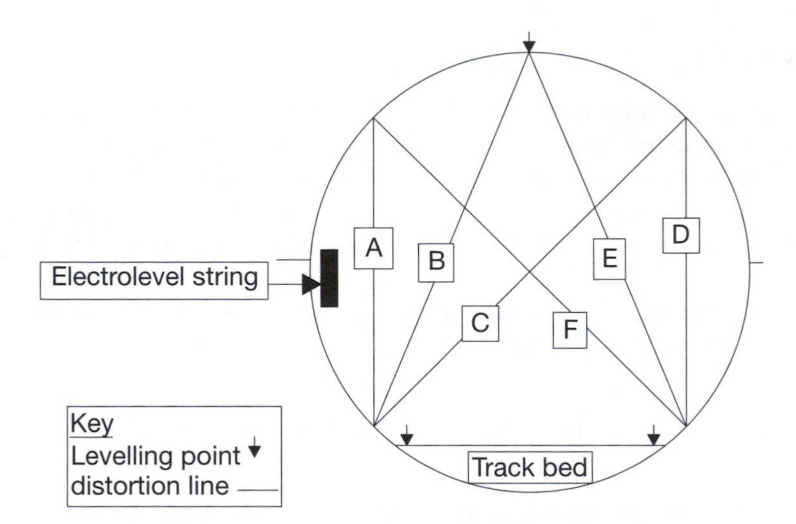

Figure 7.27 Example of monitoring within an existing tunnel in order to assess the effects of adjacent construction activities (after Cooper 2002)

8 Case studies

This chapter focuses on three different case studies. Each case study describes distinctive aspects of the tunnelling works and how some of the information described in previous chapters is applied in practice.

8.1 Eggetunnel, Germany

Unexpected invert failures of sprayed concrete linings are not unusual, despite a rigorous monitoring regime within the observational method. This case study gives an example of an invert failure, explains the failure mechanism and how to detect an invert failure by means of monitoring.

8.1.1 Project overview

The railroad line connecting Kassel and Paderborn, Germany, of which the Eggetunnel is a part, has been upgraded to allow higher speeds and to increase the route capacity. The Eggetunnel crosses the so called Egge Mountains between the towns of Willebadessen and Neuenheerse, both located in the federal state of North-Rhine-Westphalia. The tunnel became necessary because the existing railroad was affected by landslides which caused damage and delays. It was found to be safer, more efficient and less costly to build a tunnel right through the mountains rather than to stabilize the unstable slopes. The two-track railroad tunnel, length 2880 m and width 14.5 m, was constructed between 1998 and 2000, and was opened in 2003.

Construction commenced from both ends of the tunnel. From the northern portal, where hard rock formations were dominant (sand and limestone), the excavation was achieved using drill and blast, and from the south portal, where soft rock (clay) was present, excavator and side wall drifts were used. Temporary support was provided by a sprayed concrete lining, lattice girders and wire mesh reinforcement throughout the tunnel.

Some parts of the sprayed concrete-supported invert of the soft rock section experienced heavy cracking and failure, and despite a rigorous displacement monitoring regime this remained undiscovered for a long time and created some critical situations when finally detected.

The authors have come across this displacement pattern with a number of tunnels that have suffered from invert failures. However, only in rare occasions have invert failures been recorded sufficiently or even published. This is possibly because people are afraid to discuss their potential mistakes, or invert failure has not been considered a serious enough stability concern to be worth publishing. Due to the lack of knowledge, the latter is a popular fallacy among tunnel builders. Invert failure could cause a serious tunnel collapse, and to the authors' knowledge, a few have already occurred (John *et al.* 1987, Golser and Burger 2001). Even in some cases of heading collapses where the causes were not known unequivocally, a broken invert may be suspected of being one of the causes or even the main cause of the failure. The collapse of three tunnels at Heathrow airport during construction in October 1994, for example, can be put into this category (Rokahr and Mussger 2001).

The Eggetunnel, otherwise built and monitored perfectly, did not collapse; the invert failure was detected early enough to put contingency measures into effect. It has been chosen here by way of an example because, in this particular case, the invert failure – once detected – was well recorded, which should help to answer the question, why do invert failures often remain undetected for too long?

8.1.2 Invert failure of the total cross section in the Eggetunnel

It had been known from the top heading that the clay developed a significant long-term settlement due to the excavation. Therefore precautions were taken to protect the sprayed concrete invert of the full cross section. The thickness of the sprayed concrete was increased to 40 cm (which is about the maximum thickness for a regular sprayed concrete lining); a deep invert vault was built so that the shape of the tunnel was close to circular; in addition an invert monitoring device was developed and implemented as described below (see section 8.1.3).

The difficult clay zone extended over a distance of 250 m. In addition to the regular displacement monitoring it was agreed with all the parties involved to monitor the sprayed concrete invert in this area. The invert was covered with an approximately 3 m thick layer of backfill, which provided a track for the plant. In general the monitoring should not disturb or hinder the ongoing excavation. These demands were fulfilled by the so called 'Eggemouse' (see Figure 8.1a): During excavation a tube was placed every 15 m diagonally into the invert and was sprayed in place. The ends of each tube extended out of the sprayed concrete. A steel cylinder, attached to a cable, was pulled through every tube once a day from one end to the other (Figure 8.1b). The steel cylinder, the 'Eggemouse', was 70 mm long and 20 mm in diameter. In the case of a broken invert, the steel cylinder would not have passed through the tube and would have got stuck. Since this first use, the Eggemouse has been used in many other tunnels.

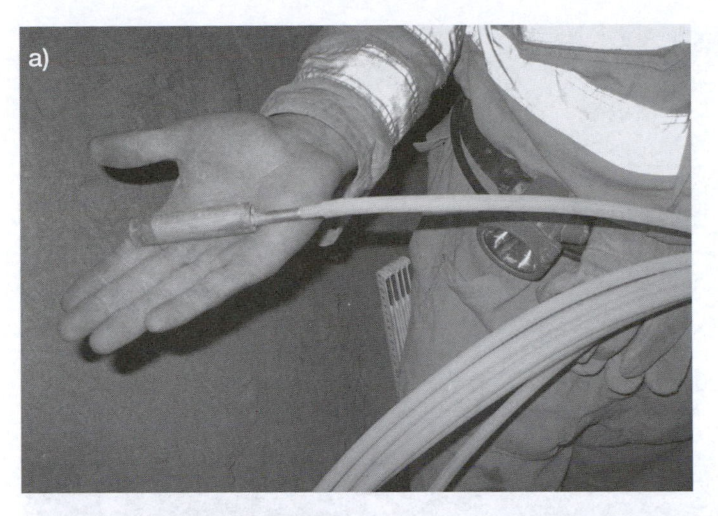

Figure 8.1 a) 'Eggemouse' and b) invert control with the 'Eggemouse' at the Lainzer Tunnel LT31, Vienna, Austria

After months of stable monitoring results, the monitoring was finally abandoned. As it turned out this was premature. After completion of the excavation, the backfill was removed to allow construction of the inner lining. While removing the backfill, the miners were surprised by a longitudinal crack, nearly 100 m long, in the middle of the invert. The displacement monitoring had not given even the smallest indication of any disturbance in the sprayed concrete lining. However, the edges of the crack began pushing

Figure 8.2 Contingency measures in the form of tree trunks at the completed cross section due to cracking after removing the backfill

over each other, with the damage to the invert getting worse and a collapse to the tunnel seemed to be a possible scenario. The movement had to be stopped immediately. But how could this be done? Replacing the backfill would have taken too long, as would stiffening the invert using massive sprayed concrete ribs. Anchoring back the invert would not have been effective enough since the displacements were mostly horizontal. The tunnel was about 13 m high, i.e. too high for setting up tree trunks. The only option was to place the tree trunks horizontally: 50 logs, each 7 to 8 m long, stopped the increasing displacements and gave enough time to reconstruct the invert with a thicker, 60 cm, sprayed concrete lining (Figure 8.2).

The invert must have failed over a longer period of time, and only the weight of the backfill provided the necessary force to maintain equilibrium; once removed, the displacements started again. But how could it happen that despite a rigorous monitoring regime, experienced staff and vigilance against any failure, massive cracks in the invert could develop without detection? This will be investigated in the following sections.

8.1.3 Sprayed concrete invert – its purpose and monitoring

In soft rock tunnels, due to low strength of the ground a sprayed concrete invert is constructed to avoid shear failure under the footings. A quick ring closure ensures the bearing capacity of both the ground and the lining. 'Quick' in this case means that the invert has to be constructed close to the leading face; a practical distance with respect to buildability is 1 to 4 m in the top heading (as part of a larger tunnel) and 4 to 10 m in a complete cross section. After construction, the invert is covered with muck and backfill. This protects the invert against damage and provides a track for the heavy plant while the excavation continues. Depending on the shape of the tunnel and the space required for manoeuvring the plant, the thickness of this cover is approximately 1 to 3 m. However, this makes the invert invisible and it is impossible either to inspect its integrity or to install any of the common optical monitoring systems.

These days, the state-of-the-art in monitoring the invert is by doing it indirectly and involves interpreting the displacement measurements of the vault (i.e. crown or crown and bench). However, the wide-spread experience on many construction sites is that a broken invert can only be detected in a very progressive state of damage – if at all. At least the Eggemouse can tell if the invert is broken, which is of paramount importance. However, it is still not known exactly how to recognize the beginning of an invert failure by interpreting vault monitoring data. Any conclusions drawn from interpreting the monitoring data of the vault with respect to the integrity of the invert cannot be proven as the invert is not visible. The behaviour of the invert is therefore open to speculation.

Based on these experiences some essential questions kept recurring:

1 Is it generally possible to get early signs of the reduced bearing capacity of the invert by interpreting the displacement measurements of the crown?
2 Also, is it possible to assess the residual bearing capacity of the broken sprayed concrete lining?

In order to answer these questions a comprehensive research project was undertaken at Hanover University, Germany (Stärk 2002). At the beginning of the research, measured data from tunnels with broken inverts were analysed. However, the data, even at the Eggetunnel, gave no indication of any problem with the invert, and since the moment of the collapse of the invert was never known, this approach was not successful. Calculations were also not helpful, as the theoretical model could not be verified due to the lack of measured data in the invert. The research was therefore advanced by using model tests (Figure 8.3). Two different shapes were used: a full cross section with a deep invert vault, and a crown section with a temporary invert. The model tunnels were made from gypsum and embedded in clay.

The load was applied by horizontal and vertical hydraulic jacks, independently controlled to achieve different coefficients of lateral earth pressure. Depending on the geometry, 12 or 14 monitoring points were distributed over the complete cross section including, of course, the invert. The location of the monitoring points in the crown and bench corresponded to the traditional monitoring positions in a real tunnel.

Figure 8.3 Model testing

Figure 8.4 shows, by way of example, the vertical displacements in the crown (monitoring points 1 to 5) versus the applied pressure of the jacks. As expected the displacements increased, in this case linearly, with increasing pressure. It seemed likely that the displacements in the crown would be affected by the cracking of the invert. However, nothing happened throughout the test in the crown even though five cracks developed in the invert. Looking at Figure 8.6a, this was quite a surprising result. However, in fact this reflects the typical behaviour of the crown perfectly.

According to widespread opinion, the sign of collapse of the invert is horizontal convergence of the footing (i.e. points 4 and 5, Figure 8.4). However, this could not be confirmed as inverts have collapsed in situations where convergence has occurred, where divergence has occurred, and where no horizontal displacements have happened at all.

The only way of achieving reliable information on what was happening was to monitor the invert itself, and especially to determine the stress-intensity-index of the lining. The stress-intensity-index is based on the monitored displacements, and according to Rokahr and Zachow (1997) is defined as the ratio of the existing stress and the permissible stress at a certain point in time. If the stress-intensity-index is 100%, the ultimate strength of the lining has been reached. With the stress-intensity-index the

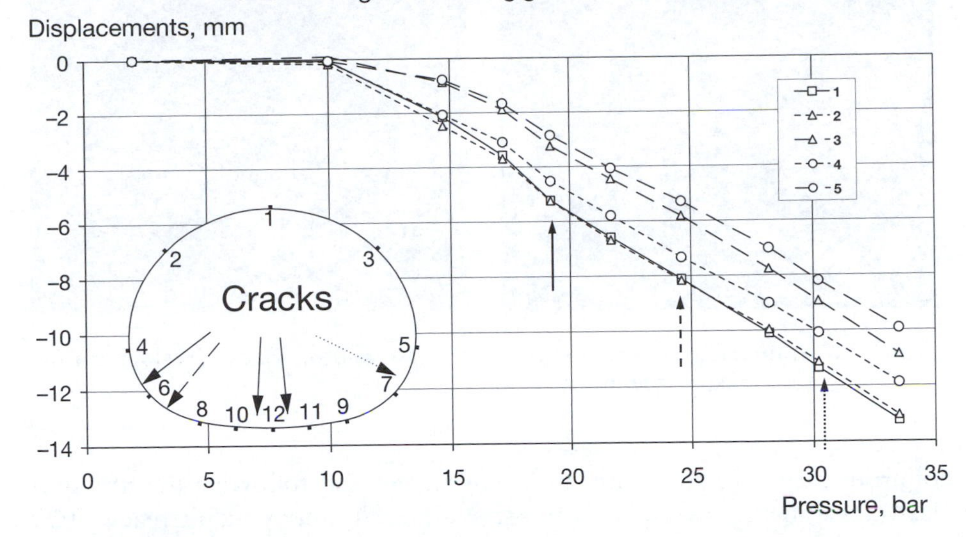

Figure 8.4 Vertical displacements at the crown and corresponding cracks

actual stress state can be determined in real time, so that there are no uncertainties about the residual bearing capacity due to increasing displacements. Figure 8.5 shows the development of the stress-intensity-index in two adjacent locations on the lining, at the crown and at the invert. The stress-intensity-index in the crown (points 1–3–5) initially reaches an untypically high value of approximately 40%. The remaining curve has typical values

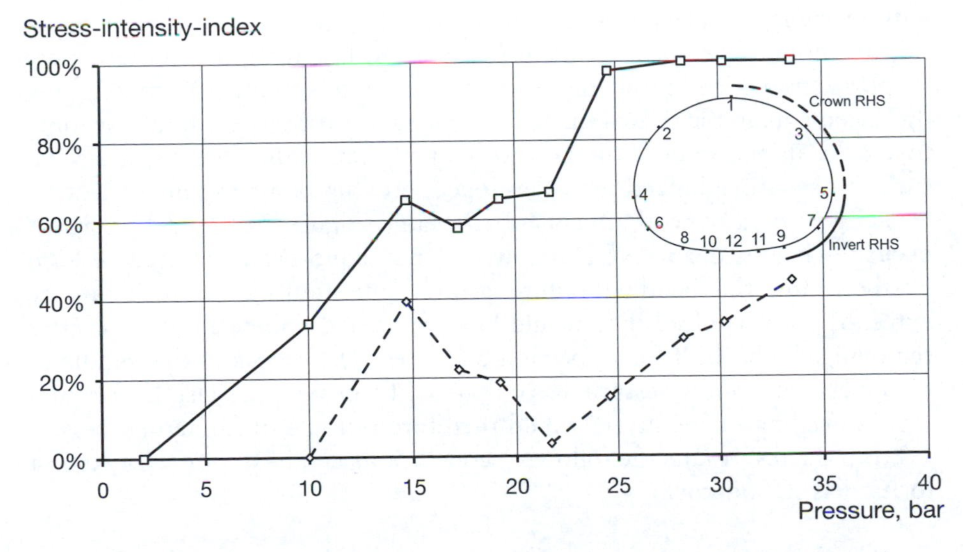

Figure 8.5 Stress-intensity-index at the crown (monitoring points 1–3–5), and invert (5–7–9)

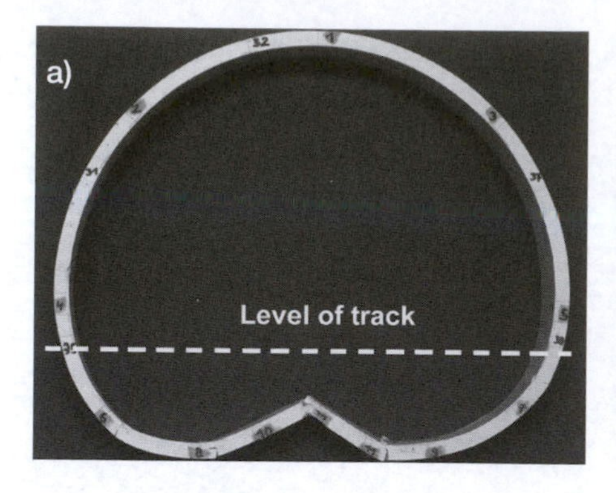

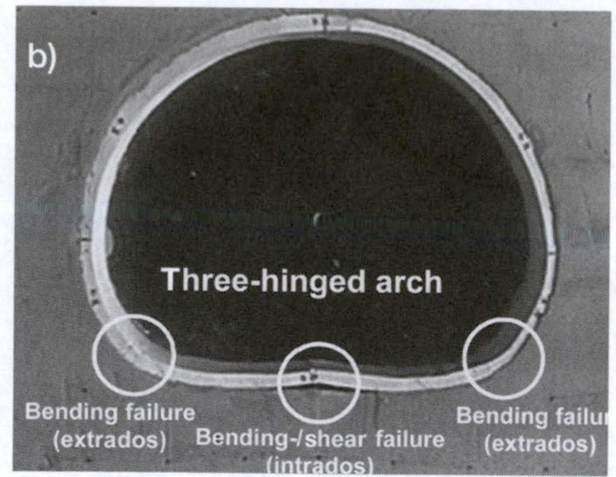

Figure 8.6 a) Invert collapse at the end of the test, and b) typical development of the cracks at the invert

of around 20% or less. Only at the end of the test, following the fifth and last crack in the invert, does the stress-intensity-index again reach 40% and peaks at an absolutely uncritical 45% as the invert finally collapses.

At the same time, the increase in load in the invert is clearly visible. The stress-intensity-index (points 5–7–9) rises from the start and indicates with a value of 100%, corresponding to the crack at point 7 (30 bar), that the load capacity of this part of the lining has been reached.

Generally, it was not predictable whether the model tunnels were going to collapse immediately after the first crack in the invert had developed or if the pressure could still be increased as shown in the test above.

Nevertheless, the invert failure mechanism can be described as follows. The vault, i.e. above the backfill, showed vertical displacements, mostly without bending. The stress remained low. The invert, i.e. below the backfill, did not follow the vertical displacements, but remained more or less in position, which is the purpose of an invert support. With ongoing displacements in the crown the load in the invert increased rapidly, leading to cracks in the middle of the invert and around the footing, a system similar to a three-hinged arch developed. At this point the invert lost its capacity to be a wide abutment for the vault (Figure 8.6b). All the cracks occurred below the backfill and would not have been detected *in situ*. Furthermore, the bending failure around the footing occurred at the extrados and therefore they would have remained undetectable even after removing the backfill for inspection purposes. The cracks in the middle of the invert were not necessarily accompanied by heaving, as visible in Figure 8.6a. Levelling of the invert would therefore only be of limited success.

From the test results the following conclusions can be drawn with respect to tunnels in soft rock:

1 Monitoring the vault only gave very late hints on a collapse of the invert, if at all.

2 Therefore, the invert should be monitored. As long as displacement monitoring of the invert is not possible, and the stress-intensity-index cannot be determined, the 'Eggemouse' is a proven and effective tool.
3 The remaining bearing capacity of the system after cracking of the invert is generally unpredictable. Just one crack in the invert can be followed by the collapse of the system. If one has knowledge about an invert failure it is recommended to repair this immediately.
4 The invert does not follow the vertical displacements of the crown causing a quicker development of the stress in the invert. This must be considered in the design. Even a temporary invert must be designed to be as robust as the vault.

8.2 London Heathrow T5, UK: construction of the Piccadilly Line Extension Junction

For complicated geometries or short tunnels, there is no alternative to using excavator and sprayed concrete support. On the London Heathrow T5 project a new tunnelling method called LaserShell™ was utilized, setting benchmarks in safety of construction and quality of SCL tunnels. This section describes arguably the most difficult and interesting part of T5's tunnelling work.

8.2.1 Project overview

London Heathrow airport has been expanded by adding a new Terminal 5 (T5), which opened in 2008 and was constructed away from the central terminal area. This had to be connected to the existing terminal buildings and to downtown London by means of a total of seven tunnels. The running tunnels were constructed by TBMs, but all the connecting constructions, such as headshunts, shafts, emergency exits, ventilation openings and cross passages were constructed using sprayed concrete lining (SCL). In total, more than 40 SCL structures with an overall length of more than 1100 m had to be built (Hilar *et al.* 2005, Williams *et al.* 2004). The tunnelling work was successfully completed in 2006. This case history focuses on the Piccadilly Line Extension (PiccEx) Junction.

London Underground provides public transport to London Heathrow via the Piccadilly Line, which had to be extended to serve the new T5. To connect the T5 Piccadilly Line Extension with the existing Piccadilly Line, the so called PiccEx Junction had been constructed in the middle of the heart of one of the busiest airports in the world (Figure 8.7).

8.2.2 The 'Box'

A 'box' was excavated, approximately 20 m deep, 50 m long and 20 m wide, to allow access to the existing tunnel systems (Figure 8.8).

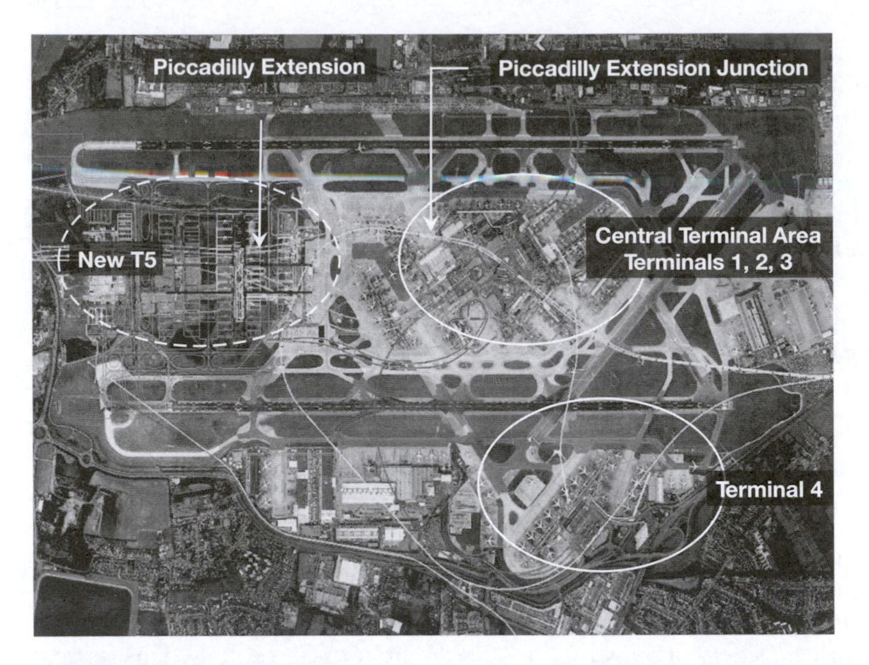

Figure 8.7 Overview of London Heathrow airport

Figure 8.8 The 'Box'

The box structure consisted of 1.2 m thick reinforced 'diaphragm' retaining walls. Half way down it was stiffened by a gallery-like intermediate slab. Two concrete pillars in the middle of the box provided additional support. All machinery, gear, equipment and material had to be lifted down the box opening by means of a crawler crane. For the first few weeks this also included the sprayed concrete supply until a permanent sprayed concrete pipe could be installed to the bottom of the box.

At the openings for the tunnels, one metre thick and heavily reinforced headwalls were cast to support the portal.

8.2.3 Construction of the sprayed concrete lining tunnels

In order to enable trains to reach the new T5, a turnout from the existing Piccadilly Line had to be built (Figure 8.9). East of the box the existing Piccadilly running tunnels needed an enlargement to provide enough space for the necessary switches for the Piccadilly Line Extension turnout, resulting in Eastbound and Westbound Turnout tunnels.

On the West side of the box the connection to the previously built Piccadilly Extension had to be completed, creating a need for Eastbound and Westbound Stub tunnels. Table 8.1 gives a brief overview of all four SCL tunnels at the PiccEx Junction.

8.2.4 Ground conditions

The only ground encountered during these works was London Clay, which appeared homogeneous, with almost no water seepage. A layer of scattered clay stones of up to 300 mm in diameter followed all four headings,

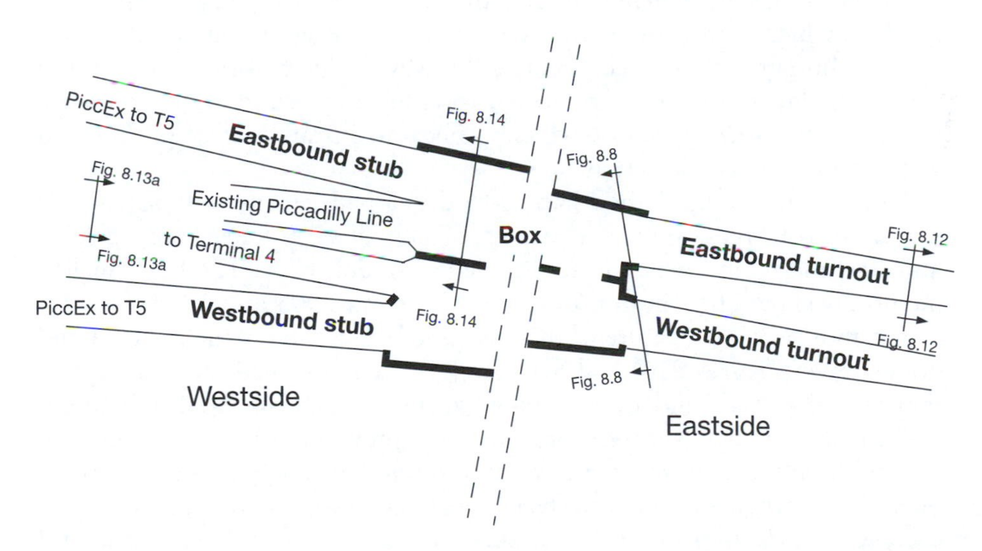

Figure 8.9 Overview of the PiccEx Junction

Table 8.1 Tunnels at the PiccEx Junction (in order of construction)

Tunnel	*Length (m)*	*Excavation diameter (m)*
Eastbound turnout	52.9	6.755–4.850
Eastbound stub	26.7	5.500/4.500
Westbound turnout	52.9	6.755–4.850
Westbound stub	30.1	5.500/4.500

approximately at tunnel axis. Occasional 'greasy backs' were not a problem since the excavation followed the principles of the LaserShell™ method (see section 8.2.5). Greasy backs are boulders likely to fall from the face. This is a particular problem in clay as water seepage in fissures reduces the friction holding the boulder in place. Greasy backs are not usually visible and can fall off the face without warning making them dangerous, which is one of the reasons why the Health and Safety Executive in the UK forbids anyone from entering the unsupported vault and face.

8.2.5 The LaserShell™ method

Until now, it has been common when using sprayed concrete support, to enter the unsupported heading to install girders and steel mesh. This is a fundamental aspect of the support system in soft ground. Girders provide immediate support and are required to fix the first layer of mesh. In addition, the girders are used to control the profile. Both mesh and girders can only be installed manually and therefore it is necessary to enter the heading, which, during this construction phase, is unsupported or only sealed with a thin layer of sprayed concrete. However, due to British health and safety regulations, no one is allowed to enter an unsupported, or even a partially supported, heading, which implies no installation of girders and hence no tunnel construction.

Thus, a tunnelling method had to be developed which could satisfy the British health and safety regulations, which does not need any girders and steel mesh, and which is applicable in London Clay. The LaserShell™ method was therefore born (Eddie and Neumann 2003 and 2004).

The most obvious feature of the LaserShell™ method is the inclined and domed face (Figures 8.10 and 8.11 Stage 1). The inclined face guarantees that nobody stands under any exposed ground at any time and only under the already hardened sprayed concrete of the previous advance. In addition, the inclination is contributing to a more stable face compared to a vertical one. As a permanent support steel fibre reinforced sprayed concrete was used with a 75 mm initial layer, and a minimum of 200 mm structural layer (Figures 8.11 Stages 2 to 5). A 50 mm finishing layer without steel

Figure 8.10 LaserShell™, inclined face

fibres covered up the steel fibres of the previously sprayed structural layer and provided a smooth finish comparable to a shuttered concrete surface. There was no inner lining.

Another less obvious feature of this method is that the structural layer is sprayed circumferential in one go, including the invert. This reduces the joints to an absolute minimum, i.e. just the radial joints between heading advances remain, with the added bonus of increased quality.

8.2.6 *TunnelBeamer™*

Another issue under the rigid British health and safety regulations is profile control. Since nobody is allowed to enter the unsupported vault, the profile can only be controlled from a distance, thus the TunnelBeamer™ became of paramount importance. The TunnelBeamer™, a laser theodolite, was connected to a laptop. The latter contained all relevant information with respect to geometry, i.e. the chainage of each advance and shape of its face (inclined and domed), and, furthermore, the geometry of the unsupported heading, as well as the geometry of every layer of sprayed concrete. During excavation of the profile of the heading and during the spraying process, the thickness of each layer of sprayed concrete was monitored continuously and easily in real time. All as-built profiles were stored and could be used for quality control purposes and later evidence as required.

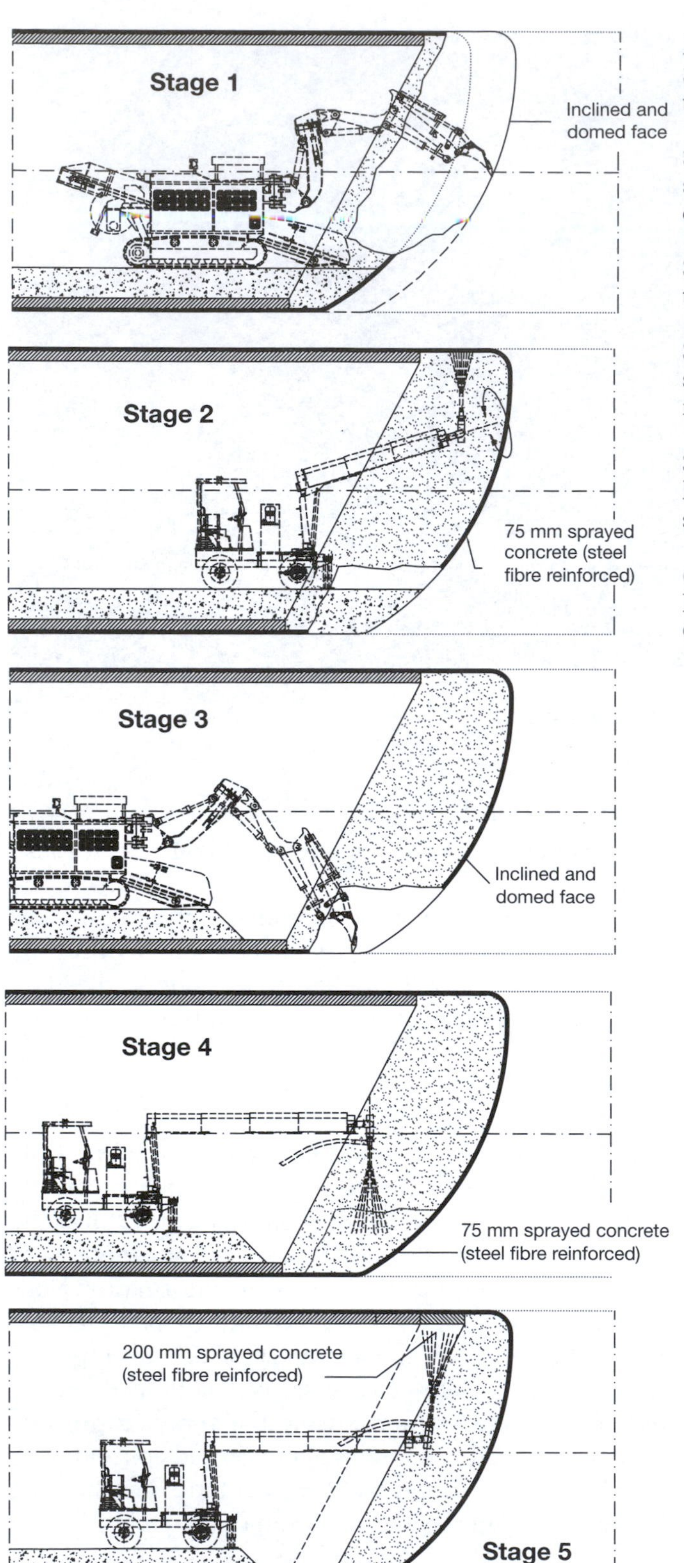

Figure 8.11
Stage 1: excavation of the crown and bench;

Stage 2: initial layer of sprayed concrete at the crown and bench;

Stage 3: excavation and trimming of the invert;

Stage 4: initial layer of sprayed concrete at the invert;

Stage 5: structural lining (permanent support)

(courtesy of ALPINE BeMo Tunnelling GmbH Innsbruck)

8.2.7 Monitoring

8.2.7.1 Existing Piccadilly Tunnel Eastside

On the Eastside of the PiccEx Junction the existing Piccadilly Line tunnels had to be enlarged over a length of 53 m (see Figure 8.9 for the location). The first half of the existing Piccadilly Eastbound was plugged with foam concrete for stability reasons during construction of the enlargement, with the second half remaining unplugged during excavation. This gave the unique opportunity to monitor the unplugged area ahead of the SCL-face once the plugged area had been left behind (Figure 8.12). The information obtained from monitoring the existing tunnel helped to determine the amount of stress re-distribution ahead of the current SCL heading and thus to assess the displacements already acting on the sprayed concrete lining before the base reading could be done.

It was possible to take readings up until the approaching heading was within just one width of a ring, i.e. 600 mm. It was discovered that there was no evidence of any deformation ahead of the face. Two conclusions could be drawn from this surprising result:

1. It confirmed previous observations that displacements in advance of the heading are small or negligible. Although it should be noted that this is not a generic statement for any tunnel construction in London Clay, and only applies to the particular conditions at the PiccEx Junction.
2. Under normal circumstances there is a delay between applying the sprayed concrete, installing the monitoring array, and taking the base reading. During this period the freshly applied sprayed concrete is already loaded and displaced by the stress re-distribution around the heading. Due to this inevitable delay in taking the base reading, this pre-displacement cannot be read and this information is lost. With the knowledge of the results discussed above, it was clear that this unreadable pre-displacement must have been very small. Therefore, the ordinary monitoring provided an almost complete picture of the total displacement and hence of the load acting upon the sprayed concrete lining. Again, it should be pointed out that this is not a general statement for any tunnel construction in London Clay, and only applies to these particular conditions at the PiccEx Junction.

8.2.7.2 Existing Piccadilly Tunnel Westside

Before this particular part of the monitoring is described, some general information on the existing Piccadilly Line tunnels should be given. The lining of the existing Piccadilly Eastbound tunnel was made of spheroidal graphite (cast) iron segments (SGI-segments). The joints between the segments were filled with plywood, or similar, to cope with curves and changes in gradient.

Figure 8.12 View into the unplugged existing Piccadilly Line Eastbound Turnout tunnel

No gaskets had been used to prevent water from infiltrating through the joints. The London Clay at T5 in general is considered to be practically impermeable, however water did somehow find its way down to the tunnels and through the joints. As a result plenty of stalactites were growing on the lining and rusting of the iron segments was also visible (see Figure 8.13a, left-hand side and Figure 8.13b for the detail).

The distance between the existing Piccadilly tunnels and the SCL Stub tunnels was very small (Figure 8.9, Figure 8.14). The remaining ground pillar between the tunnels varied from 0.4 m at the box to approximately 3 m at the end.

The integrity of the existing structures during the SCL work had to be ensured, therefore monitoring of the existing Piccadilly Line tunnels was essential. Of particular interest was the performance of the existing Piccadilly Line Eastbound Tunnel during the excavation of the SCL Eastbound Stub Tunnel, which was the first of the stub tunnels to be constructed.

The existing Piccadilly Line Eastbound was approximately 20 m long (before it merged with the existing Piccadilly Line Westbound, Figures 8.9 and 8.13a), of which the first 8 m, measured from the box, were plugged with foam concrete to enhance the stability during the excavation of the adjacent SCL Eastbound Stub. In the remaining unplugged part four monitoring arrays were installed. Access was possible through the neighbouring Piccadilly Line Westbound Tunnel.

The existing Piccadilly Line Eastbound Tunnel experienced 3 to 5 mm vertical displacement in the crown, and 5 to 7 mm horizontal displacement at axis level directed towards the SCL Eastbound Stub, indicating a clear ovalization. All displacements were limited to the crown and the side of the lining where the SCL Eastbound Stub passed by (Figure 8.15).

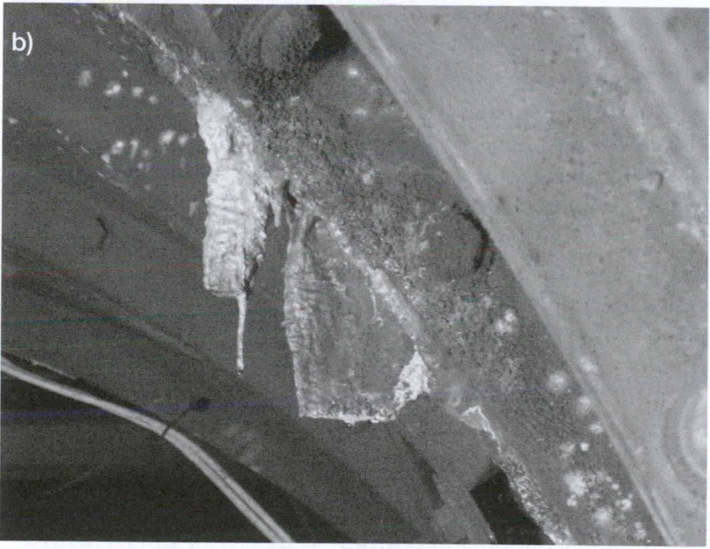

Figure 8.13 a) Existing Piccadilly Line tunnels, SGI-Segments (LHS), Concrete Segments (RHS), b) existing Piccadilly Line Eastbound Tunnel, stalactites on the SGI-Segments

Figure 8.14 (LHS) existing Piccadilly Line Tunnel, (RHS) SCL Eastbound Stub Tunnel

Significant displacements could only be observed approximately 2 m ahead of the current SCL face and continued 4 to 5 m after the SCL face had passed the relevant monitoring array (equivalent to half a tunnel diameter in advance and one tunnel diameter after passing the monitoring array, or in terms of time equivalent to approximately three days altogether).

Although the overall displacements in the existing Piccadilly Line Eastbound Tunnel were 2 mm smaller than anticipated in the design (Jäger and Stärk 2007), the gradient appeared steeper, creating a larger longitudinal 'bending' of the SGI lining towards the SCL Stub Tunnel. There was no concern for the stability of the SGI lining as all the segments were bolted together to form rigid and robust rings. However, this information was important with respect to the stability of the adjacent tunnel, the existing Piccadilly Line Westbound Tunnel, during the upcoming excavation of the SCL Westbound Stub. As shown in Figure 8.13a the existing Piccadilly Line Westbound Tunnel was supported by expanded lining. Precast concrete segments form rings which were not bolted, but were held in place only by radial forces generated by the keystone and ground pressure, and longitudinal forces generated by the TBM rams during tunnel construction. This technique normally works perfectly well. However, for whatever reason in this case

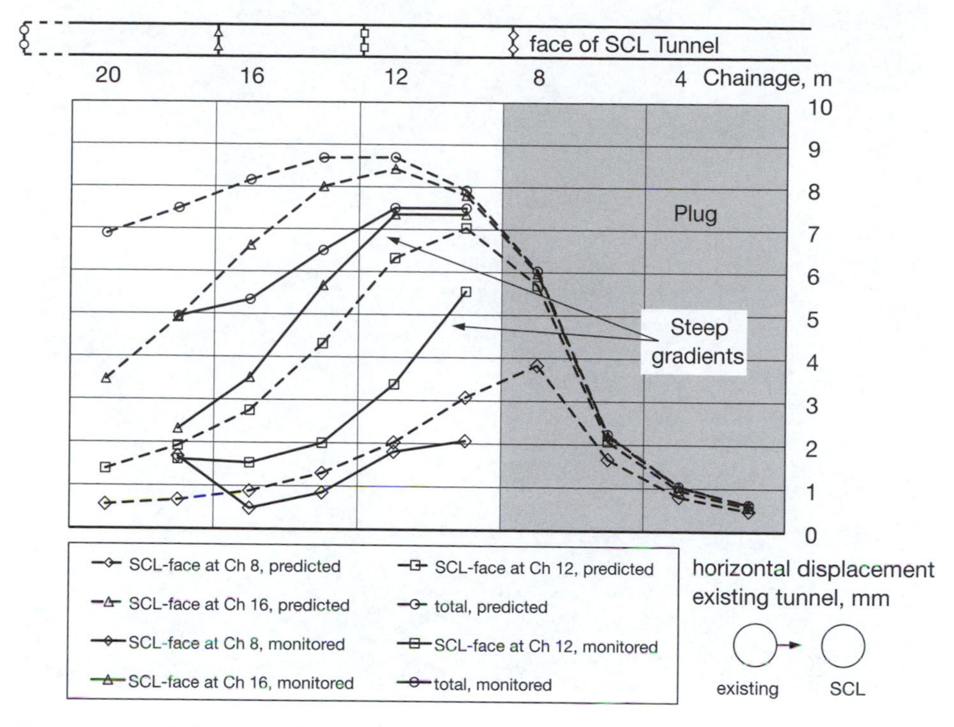

Figure 8.15 Predicted and measured horizontal displacements in the existing Piccadilly Line Eastbound Tunnel

the precast concrete segments did not form an even surface, but showed irregular steps between the joints of the precast concrete segments of up to approximately 60 mm (Figure 8.16).

Knowing the future development of the displacements as shown in Figure 8.15, it was quite clear that the concrete segments would experience additional forces during construction of the SCL Westbound Stub causing the joints to open wider. The question was how this would affect the stability of the concrete segments? In fact, nobody wanted to find out and it was decided to extend the foam concrete plug in the existing Piccadilly Line Westbound Tunnel to the full length (instead of only 8 m). Thus any additional movement of the expanded lining and any further opening of the joints were prevented. This example shows that the overall magnitude of relatively small displacements is not important, but it is their effect on the ground and structures.

The civil engineering and tunnelling works at PiccEx Junction were completed successfully in 2006. The Laser Shell™ tunnelling method is now commonly used in the UK. (Note: LaserShell™ and TunnelBeamer™ are registered trademarks by Beton- und Monierbau and Morgan=Est. The TunnelBeamer™ has been patented by Beton- und Monierbau and Morgan=Est.)

Figure 8.16 Existing Piccadilly Line Westbound Tunnel showing steps of up to 60 mm between concrete segments

8.3 Lainzer Tunnel LT31, Vienna, Austria

The Lainzer Tunnel Lot 31 (LT31) is a large railway tunnel in shallow soft ground beneath a densely built up urban area. This section describes the construction of 3 km of side wall drift, and highlights fundamental design issues as well as essential aspects of the complex monitoring regime.

8.3.1 Project overview

Today's rail traffic runs overground through Vienna, and rail traffic is quite heavy. The noisy freight trains are especially likely to disturb people's sleep at night in this densely built up urban area. This will change when most of the trains start running through the Lainzer Tunnel, which will be opened at the end of 2012 by the Federal Austrian Railroad.

The Lainzer Tunnel is 12.3 km long. Due to its length and changing geological conditions, the project has been divided into different 'lots'. LT31 forms, together with the neighbouring LT33, the core of the Lainzer Tunnel, i.e. the 6.5 km long Connection Tunnel. A significant change from soft ground to hard rock divides the Connection Tunnel into two sections of nearly the same length, which separated LT31 (soft ground) from LT33

Table 8.2 Overview of headings

Shaft	*Section*	*Length*	*Excavation method*	*Geology*	*Direction of excavation*
Lainzer Str.	W_{new}	595 m	Crown/Bench/Inv.	Hard rock	LT33
	W	790 m	Side wall drift	Soft ground	LT33
	P	596 m	Side wall drift	Soft ground	Klimtgasse
Klimtgasse	M	593 m	Side wall drift	Soft ground	Lainzer Str.
	S	1051 m	Side wall drift	Soft ground	LT44

(hard rock). In order to ensure a coordinated date for the opening of the Lainzer Tunnel in connection with the new railway line from Vienna to St. Pölten, the lot boundary was moved for the benefit of LT31 by 595 m. In addition to the original 3.05 km long soft ground section, LT31 was extended to include a 595 m long hard rock section. The soft ground was excavated completely by means of side wall drifts, which was possibly the longest side wall drift in the world at that time. The remaining 595 m in hard rock were excavated conventionally with crown/bench/invert using a roof pipe umbrella and drill and blast, respectively. This case history will focus on the construction and monitoring of the side wall drift section.

Tunnelling started in October 2006 from two 30 m deep mucking and delivery shafts, 'Lainzer Straße' and 'Klimtgasse', in two directions, each resulting in four headings excavated simultaneously. Section 'S' connected LT31 to LT44 in the East, section 'W' and its hard rock extension 'W_{new}' connected to LT33 in the West, and sections 'M' and 'P' met in the middle. Table 8.2 gives an overview of this.

The breakthrough between sections P and M was in September 2008, with the breakthrough to LT44 in December 2008, and the excavation of section W_{new} completed in May 2009. Figure 8.17 depicts a bird's eye view of LT31 with direction of sight to the West.

The Lainzer Tunnel is designed with emergency exits approximately every 500 m, seven of which are within LT31 (see Table 8.3 for details). Both of the existing mucking and delivery shafts were to be converted into emergency exits, and the remaining five had to be newly constructed. At each emergency exit a shaft provides a vertical access down to the main tunnel level. Galleries connect the shafts with the main tunnel. The shafts have a diameter of 9.4 m and a depth ranging from 20 m to 55 m. The connecting galleries have cross sections of 25 m^2 to 30 m^2 and they are 20 m to 258 m long. On the opposite side of the five new emergency exits 8 m long transformer niches had to be constructed. Furthermore, at the lowest level of the tunnel an 8 m deep sump was excavated to collect water flowing into the main tunnel during operation. This water will be pumped out through the nearest emergency exit, Jagdschlossgasse.

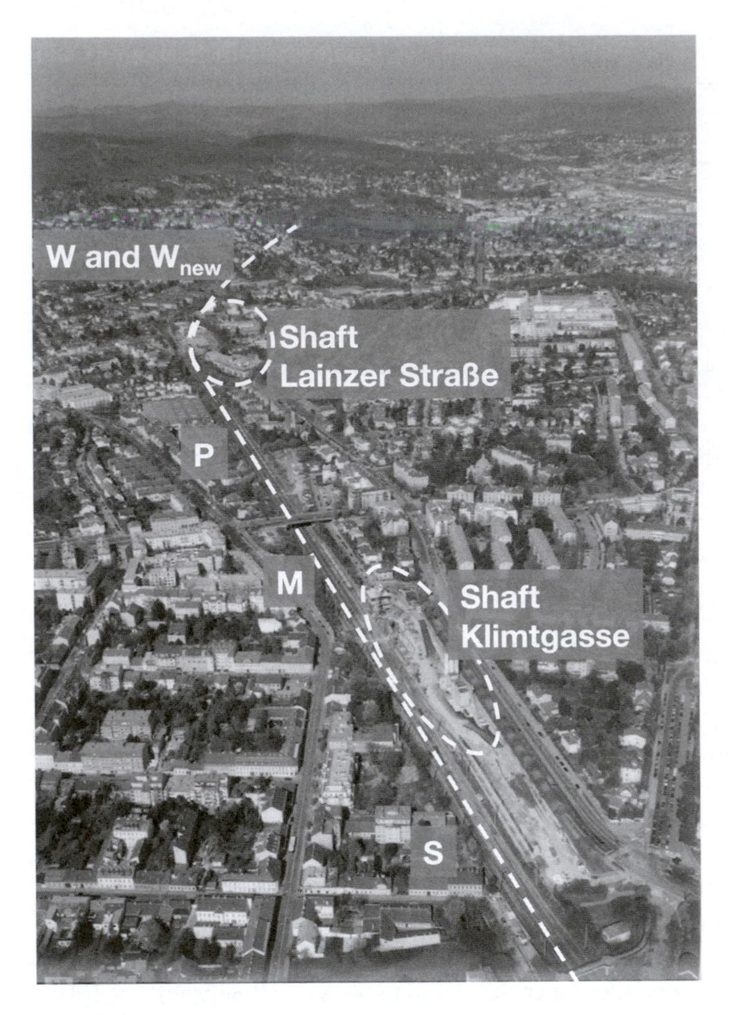

Figure 8.17 Project overview of LT31

Table 8.3 Overview of emergency exits

Emergency exit	*Section*	*Shaft depth*	*Gallery length*	*Geology*
Veittingergasse	W_{new}	55 m	258 m	Hard rock
Jagdschlossgasse	W	35 m	67 m	Soft ground
Lainzer Straße	–	30 m	Conversion of shaft	
Himmelbaurgasse	P	30 m	27 m	Soft ground
Schönbachstraße	M	32 m	32 m	Soft ground
Klimtgasse	–	30 m	Conversion of shaft	
Schlöglgasse	S	23 m	50 m	Soft ground
5 transformer niches	Same as exit	–	8 m	Same as exit
Sump Waldvogelstraße	W	8 m	11 m	Soft ground

8.3.2 Geology

The soft ground (sections W, P, M, S) was dominated by alternating layers of silt/clay, sand, gravel, and wide graded sediments. Except for the silt/clay layers the ground is permeable with the water table above the tunnel roof. Groundwater lowering was necessary by means of wells from the ground surface well ahead of the leading excavation face to avoid stability problems at the face. The groundwater layer was often enclosed by silt/clay layers resulting in confined groundwater aquifers. Rigid conglomerates and layers of hard sandstone, up to 3 m thick, were embedded between the soft ground layers. The hard rock formations 'Flysch' (section W_{new}) were of varying quality from extremely poor to fair. At the transition from soft ground to hard rock a 300 m long section of lower ground quality was stabilized by roof pipe umbrellas. The overburden extended from 6 to 26 m in the soft ground, and from 26 to 66 m in the hard rock formation.

8.3.3 Starting construction from the shafts

Both shafts, 'Lainzer Straße' and 'Klimtgasse', consisted of an open oval section supported by sprayed concrete and a rectangular section with the live railway on top. The open oval section was used for mucking and delivery, while the excavation started from the rectangular section. The rectangular section was supported by a bored pile wall and four levels of bracing, the lowest of which ran right through the tunnel profile (Figures 8.18a and 8.21). The bracing was not allowed to be dismantled all at once, but only in sections with the advancing excavation. Since a tunnel construction starts with the top heading this was a problem, because it was not known exactly how the excavator could reach the top heading while the bracing was still in place. It was decided to set up a platform above the lowermost bracing. This kind of platform was a challenge because there was no reference on how to design a platform for dynamic loads of heavy excavators. Shoring towers used for formwork seemed to be the most robust support for the platform. The shoring towers were rigidly attached to the walls. On top of the shoring towers solid web girders were laid close together, which were covered by a two layer crisscross nailed up planking (Figure 8.18b). The planking protected the girders against damage (surface wear and tear) and acted statically like a rigid disc, helping to distribute the load equally onto the girders and towers.

From the platform, the crown sections were excavated to a length of 5 m to both sides of the rectangular shaft, and then the platform was dismantled (Figure 8.19). The excavation of the bench and invert of the leading site wall drift could also start after the bracing was partly dismantled. Figure 8.20 shows the tunnel after the total cross section was constructed over a length of approximately 30 m, giving space in the shaft area to manoeuvre the plant.

8.3.4 Side wall drift section: excavation sequence and cross section

Approximately two-thirds of the 3.05 km side wall drift section was excavated beneath the existing railroad, which was still subjected to the regular heavy rail traffic; the rest was located under buildings and streets. Therefore the need for a robust excavation method with low subsidence

Figure 8.18
a) Shaft 'Klimtgasse'. Excavation platform with planking nearly finished, b) excavation platform. Shoring towers in place, web girders are laid close together and covered in planking

Figure 8.19 Crown excavated, and enlargement of leading side wall drift started. Remaining bracing still in place

Figure 8.20 30 m of total cross section (on the RHS, the open oval section, the remainder of the lower most bracing can be seen still in place)

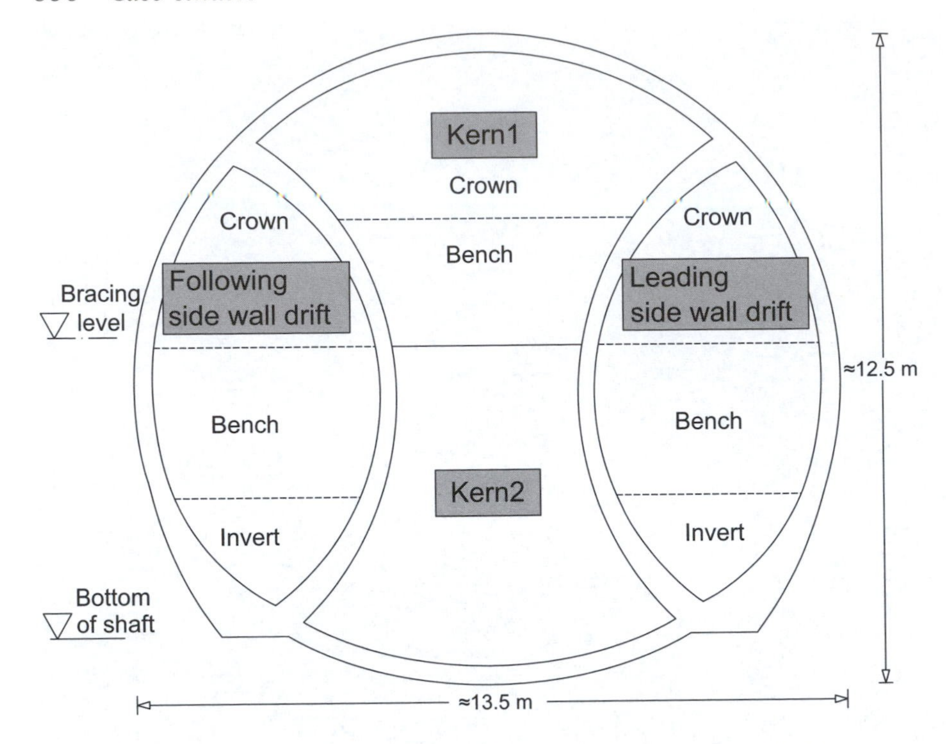

Figure 8.21 Cross section of the tunnel, including the two side wall drifts

was important. Figure 8.21 shows the cross section of the area with side wall drift excavation.

There were several reasons for choosing this design:

1. The soft ground has limited bearing capacity. With respect to railroad, buildings and service structures above the tunnel a stiff support was a paramount issue to control ground loss at the face, to avoid face or roof instability, and finally to minimize settlement at the ground surface. This could be achieved by dividing the tunnel into smaller headings and limiting the cross section of each heading.
2. Most disturbance of the original stress level in the ground had been expected when excavating the remaining top heading. To mitigate this effect, the roof of the side wall drifts were positioned so as to reduce the span of the remaining top heading.
3. Big footings at the invert of the side wall drifts, similar to elephants' feet, helped to avoid settlement and any inward orientated movement during excavation of the remaining top heading.
4. With respect to buildability, quality of construction joints and safety of construction, the external walls of the side wall drifts were integrated into the permanent sprayed concrete lining as much as possible, resulting in a high and slender shape.

5 The shape of the side wall drifts provided enough space for plant to excavate the remaining top heading and bench.
6 The high and slender side wall drifts were more sensitive to high horizontal loads (rather than more circular side wall drifts). This had to be considered in the design process and resulted in a sprayed concrete thickness of 30 cm for the internal wall and 35 cm of sprayed concrete for the circumferential outer walls, all girder supported and rebar reinforced.
7 The side wall drifts, approximately 9 m high, were divided into crown, bench and invert with a short distance for ring closure of at most 10 m. The remaining top heading was also divided into crown and bench sections.
8 The crown had to be opened in up to four sub cross sections and the bench in up to two sub cross sections.
9 The crown and bench support was accompanied by forepoling and face anchors.

Figure 8.22 shows a plan view of the excavation sequence. The minimum distance of each heading had to be 10 days or 20 m, respectively. The first reason for this was to let the sprayed concrete gain enough strength to cover the additional load of the following headings. The second reason was to allow the displacements of each heading to come to a halt before the following heading passed the relevant area. This was necessary to fully control the ground surface settlements with respect to third party structures. In addition, the leading side wall drift allowed for dewatering of any residual groundwater making the excavation safer and easier for the following headings.

With ongoing excavation it could be proven that stress redistribution within the less cohesive/non-cohesive soil (gravel and sandy layers) was only 10 m. In combination with an achieved high compressive strength for the sprayed concrete (the required 28-days-values were already reached after seven days), the distance between heading faces could be reduced to five

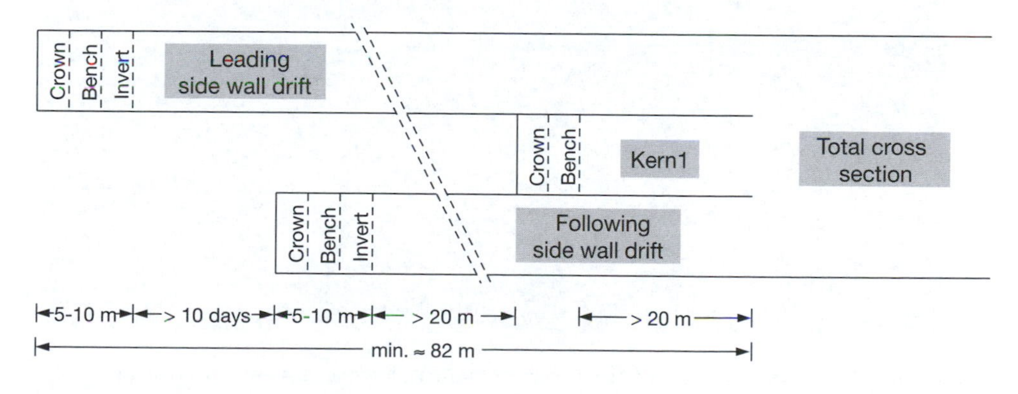

Figure 8.22 Plan view of the excavation sequence

days or 10 m, respectively. The stress redistribution in the cohesive silt/clay layers took longer due to creep effects; so they had to adhere to the original designed sequence.

Originally, a simultaneous excavation of the side wall drifts and Kern1 was not allowed for safety reasons. The risk assessment identified an overstressing of the inner walls of the side wall drifts during the excavation of Kern1 as a possible hazard. This could result in a collapse of the side wall drifts with the miners trapped at the face. The analysis of in-tunnel monitoring during construction, however, proved no adverse effects on the stability of the side walls. After a reasonable observation period of approximately six months the simultaneous excavation of both side wall drifts and Kern1 was permitted. As a precaution the in-tunnel monitoring was intensified and the face had to be opened in smaller sub-areas, respectively. Altogether the excavation was now much quicker. In-tunnel monitoring is usually used to identify adverse developments. However, in this case it helped to improve the performance, while keeping the safety to the same high level.

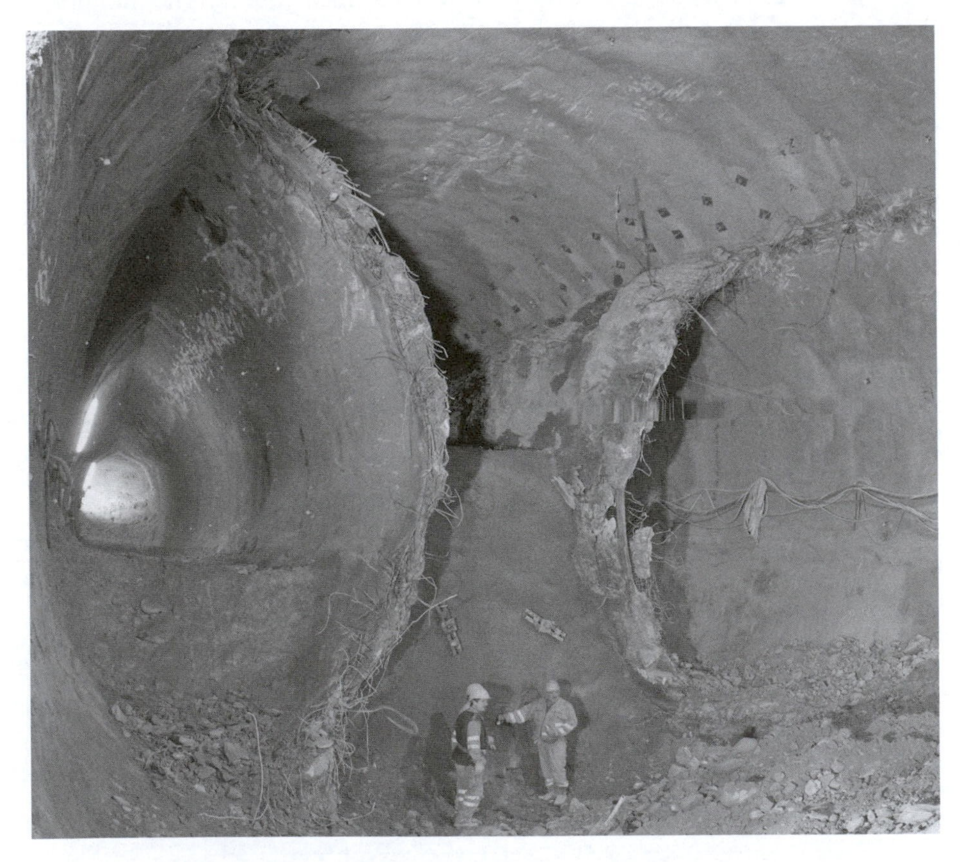

Figure 8.23 Construction of the invert at the total cross section (excavation Kern2 and demolition of inner walls of side wall drift)

A simultaneous excavation of the side wall drifts and Kern2 was still not allowed, but this was never the aim for buildability reasons. The excavation of Kern2 included the ring closure at the total cross section and the dismantling of the inner walls of the side wall drifts. During this process access to Kern1 was not possible and only with difficulty to the side wall drifts (see Figure 8.23). Therefore, these excavations had to be suspended. From a practical point of view, a compromise had to be found in such a way that on the one hand driving cycles into the side wall drifts would not increase too much, and on the other hand that the time consuming preparation for the excavation of Kern2 was kept to a minimum. It turned out that changing to Kern2 every 30 m to 60 m was the best option.

8.3.5 Monitoring of the sprayed concrete lining of the side wall drift section

During the design process a geotechnical safety management concept was established, which was a live document that was continuously revised during construction (Heissenberger *et al.* 2008). According to this concept the regular distance of monitoring cross sections was 10 m throughout LT31 and if necessary this was reduced to 5 m. Readings had to be taken 20 m ahead and 30 m behind the face on a daily basis. In consideration of the distance between faces, as shown in Figure 8.22, an area of 100 m to 140 m in each of the sections S, M, P, and W had to be monitored by means of displacement measurements. Additionally, measurements had to be taken during construction of all emergency shafts and galleries as well as the adjacent areas of the main tunnel. Since LT31 was situated almost completely under a live railway and other urban infrastructure a large number of surface surveying points had to be monitored. This section highlights some special features, however further information on the in-tunnel monitoring can be found in Moritz *et al.* (2008).

8.3.6 Cracks in the sprayed concrete lining

Cracks were detected for the first time in section W between Chainage (Ch.) 50 and 60 at the inner walls of both side wall drifts. The horizontal cracks occurred at the intrados approximately 0.5 m to 1.0 m above the intersection between the bench and invert (Figure 8.24). The ring closure of the total cross section, including dismantling of the inner walls of the side wall drifts, was completed up to Ch. 50, i.e. the cracks ran in the remaining inner walls in the direction of the face. The width of the cracks was up to several millimetres in some areas. From the experience of former projects (e.g. Eggetunnel, railway line Kassel-Dortmund) the development of cracks had been expected, but at the extrados in the area of the crown. The location and the extent of the cracks were therefore irritating and due to the sensitive urban area a detailed investigation was done. In order to

Figure 8.24 Crack in the inner wall of the side wall drift, section W, Chainage 52–60

check the integrity of the sprayed concrete lining, three cores were taken out of the inner walls of section W, Ch. 52 to 54, and these are shown in Figure 8.25. The cores had a diameter of 160 mm and a length of 37 cm to 42 cm. Only with specimen No. 1 was the sprayed concrete lining drilled through completely, i.e. the overall thickness of the inner walls was comfortingly greater than the required 30 cm. Specimens No. 1 and No. 2 were broken at the construction joint between the first and second layer of sprayed concrete; based on this fact the manufacturing of the second layer of sprayed concrete was improved immediately. Specimen No. 3 got jammed in the core-barrel and had to be drilled out, leading to it breaking in a couple of places. Nevertheless, the original crack in this specimen could still be identified. The straight nature and the opening of the cracks towards the intrados indicated flexural tension as the most possible cause of all the cracks.

The monitoring data confirmed the visual observations. During excavation of the side wall drifts the most significant displacement was a convergence between monitoring points 10 and 4, and 11 and 5 in the crown of the side wall drifts. During excavation of the following Kern1 the direction of the displacements changed in crown-points 10 and 11 resulting in a clear divergence developing between points 10 and 4, and 11 and 5

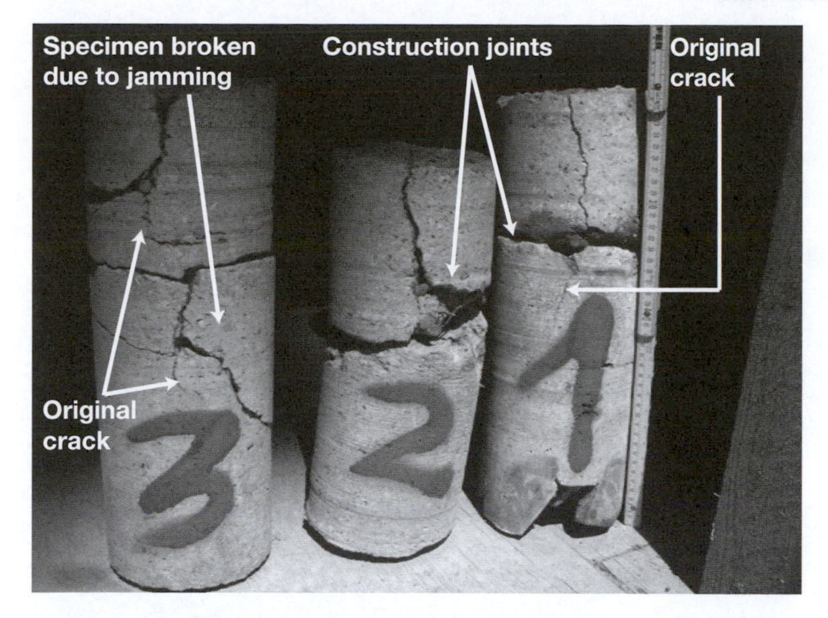

Figure 8.25 Cores, section W, Chainage 52–54

(Figure 8.26). The explanation is quite clear: with the excavation of Kern1 the bedding of the inner walls of the side wall drifts was taken away and the inner walls moved in the freshly excavated open space. This behaviour, although not in this magnitude, was well known from previous projects as mentioned above, with only the expected cracks at the extrados around

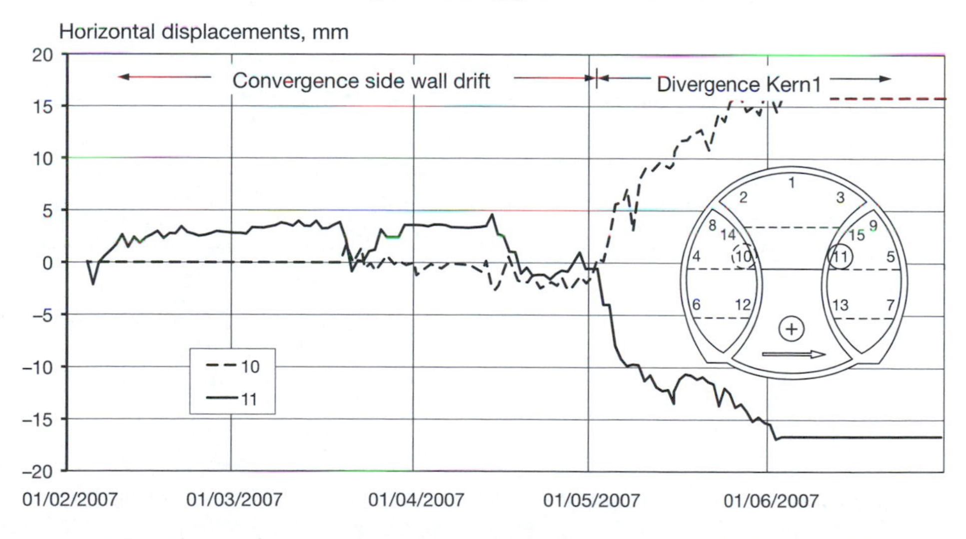

Figure 8.26 Change from convergence to divergence, section W, Chainage 60

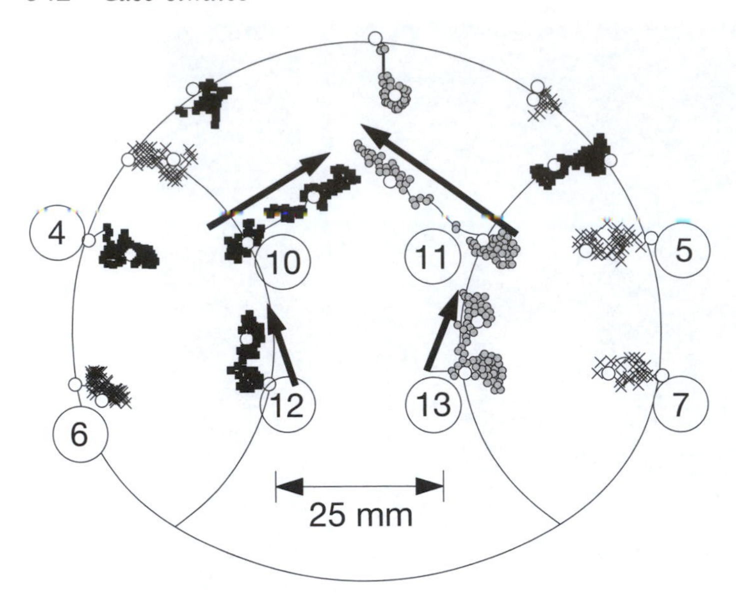

Figure 8.27 Displacements of side wall drift during excavation of Kern1, section W, Chainage 60

points 10 and 11 not being found. Instead, cracks at the intrados developed (as shown in Figure 8.24). The reason for this was that points 10 and 11 also showed a heave, which caused an unexpected heave of approximately the same amount in the bench-points 12 and 13. Figure 8.27 shows qualitatively how the points moved. The movement generated a negative

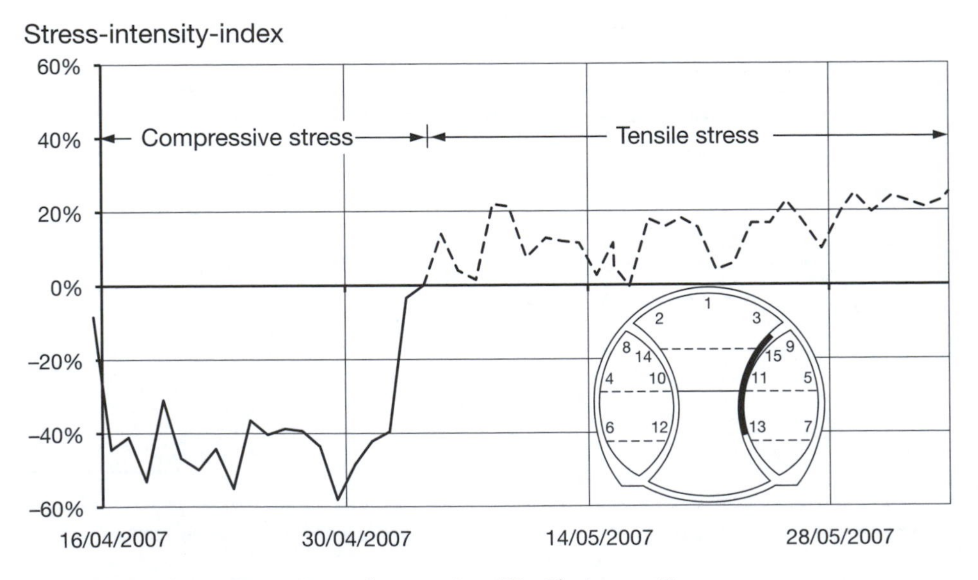

Figure 8.28 Stress-intensity-index, section W, Chainage 60

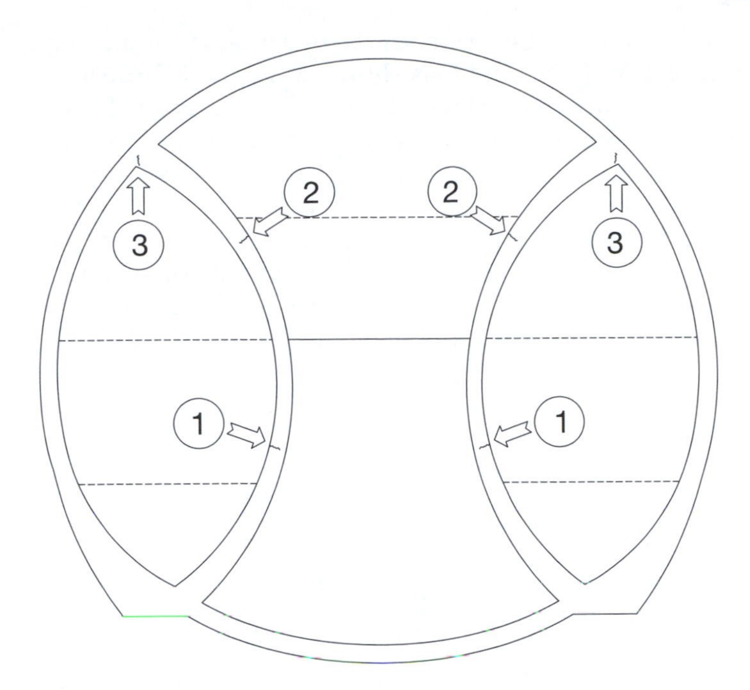

Figure 8.29 Crack pattern occurring in the walls of the side drifts

bending moment around points 10 and 11 (tensile stress at the extrados), and a positive bending moment/elongation around points 12 and 13. The sprayed concrete was, at this time, some weeks old and already hardened. Stress redistribution inside the sprayed concrete, e.g. due to creeping effects, was negligible. The movements therefore immediately caused an increase in stress inside the sprayed concrete, which was made visible by the cracks.

This was confirmed by the stress-intensity-index (see section 7.3.4.6) as shown in Figure 8.28 for the inner wall of the right-hand side wall drift at Chainage 60 (see marking in Figure 8.28 between points 13–11–15). With the face of Kern1 passing the monitoring cross section at Chainage 60, the stress changes from a compressive stress (negative sign) to a tensile stress (positive sign) and leads to the cracks. The stress-intensity-index shows a tensile stress of only about 20% after the development of the cracks. The safety factor against failure (η = 100%) was still around five. This gave the certainty that the tunnel was in a very stable situation. With the knowledge of the stress-intensity-index another conclusion could be made: Due to the cracks and the low stress level in the inner wall, most of the load had been redistributed into the outer walls of the side wall drifts. This effect was considered desirable with respect to the later demolition of the inner walls. Demolition was safer and easier with unloaded inner walls. Overall, the development of cracks in the sprayed concrete lining was beneficial in this case.

During further excavation, a uniform crack pattern developed at the same time in all four sections W, P, M, and S as shown in Figure 8.29. All the cracks ran towards the face over the complete length of the excavated Kern1. This confirmed the experience from earlier projects, although only the cracks at the extrados, marked with '2', had been expected. However, the other unexpected cracks at the intrados, marked with '1' and '3' were a logical consequence of the displacements according to Figures 8.26 and 8.27. Due to the rough surface of the sprayed concrete lots of soil stuck to it and the expected cracks '2' were very hard to detect even for experienced eyes, leading to the previously mentioned issue.

Appendix A

Further information on rock mass classification systems

A.1 Rock Mass Rating

Brief details of this rock mass classification system are provided in section 2.4.4.2, with further information provided in this appendix. Table A.1 shows the classification parameters used.

In section A of Table A.1, five parameters are grouped into five ranges of values. As these parameters are not equally important for the overall classification of a rock mass, importance ratings are allocated to the different value ranges of the parameters. A higher rating indicates a better rock mass condition. The ratings for the strength of the intact rock, RQD and discontinuity spacing can be interpolated between the values indicated in the Table A.1 and these are shown in Figures A.1a to c, respectively. If either RQD or discontinuity data are lacking, then Figure A.1d can be used.

Once the ratings for the five parameters in section A of Table A.1 have been established, these are summed to provide the basic Rock Mass Rating (RMR) for the area of the rock mass being considered. The next stage is to include the sixth parameter, i.e. the orientation of the discontinuities, by adjusting the basic RMR according to section B of Table A.1. With regard to tunnelling projects, further information on this section can be found in section F of Table A.1. After adjustment for discontinuity orientation, the rock mass is classified using section C of Table A.1, which groups the final (adjusted) RMR into five rock mass classes. This value varies from 0 to 100. Subsequently, section D of Table A.1 provides practical meaning to each rock mass class as it relates this to specific engineering problems. Section E of Table A.1 provides guidelines for classifying the discontinuity conditions.

Davis (2006) states that one of the important aspects of this method is the way the various parameters are derived. It is advisable to choose a 'best estimate' and a 'worst credible' case and assess these for each parameter. Davis (2006) has the following advice on deriving the various parameters:

- *Uniaxial compressive strength of the rock material* – This can be obtained from laboratory UCS testing or point load strength testing of samples. Descriptions of the borehole logs can be used if no test data are available.

Table A.1 Rock Mass Rating system (after Bieniawski 1989)

	Parameter		*Range of values*						
A. CLASSIFICATION PARAMETERS AND THEIR RATINGS									
1	Strength of intact rock material	Point-load strength index	> 10 MPa	4–10 MPa	2–4 MPa	1–2 MPa	For this low range – uniaxial compressive test is preferred		
		Uniaxial comp. strength	> 250 MPa	100–200 MPa	50–100 MPa	25–50 MPa	5–25 MPa	1–5 MPa	< 1 MPa
		Rating	15	12	7	4	2	1	0
2	Drill core quality	*RQD*	90–100%	75–90%	50–75%	25–50%	< 25%		
		Rating	20	17	13	8	3		
3	Spacing of discontinuities		> 2 m	0.6–2 m	200–600 mm	60–200 mm	< 60 mm		
		Rating	20	15	10	8	5		
4	Condition of discontinuities (see E)		Very rough surfaces; Not continuous; No separation; Unweathered wall rock	Slightly rough surfaces; Separation < 1 mm; Slightly weathered walls	Slightly rough surfaces; Separation < 1 mm; Highly weathered walls	Slickensided surfaces or Gouge <5mm thick or Separation 1–5 mm; Continuous	Soft gouge >5 mm thick or separation > 5 mm Continuous		
		Rating	30	25	20	10	0		
5	Ground-water	Inflow per 10 m tunnel length (l/m)	None	< 10	10–25	25–125	> 125		
		Joint water press/(Major principal σ)	0	< 0.1	0.1–0.2	0.2–0.5	> 0.5		
		General conditions	Completely dry	Damp	Wet	Dripping	Flowing		
		Rating	15	10	7	4	0		

B. RATING ADJUSTMENT FOR DISCONTINUITY ORIENTATIONS (see F)

Strike and dip orientations		Very favourable	Favourable	Fair	Unfavourable	Very unfavourable
Ratings	Tunnels and mines	0	–2	–5	–10	–12
	Foundations	0	–2	–7	–15	–25
	Slopes	0	–5	–25	–50	

Table A.1 *(continued)*

Parameter	*Range of values*				
C. ROCK MASS CLASSES DETERMINED FROM TOTAL RATINGS					
Rating	100 ← 81	80 ← 61	60 ← 41	40 ← 21	< 21
Class number	I	II	III	IV	V
Description	Very good rock	Good rock	Fair rock	Poor rock	Very poor rock
D. MEANING OF ROCK CLASSES					
Class number	I	II	III	IV	V
Average stand-up time	20 yrs for 15 m span	1 year for 10 m span	1 week for 5 m span	10 hrs for 2.5 m span	30 min for 1 m span
Cohesion of rock mass (kPa)	> 400	300–400	200–300	100–200	< 100
Friction angle of rock mass (deg)	> 45	35–45	25–35	15–25	< 15
E. GUIDELINES FOR CLASSIFICATION OF DISCONTINUITY CONDITIONS[a]					
Discontinuity length (persistence)	< 1 m	1–3 m	3–10 m	10–20 m	> 20 m
Rating	6	4	2	1	0
Separation (aperture) in (mm)	None	< 0.01	0.1–1.0	1–5	> 5
Rating	6	5	4	1	0
Roughness	Very rough	Rough	Slightly rough	Smooth	Slickensided
Rating	6	5	3	1	0
Infilling (gouge) in (mm)	None	Hard filling < 5	Hard filling >5	Soft filling < 5	Soft filling > 5
Rating	6	4	2	2	0
Weathering	Unweathered	Slightly weathered	Moderately weathered	Highly weathered	Decomposed
Rating	6	5	3	2	0

F. EFFECT OF DISCONTINUITY STRIKE AND DIP ORIENTATION IN TUNNELLING[b]

Strike perpendicular to tunnel axis		Strike parallel to tunnel axis	
Drive with dip Dip 45°–90° Very favourable	Drive with dip Dip 20°–45° Favourable	Dip 45°–90° Very unfavourable	Dip 20° –45° Fair
Drive against dip Dip 45°–90° Fair	Drive against dip Dip 20°–45° Unfavourable	Dip 0°–20° – Irrespective of strike Fair	

Notes: (a) Some conditions are mutually exclusive. For example, if infilling is present, the roughness of the surface will be overshadowed by the influence of the gouge. In such cases use A4 of this table directly. (b) Modified after Wickham *et al.* (1972).

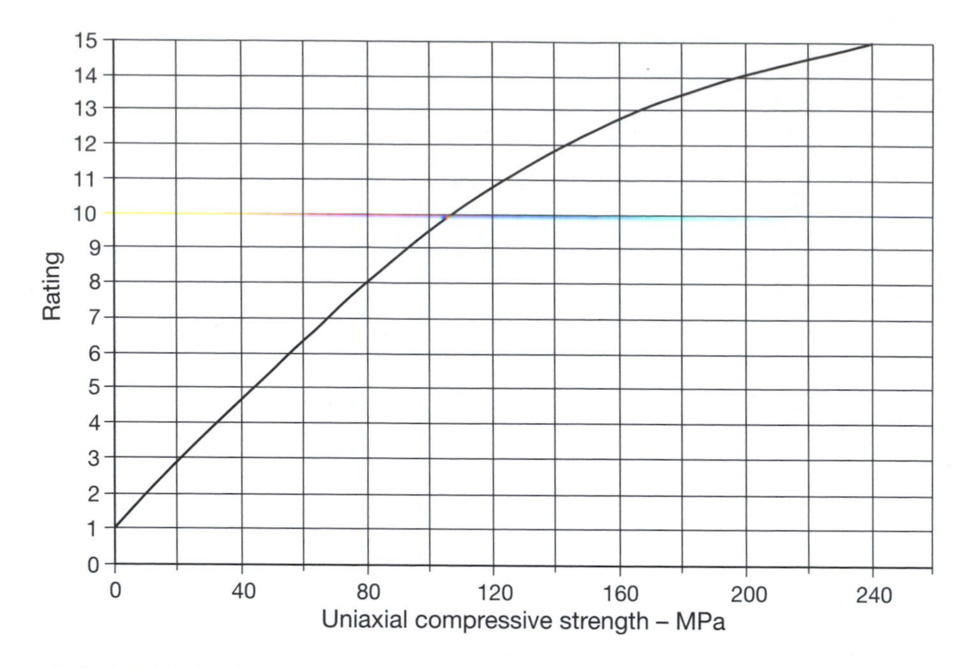

a) **CHART A Ratings for strength of intact rock**

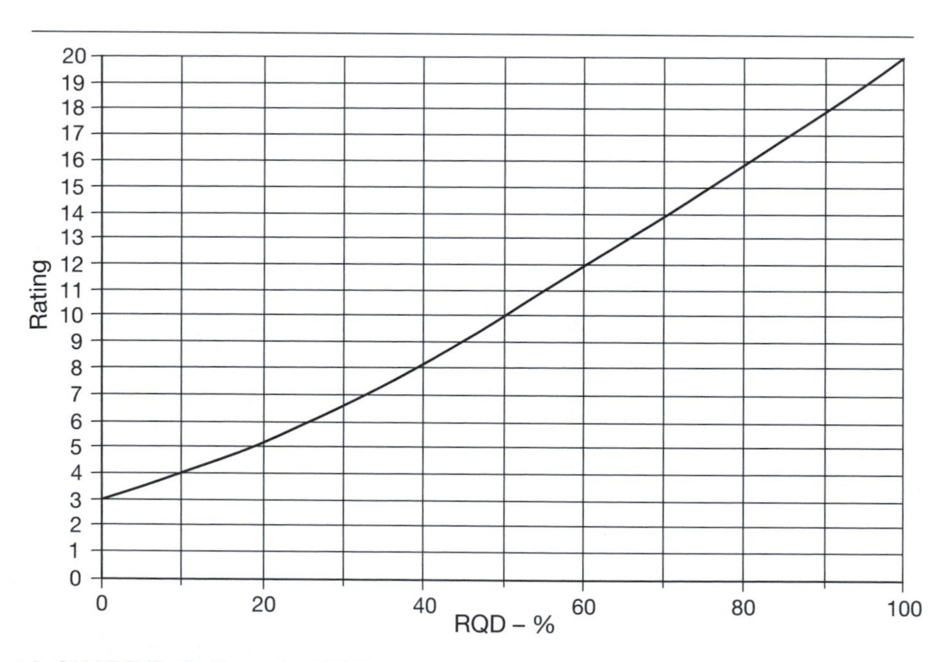

b) **CHART B Ratings for RQD**

Figure A.1 a) to d) Charts for various RMR ratings (after Bieniawski 1989)

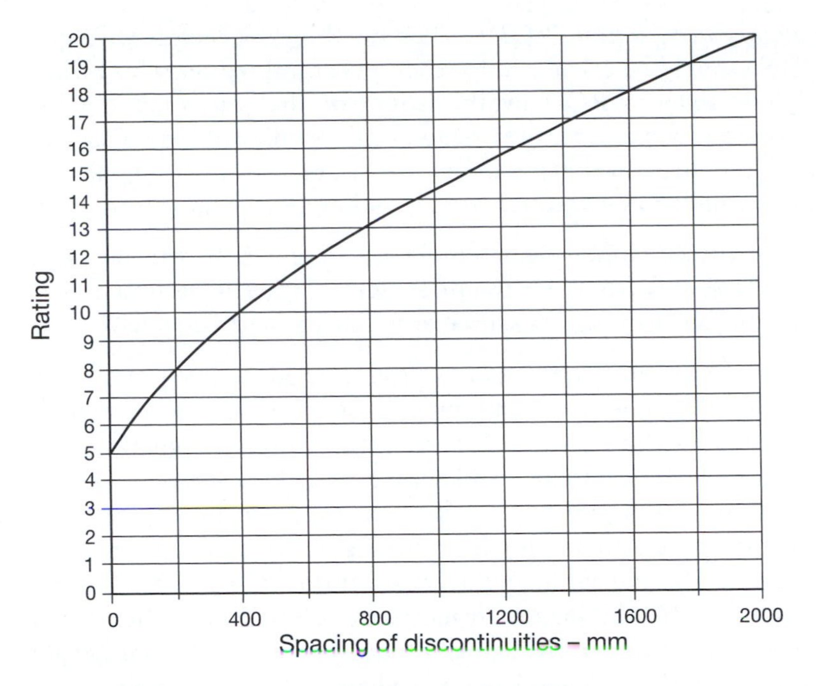

c) CHART C Ratings for discontinuity spacing

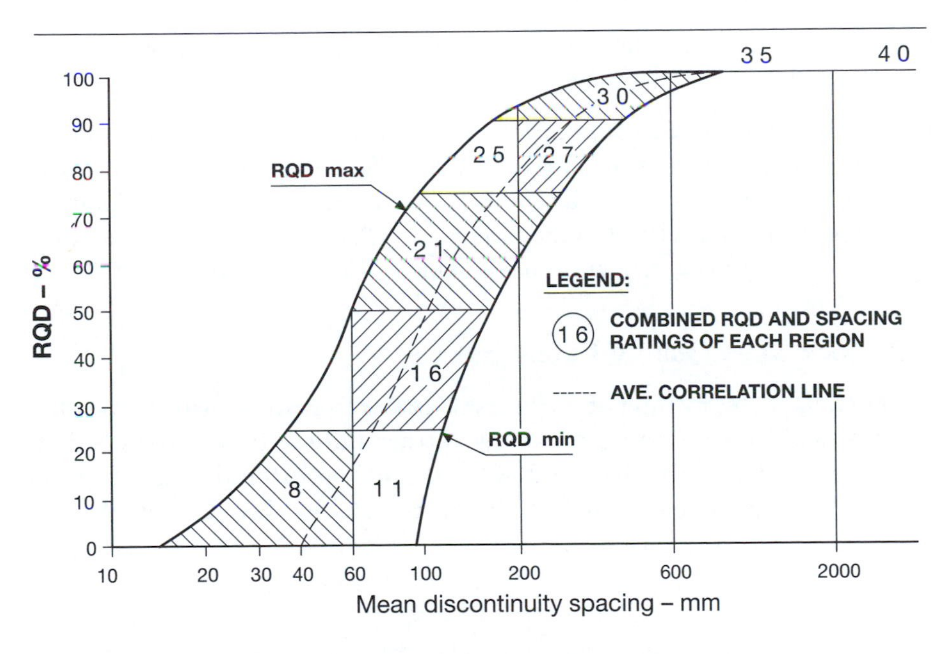

d) CHART D Correlation between RQD and discontinuity spacing

Figure A.1 (continued)

- *Rock quality designation (RQD)* – For boreholes, a length weighted mean RQD should be calculated for each structural region. This means multiplying each run RQD by the length of the run, summing the results for the whole structural region and dividing the sum by the length of the structural region. For exposure, or face logging, an assessment can be made directly from scan lines, or using Figure A.1d.
- *Spacing of discontinuities* – For boreholes, Figure A.1d can be used, or can be assessed from the fracture index if this is recorded. For exposures or face logging, measurements can be made directly.
- *Condition of discontinuities* – Davis (2006) suggests five parameters to assess the condition of discontinuities. These are persistence (length of the discontinuity in exposure), aperture (discontinuity separation or openness), roughness, infilling and weathering. Persistence – can be obtained from exposures, but not from cores. Aperture – cannot be obtained from cores, although where infill is present, the aperture can be assumed to be the infilling thickness. With both of these, if no information can be obtained then judgement should be made on the significance on the design. Roughness – can be obtained from logged discontinuities for a structural region, where these are not available, but summary descriptions are available for each joint set, the summary term can be used to derive a rating.
- *Groundwater conditions* – This can be obtained from piezometer data along the tunnel alignment.
- *Orientation of discontinuities* – Possibly available before tunnel construction via an orientated core, downhole logging or exposure logging. The dip direction of discontinuities can be plotted on stereographic projections and related to the tunnel axis. Where no data are available, Davis (2006) suggests adopting a 'fair' rating as a best estimate.

A.2 Rock Mass Quality Rating (Q)

Brief details of this method of rock mass classification are provided in section 2.4.4.3, with this appendix providing further information. Tables A.2 to A.7 (Barton *et al.* 1974, Barton 2000 and 2002) provide the classification of individual parameters used to obtain the Rock Mass Quality Rating value, *Q*, for a rock mass.

Table A.2 Rock quality designation (after Barton 2002)

		RQD (%)
A	Very poor	0–25
B	Poor	25–50
C	Fair	50–75
D	Good	75–90
E	Excellent	90–100

Notes: (i) Where RQD is reported or measured as ≤10 (including 0), a nominal value of 10 is used to evaluate Q. (ii) RQD intervals of 5, i.e.100, 95, 90, etc, are sufficiently accurate.

Table A.3 Joint set number (after Barton 2002)

		J_n
A	Massive, no or few joints	0.5–1
B	One joint set	2
C	One joint set plus random joints	3
D	Two joint sets	4
E	Two joint sets plus random joints	6
F	Three joint sets	9
G	Three joints sets plus random joints	12
H	Four or more joint sets, random, heavily jointed, 'sugar cube', etc	15
J	Crushed rock, earthlike	20

Notes: (i) For tunnel intersections, use ($3.0 \times J_n$). (ii) For portals use ($2.0 \times J_n$).

Table A.4 Joint roughness number (after Barton 2002)

		J_r
(a) Rock-wall contact, and (b) rock-wall contact before 10 cm shear		
A	Discontinuous joints	4.0
B	Rough or irregular, undulating	3.0
C	Smooth, undulating	2.0
D	Slickensided, undulating	1.5
E	Rough or irregular, planar	1.5
F	Smooth, planar	1.0
G	Slickensided, planar	0.5
(c) *No rock-wall contact when sheared*		
H	Zone containing clay minerals thick enough to prevent rock-wall contact	1.0
J	Sandy, gravely or crushed zone thick enough to prevent rock-wall contact	1.0

Notes: (i) Descriptions refer to small-scale features and intermediate-scale features, in that order. (ii) Add 1.0 if the mean spacing of the relevant joint set is greater than 3 m. (iii) J_r = 0.5 can be used for planar, slickensided joints having lineations, provided that the lineations are orientated for minimum strength. (iv) J_r and J_a classification is applied to the joint set or discontinuity that is least favourable for stability both from the point of view of orientation and shear resistance, τ, where $\tau \approx \sigma_n \tan^{-1}(J_r/J_a)$.

Table A.5 Joint alteration number (after Barton 2002)

		ϕ_r *approx. (deg)*	J_a
(a) Rock-wall contact (no mineral fillings, only coatings)			
A	Tightly healed, hard, non-softening, impermeable filling, i.e. quartz or epidote	—	0.75
B	Unaltered joint walls, surface staining only	25–35	1.0
C	Slightly altered joint walls, non-softening, mineral coatings, sandy particles, clay-free disintegrated rock, etc	25–30	2.0
D	Silty- or sandy-clay coatings, small clay fraction (non-softening)	20–25	3.0
E	Softening or low friction clay mineral coatings, i.e. kaolinite or mica. Also chlorite, talc, gypsum, graphite, etc., and small quantities of swelling clays	8–16	4.0
(b) Rock-wall contact before 10 cm shear (thin mineral fillings)			
F	Sandy particles, clay-free disintegrating rock, etc	25–30	4.0
G	Strongly over-consolidated non-softening clay mineral fillings (continuous but < 5 mm thickness)	16–24	6.0
H	Medium or low over-consolidation, softening, clay mineral fillings (continuous, but < 5 mm thickness)	12–16	8.0
J	Swelling-clay fillings, i.e. montmorillonite (continuous but > 5 mm thickness. Value of J_a depends on percent of swelling clay-sized particles, and access to water, etc.	6–12	8–12
(c) No rock-wall contact when sheared (thick mineral fillings)			
KLM	Zones or bands of disintegrated or crushed rock and clay	6–24	6, 8, or 8–12
N	Zones or bands of silty- or sandy-clay, small clay fraction (non-softening)	—	5.0
OPR	Thick, continuous zones or bands of clay (see G, H, J, for description of clay condition)	6–24	10, 13 or 13–20

Table A.6 Joint water reduction factor (after Barton 2002)

		Approx. water pressure (kg/cm²)	J_w
A	Dry excavations or minor inflow	< 1	1.0
B	Medium inflow or pressure, occasional outwash of joint fillings	1–2.5	0.66
C	Large inflow or high pressure in competent rock with unfilled joints	2.5–10	0.5
D	Large inflow or high pressure, considerable outwash of joint fillings	2.5–10	0.33
E	Exceptionally high inflow or water pressure at blasting, decaying with time	> 10	0.2–0.1
F	Exceptionally high inflow or water pressure continuing without noticeable decay	> 10	0.1–0.05

Notes: (i) Factors C to F are crude estimates. Increase J_w if drainage measures are installed. (ii) Special problems caused by ice formation are not considered. (iii) For general characterisation of rock masses distant from excavation influences, the use of J_w = 1.0, 0.66, 0.5, 0.33, etc as depth increases from say 0–5, 5–25, 25–250 to > 250 m is recommended, assuming that RQD/J_n is low enough for good hydraulic connectivity. This will help to adjust *Q* for some of the effective stress and water softening effects, in combination with appropriate characterisation values of the Stress Reduction Factor. Correlations with depth-dependent static deformation modulus and seismic velocity will then follow the practice used when these were developed.

Table A.7 Stress Reduction Factor (SRF) (after Barton 2002)

		σ_c/σ_1	σ_θ/σ_c	*SRF*
(a) Weakness zones intersecting excavation, which may cause loosening of rock mass when tunnel is excavated				
A	Multiple occurrences of weakness zones containing clay or chemically disintegrated rock, very loose surrounding rock (any depth)			10
B	Single weakness zones containing clay or chemically disintegrated rock (depth of excavation ≤ 50 m)			5
C	Single weakness zones containing clay or chemically disintegrated rock (depth of excavation > 50 m)			2.5
D	Multiple shear zones in competent rock (clay-free), loose surrounding rock (any depth)			7.5
E	Single shear zones in competent rock (clay-free), (depth of excavation ≤ 50 m)			5
F	Single shear zones in competent rock (clay-free), (depth of excavation > 50 m)			2.5
G	Loose, open joints, heavily jointed or 'sugar-cube', etc. (any depth)			5
(b) *Competent rock, rock stress problems*				
H	Low stress, near surface, open joints	> 200	< 0.01	2.5
J	Medium stress, favourable stress condition	200–10	0.01–0.3	1
K	High stress, very tight structure. Usually favourable to stability, maybe unfavourable for wall stability	10–5	0.3–0.4	0.5–2
L	Moderate slabbing after > 1 h in massive rock	5–3	0.5–0.65	5–50
M	Slabbing and rock burst after a few minutes in massive rock	3–2	0.65–1	50–200
N	Heavy rock burst (strain-burst) and immediate dynamic deformations in massive rock	< 2	> 1	200–400
(c) *Squeezing rock: plastic flow of incompetent rock under the influence of high rock pressure*				
O	Mild squeezing rock pressure		1–5	5–10
P	Heavy squeezing rock pressure		> 5	10–20
(d) *Swelling rock: chemical swelling activity depending on the presence of water*				
R	Mild swelling rock pressure			5–10
S	Heavy swelling rock pressure			10–15

Notes: (i) Reduce these values of SRF by 25–50% if the relevant shear zones only influence but do not intersect the excavation. This will also be relevant for characterisation. (ii) For strongly anisotropic virgin stress field (if measured); when $5 \leq \sigma_1/\sigma_3 \leq 10$, reduce σ_c to $0.75\ \sigma_c$. When $\sigma_1/\sigma_3 > 10$, reduce σ_c to $0.5\ \sigma_c$, where σ_c is the unconfined compression strength, σ_1 and σ_3 are the major and minor principal stresses, and σ_θ the maximum tangential stress (estimated from elastic theory). (iii) Few case records available where depth of crown below surface is less than span width, suggest an SRF increase from 2.5 to 5 for such cases (see H). (iv) Cases L, M, and N are usually most relevant for support design of deep tunnel excavations in hard massive rock masses, with RQD/J_n ratios from about 50–200. (v) For general characterisation of rock masses distant from excavation influences, the use of SRF = 5, 2.5, 1.0, and 0.5 is recommended as depth increases from say 0–5, 5–25, 25–250 to >250 m. This will help to adjust Q for some of the effective stress effects, in combination with the appropriate characterisation values of J_w. Correlations with depth-dependent static deformation modulus and seismic velocity will then follow the practice used when these were developed. (vi) Cases of squeezing rock may occur for depth $H > 350Q^{1/3}$. Rock mass compression strength can be estimated from $SIGMA_{cm} \approx 5\gamma Q_c^{1/3}$ (MPa) where γ is the rock density in t/m³, and $Q_c = Q \times \sigma_c/100$.

A.2.1 Use of the Q-method for predicting TBM performance

Barton (1999), with further explanation in Barton (2000), developed a method for predicting the penetration rate and advance rate for TBM tunnelling. This method is based on an expanded Q-method of rock mass classification and average cutter force in relation to the appropriate rock mass strength. The parameter Q_{TBM} can be estimated during feasibility studies, and can also be back calculated from TBM performance during tunnelling. Equation A.1 shows the expression used to calculate Q_{TBM} and is based on equation 2.10 presented for the standard Q-system.

$$Q_{TBM} = \frac{RQD_0}{J_n} \times \frac{J_r}{J_a} \times \frac{J_w}{SRF} \times \frac{SIGMA}{F^{10}/20^9} \times \frac{20}{CLI} \times \frac{q}{20} \times \frac{\sigma_\theta}{5} \tag{A.1}$$

where RQD_0 = RQD (%) interpreted in the tunnelling direction. J_n, J_r, J_a, J_w and SRF are unchanged, except that J_r and J_a should refer to the joint set that most assists (or hinders) boring. F is the average cutter load (tnf) through the same zone, normalized by 20 tnf. SIGMA is the rock mass strength estimate (MPa) in the same zone. CLI is the cutter life index (for example 4 for quartzite and 90 for limestone). q is the quartz content in percentage terms and σ_θ is the induced biaxial stress on the tunnel face (approx. MPa) in the same zone, normalized to an approximate depth of 100 m. SIGMA incorporates the Q-value. The choice between $SIGMA_{cm}$ and $SIGMA_{tm}$ (equations A.2 and A.3) will depend on orientation (Barton 2000).

$$SIGMA_{cm} = 5\gamma Q_c^{1/3} \tag{A.2}$$

$$SIGMA_{tm} = 5\gamma Q_t^{1/3} \tag{A.3}$$

where $Q_c = Q\sigma_c/100$, $Q_t = Q.I_{50}/4$, γ = density (g/cm^3), σ_c is the uniaxial strength, I_{50} is the point load strength.

Based on empirical data, Barton (1999) suggested an approximate relationship between penetration rate (PR) and Q_{TBM} as shown in equation A.4.

$$PR \approx 5(Q_{TBM})^{-0.2} \tag{A.4}$$

and advance rate (AR) as shown in equation A.5.

$$AR \approx 5(Q_{TBM})^{-0.2} \times T^m \tag{A.5}$$

where T is total time in hours (24/day, 168/week, etc.) and m is defined from the empirical data as follows:

Best performance	m ~ –0.13 to –0.17 (variable)
Good	m ~ –0.17
Fair	m ~ –0.19
Poor	m ~ –0.21
Exceptionally poor	m ~ –0.25

m can be further refined based on the diameter of the tunnel D, CLI, q and n using equation A.6.

$$m \approx m_1 \times \left(\frac{D}{5}\right)^{0.20} \times \left(\frac{20}{CLI}\right)^{0.15} \times \left(\frac{q}{20}\right)^{0.10} \times \left(\frac{n}{2}\right)^{0.05} \tag{A.6}$$

where n = porosity (%). Some case history data using Q_{TBM} were reported by Sapigni *et al.* (2002) and Figure A.2 is reproduced from Palmström and Broch (2006).

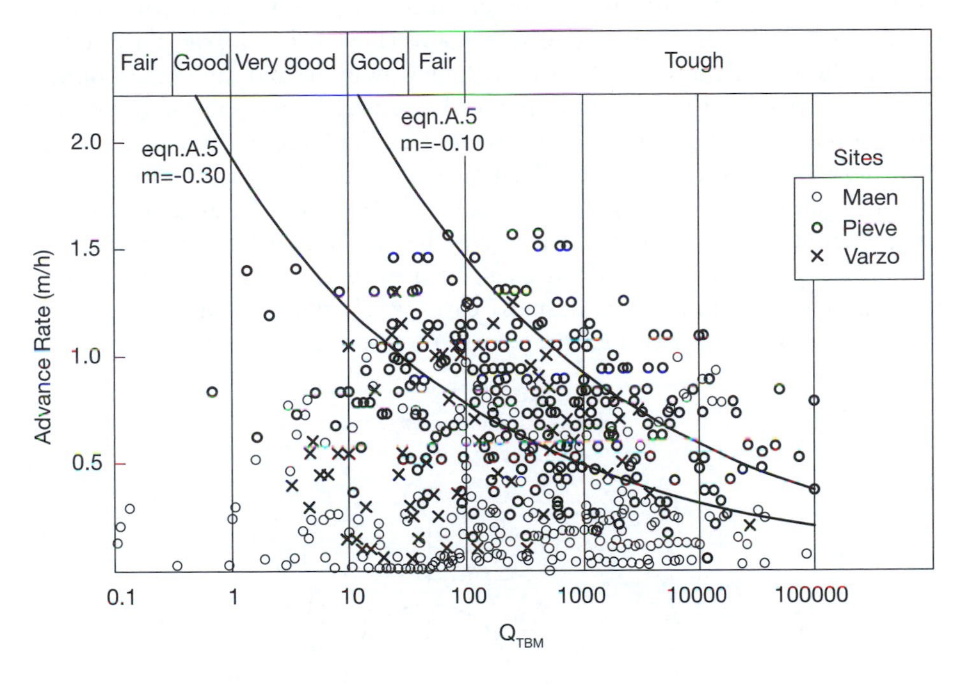

Figure A.2 Advance rate for three TBM tunnels plotted against Q_{TBM} (Sapigni *et al.* 2002, reproduced from Palmström and Broch 2006).

Appendix B

Analytical calculation of a sprayed concrete lining using the continuum method

B.1 Introduction

There are different analytical methods to estimate the internal forces in a tunnel lining and give an indication on the type of support needed (see section 3.5). In this section the focus is on tunnels which have a large overburden ($h \geq D$). This allows the ground to be treated as a continuum, i.e. a plate with deformations in one plane (Figure B.1). The plate has a circular hole (the tunnel), which is stiffened by a circular ring (the lining). It can be assumed that the area above the tunnel is not softened and can carry some load. The primary stresses can be calculated without the associated deformations and the lateral coefficient of earth pressure is K_0.

For the approach of a rigid interconnection between the ground and the tunnel lining it is important to note whether the tangential component of the stresses from the earth pressure can be transferred into the tunnel lining for example through friction. In many cases it is better to assume tangential slippage between the ground and the tunnel lining in order to be on the

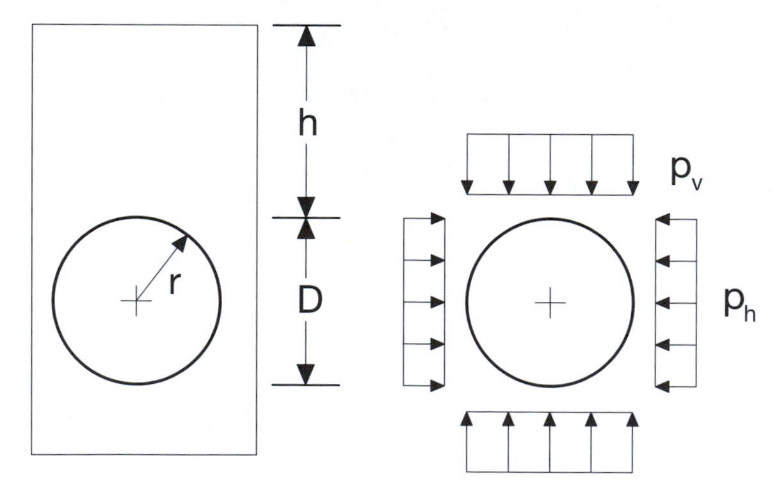

Figure B.1 a) Analytical model for deep tunnels and b) primary loads (Ahrens *et al.* 1982)

safe side. This can sometimes also be supported from a construction point of view by the type of tunnel construction for example for a shield driven tunnel or a tunnel with a membrane layer between the lining and the ground. The earth pressure approach displayed in Figure B.1 has first been suggested by DGGT (1980) and is valid independently of the depth of the tunnel and the chosen analytical model. Earlier analytical models assumed different approaches for the earth pressure using tables and diagrams for the simple determination of internal loads.

B.2 Analytical model using Ahrens *et al.* (1982)

MAIN ASSUMPTIONS AND REQUIREMENTS

- Straight tunnel.
- Load, ground parameters and cross sectional area remain constant along the tunnel.
- The tunnel construction is completed.
- Primary stress condition $\sigma_v^p = -\gamma \times h$; $\sigma_h^p = -K_0 \times \gamma \times h$
- Circular tunnel cross section.
- Homogeneous, isotropic and ideal-elastic material behaviour for the ground and the lining.
- Thin tunnel lining.
- Constant area and constant second moment of area in the φ-direction.

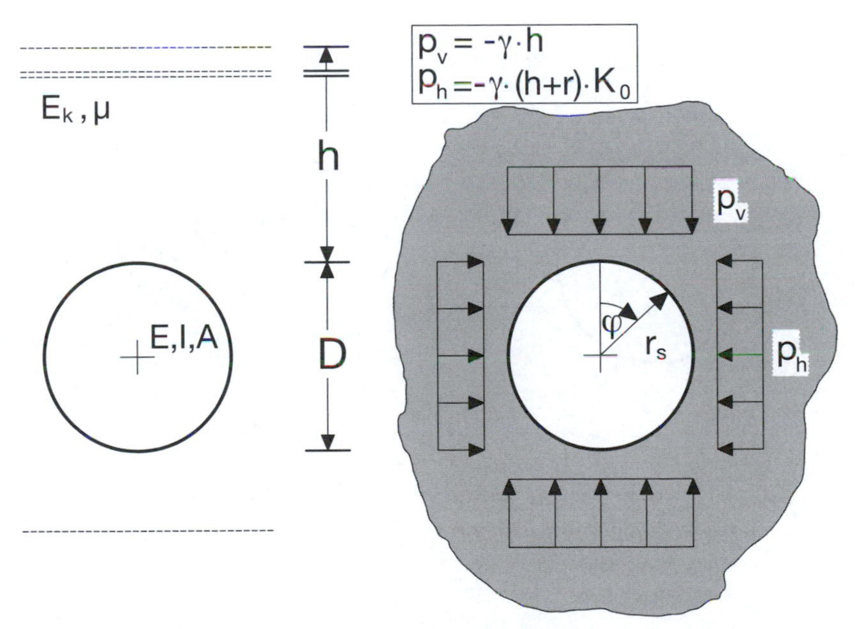

Figure B.2 a) Analytical model for deep tunnels and b) primary loads (Ahrens *et al.* 1982)

If segmental linings are used, the following additional assumptions are made in the analysis:

- Pre-deformations of the segmental linings resulting from the erection of the segments are proportional to the elastic deformations.
- The annular gap between the ground and the lining is completely grouted.
- Linearized theory of second order (small strain, large deformations) can be applied.

B.3 Required equations and calculation process

The calculation of the internal forces of the analytical model is carried out using the displacement method. This requires that the primary stress situation or the stresses determined from the earth pressure approach is used as a load displacement condition, where the, as yet unknown, deformations (in this case the deformation of the tunnel contour) are assumed to be zero.

However, as additional limitations of the deformations do not exist, the forces due to the earth pressure are not in equilibrium around the tunnel contour. Furthermore, the deformations along the tunnel contour are not equal to zero. Instead, these can be calculated taking into account the appropriate forces – the transition condition between the perforated disc

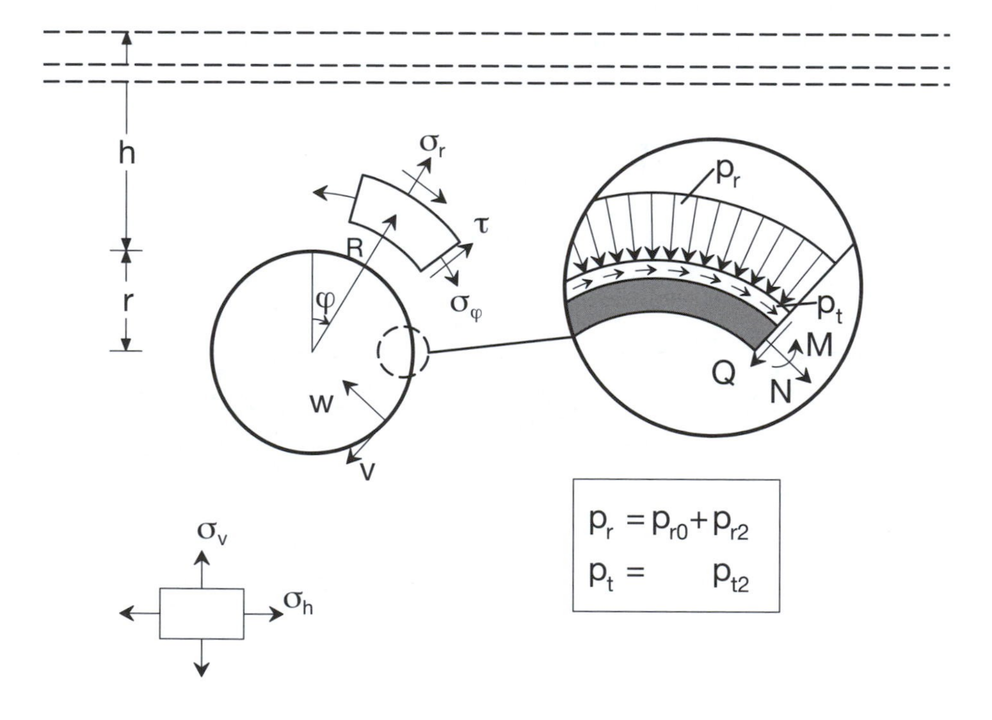

Figure B.3 Definitions of the parameters

and the circular ring – with the help of the unity deformation condition from the equilibrium conditions along the tunnel contour. The consideration of the equilibrium conditions for the individual components of the Fourier series leads to an equilibrium system, which allows for a more or less easy calculation of the unknown deformations, resulting in the internal forces. For the case of a tunnel support with infinite axial stiffness, the equations of the systems can be decoupled so that explicit equations can be given for the calculation of the deformations.

The following are the equations, using first order theory, for the case of a rigid bond between the tunnel support and the ground assuming infinite axial stiffness. If using segmental lining, the pre-deformations of the segments as a result of the installation can be considered using second order theory (Ahrens *et al.* 1982). It is assumed that the tunnel lining has an infinite axial stiffness.

The following equations are from Ahrens *et al.* (1982) and further explanations can be found in this reference.

Earth pressure:

$$p_v = -\gamma \times h \tag{B.1a}$$

$$p_h = -K_0 \times \gamma \times (h + r) \tag{B.1b}$$

Transformation into polar coordinates:

$$p_r = \overline{p}_{r0} + \overline{p}_{r2} \times \cos 2\varphi \tag{B.2a}$$

$$p_t = \overline{p}_{r2} \times \sin 2\varphi \tag{B.2b}$$

With

$$\overline{p}_{r0} = 0.5 \times \gamma \times [h + (h + r) \times K_0] \tag{B.3a}$$

$$\overline{p}_{r2} = \overline{p}_{tr2} = 0.5 \times \gamma \times [h - (h + r) \times K_0] \tag{B.3b}$$

The load and deformation variables of the plate are denoted with a superscript 'D', while the load and deformation variables associated with the circular ring frame are denoted with a superscript 'R'.

Deformations:

$$w(\varphi) = \overline{w}_0 + \overline{w}_2 \cdot \cos 2\varphi \tag{B.4}$$

$$v(\varphi) = + \overline{v}_2 \cdot \sin 2\varphi \tag{B.5}$$

With $(EA \to \infty)$

$$\bar{w}_0 = 0 \tag{B.6}$$

$$\bar{w}_2 = \frac{\bar{p}_{r2} + 0.5 \times \bar{p}_{t2}}{\frac{1}{(3-\mu-4\mu^2)} \times (2.25 - 1.5\mu) \times \frac{E_k}{r} + \frac{9EI}{r^4}} \tag{B.7}$$

$$\bar{v}_2 = 0.5 \times \bar{w}_2 \tag{B.8}$$

The proportion of the earth pressure acting on the circular frame and the continuum is equivalent to their relative stiffnesses. The circular frame load is:

$$\bar{p}_{r0}^R = \bar{p}_{r0} - \bar{p}_{r0}^D$$

$EA \to \infty : \bar{p}_{r0}^D = 0$ (as a result of the infinite axial stiffness, the complete constant load is supported by the circular ring frame)

$$\bar{p}_{r0}^R = \bar{p}_{r0}$$

$$\bar{p}_{r2}^R = \bar{p}_{r2} - \bar{p}_{r2}^D$$

$$\bar{p}_{r2}^D = \frac{E_c}{r} \times \frac{1}{(3-\mu-4\mu^2)} \times \left[(5-6\mu) \times \bar{w}_2^D + (-4+6\mu)\bar{v}_2^D\right] \tag{B.9a}$$

$$\bar{p}_{t2}^R = \bar{p}_{t2} - \bar{p}_{t2}^D$$

$$\bar{p}_{t2}^D = \frac{E_c}{r} \times \frac{1}{(3-\mu-4\mu^2)} \times \left[(-4+6\mu) \times \bar{w}_2^D + (5-6\mu)\bar{v}_2^D\right] \tag{B.9b}$$

Proportional internal force parameter:

Load part $\bar{p}_{r0}^R$:

$$N_0 = -r \times \bar{p}_{r0}^R \tag{B.10a}$$

$$Q_0 = 0 \tag{B.10b}$$

$$M_0 = 0 \tag{B.10c}$$

Load part $\bar{p}_{r2}^R$, $\bar{p}_{t2}^R$:

$$N_2 = \frac{r}{3} \times \left(2 \times \bar{p}_{t2}^R + \bar{p}_{r2}^R\right) \times \cos 2\varphi \tag{B.11a}$$

$$Q_2 = -\frac{r}{3} \times \left(\overline{p}_{t2}^R + 2 \times \overline{p}_{r2}^R\right) \times \sin 2\varphi \tag{B.11b}$$

$$M_2 = \frac{r^2}{6} \times \left(\overline{p}_{t2}^R + 2 \times \overline{p}_{r2}^R\right) \times \cos 2\varphi \tag{B.11c}$$

Or simpler:

$$Q_2 = -\frac{6EI}{r^3} \times \overline{w}_2^R \times \sin 2\varphi \tag{B.11d}$$

$$M_2 = \frac{3EI}{r^2} \times \overline{w}_2^R \times \cos 2\varphi \tag{B.11e}$$

Final internal parameters and deformations:

$$N = N_0 + N_2 \tag{B.12}$$

$$Q = Q_2 \tag{B.13}$$

$$M = M_2 \tag{B.14}$$

$$w = w_2 \tag{B.15}$$

B.4 Example for a tunnel at King's Cross Station, London

Figure B.4 shows the schematic of the geology associated with a tunnel at King's Cross Station in London, UK. The overburden is assumed to be h = 11.0 m.

A) GROUND PARAMETERS

Density	ρ	= 2000 kg/m^3
Stiffness	E_S	= 87 MN/m^2 ($E_c = f(E_S)$)
Poisson's ratio	μ	= 0.15
Coefficient of Lateral Earth Pressure	K_0	= 1.2

The ground is idealized as a homogenous continuum assuming an average density for the ground of 2000 kg/m^3.

B) STRUCTURAL SYSTEM, LOADS AND PARAMETERS

In order to simplify the calculation, it is assumed that the construction has been completed. The effects of groundwater are neglected as the tunnel was constructed in London Clay. No dead load is considered at the ground

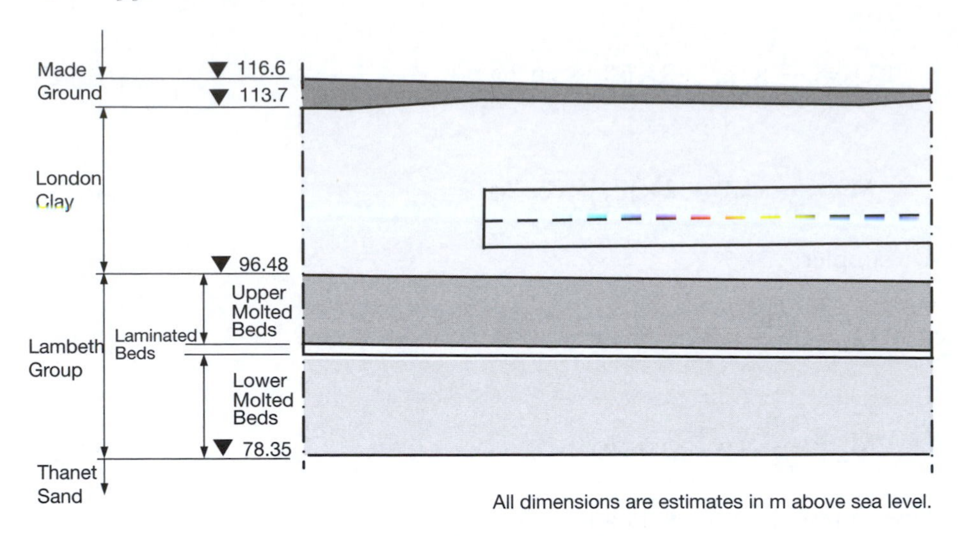

Figure B.4 Schematic showing the geology around the tunnel

surface now or in the future. A rigid bond exists between the lining and the ground.

Weight of the ground:	$\gamma = 20\ \mathrm{kN/m^3}$	The tunnel weight is neglected.
Radius of the system axis:	$r_S = 3.05\ \mathrm{m}$	
Overburden:	$h_O = h = 11.0\ \mathrm{m}$	

$p_v, p_h \rightarrow p_r, p_t$

$$\bar{p}_{r0} = 0.5 \times \gamma \times [h + (h + r) \times K_0]$$
$$= 0.5 \times 20 \times [11.0 + (11.0 + 3.05) \times 1.2] = 278.6\ \mathrm{kN/m^2}$$

$$\bar{p}_{r2} = 0.5 \times \gamma \times [h - (h + r) \times K_0]$$
$$= 0.5 \times 20 \times [11.0 - (11.0 + 3.05) \times 1.2] = -58.6\ \mathrm{kN/m^2}$$

C) TUNNEL SUPPORT PARAMETERS

The material is sprayed concrete (C20/25) with a Young's modulus of $E = 2.88 \times 10^4$ MPa. The key parameters are:

Profile parameters:

$$A = 0.175\ \mathrm{m^2/m}$$
$$I = 4.466 \times 10^{-4}\ \mathrm{m^4/m}$$
$$W_o = 5.104 \times 10^{-3}\ \mathrm{m^3/m}$$
$$W_i = 5.104 \times 10^{-3}\ \mathrm{m^3/m}$$

Allowable compressive stress:

$\sigma_{c,c}$ = 11.3 MPa (includes factor of safety and long-term influences)

D) STIFFNESS PARAMETERS

Ground stiffness: E_S = 87 MPa (stiffness parameter)

In general, the stiffness parameter E_S is used as the deformation parameter for soft ground and is determined in laboratory experiments using a compression test with restricted strain. This parameter cannot simply be used as an Elasticity modulus E_C when applying the continuity calculation. In this case, the Elasticity modulus for the three-dimensional continuum can be determined with a Poisson's ratio of $\mu = 0.15$ using Equation B.16:

$$E_C = \frac{(1+\mu)(1-2\mu)}{(1-\mu)} E_S \tag{B.16}$$

A further conversion of the Elasticity modulus of a disc like structure in a plane strain state is not required, as these specific modifications for the model are already included in the following equations. With the given parameters the Elasticity modulus can be calculated

$$E_C = \frac{(1+0.15)(1-2\times 0.15)}{(1-0.15)} 87 = 82.4 \text{ MPa} \tag{B.17}$$

Tunnel support stiffness: $EI = 2.88\ 10^4 \times 4.466\ 10^{-4} = 12.86\ \text{MNm}^2/\text{m}$

E) DETERMINATION OF THE INTERNAL FORCES

Assumption: Tunnel lining with an infinite axial stiffness; 1st Order Theory
Assumption: $EA \to \infty$
Radial displacement $\bar{w}_0 = 0$
Radial displacement $\bar{w}_2$

$$\bar{w}_2 = \frac{\bar{p}_{r2} + 0.5 \times \bar{p}_{t2}}{\frac{1}{(3-\mu-4\mu^2)} \times (2.25 - 1.5\mu) \times \frac{E_C}{r} + \frac{9EI}{r^4}}$$

$$\bar{w}_2 = \frac{-58.6 + 0.5 \times (-58.6)}{\frac{1}{(3-0.15-4\times 0.15^2)} \times (2.25 - 1.5 \times 0.15) \times \frac{82000}{3.05} + \frac{9 \times 12860}{3.05^4}}$$

$$= -0.0042 \text{ m}$$

The tangential displacement

$$\bar{v}_2 = 0.5 \times \bar{w}_2 = 0.5 \times (-0.0042) = -0.0021 \text{ m}$$

Partial load $\bar{p}_{r0}^{R}$, (the total constant partial load $\bar{p}_{r0}$ is carried by the circular ring system ($EA \rightarrow \infty$)):

$$\bar{p}_{r0}^{R} = 278.6 \text{ kN/m}^2$$

Load parts $\bar{p}_{r2}^{R}$, $\bar{p}_{t2}^{R}$

The load part acting on the circular ring support derived from the load parts $\bar{p}_{r2}$ and $\bar{p}_{t2}$ can only be calculated indirectly from the difference between the total load and the portion of the load acting on the plate due to $EA \rightarrow \infty$:

$$\bar{p}_{r2}^{D} = \frac{E_c}{r} \times \frac{1}{(3-\mu-4\mu^2)} \times \left[(5-6\mu) \times \bar{w}_2^{D} + (-4+6\mu)\bar{v}_2^{D})\right]$$

$$\bar{p}_{r2}^{D} = \frac{82000}{3.05} \times \frac{1}{(3-0.15-4\times 0.15^2)} \times$$
$$\left[(5-6\times 0.15)\times(-0.0042)+(-4+6\times 0.15)\times(-0.0021)\right]$$
$$= 9.787 \times \left[-17.03+6.44\right] = -103.7 \text{ kN/m}^2$$

$$\bar{p}_{t2}^{D} = \frac{E_c}{r} \times \frac{1}{(3-\mu-4\mu^2)} \times \left[(-4+6\mu) \times \bar{w}_2^{D} + (5-6\mu)\bar{v}_2^{D}\right]$$

$$\bar{p}_{t2}^{D} = \frac{82000}{3.05} \times \frac{1}{(3-0.15-4\times 0.15^2)} \times$$
$$\left[(-4+6\times 0.15)\times\left(-0.0042\right)+(5-6\times 0.15)\times\left(-0.0021\right)\right]$$
$$= 9.788 \times \left[12.87-8.51\right] = 42.7 \text{ kN/m}^2$$

$$\rightarrow \bar{p}_{r2}^{R} = \bar{p}_{r2} - \bar{p}_{r2}^{D} = -58.6 - (-103.7) = 45.1 \text{ kN/m}^2$$

$$\rightarrow \bar{p}_{t2}^{R} = \bar{p}_{t2} - \bar{p}_{t2}^{D} = -58.6 - 42.7 = -101.3 \text{ kN/m}^2$$

Internal loads:

Load portion $\overline{p}_{r0}^{R}$

$$N_0 = -r \times \overline{p}_{r0}^{R} = -3.05 \times 278.6 = -849.7 \text{ kN/m}$$

$$Q_0 = 0 \text{ kN/m}$$

$$M_0 = 0 \text{ kNm/m}$$

Load part $\overline{p}_{r2}^{R}$, $\overline{p}_{t2}^{R}$:

$$N_2 = \frac{r}{3} \times \left(2 \times \overline{p}_{t2}^{R} + \overline{p}_{t2}^{R}\right) \times \cos 2\varphi = \frac{3.05}{3} \times \left(2 \times (-101.3) + 45.1\right) \times \cos 2\varphi$$
$$= -160.3 \times \cos 2\varphi \text{ kN/m}$$

$$Q_2 = -\frac{r}{3} \times \left(\overline{p}_{t2}^{R} + 2 \times \overline{p}_{r2}^{R}\right) \times \sin 2\varphi = -\frac{3.05}{3} \times \left(-101.3 + 2 \times 45.1\right) \times \sin 2\varphi$$
$$= 11.3 \times \sin 2\varphi \text{ kN/m}$$

$$M_2 = \frac{r^2}{6} \times \left(\overline{p}_{t2}^{R} + 2 \times \overline{p}_{r2}^{R}\right) \times \cos 2\varphi = \frac{3.05^2}{6} \times \left(-101.3 + 2 \times 45.1\right) \times \cos 2\varphi$$
$$= -17.2 \times \cos 2\varphi \text{ kNm/m}$$

Calculation to check Q_2 and M_2 from the radial displacement $\overline{w}_2^{R}$:

$$Q_2 = -\frac{6EI}{r^3} \times \overline{w}_2^{R} \times \sin 2\varphi = -\frac{6 \times 12860}{3.05^3} \times \left(-0.0042\right) \times \sin 2\varphi$$
$$= 11.3 \times \sin 2\varphi \text{ kN/m}$$

$$M_2 = \frac{3EI}{r^2} \times \overline{w}_2^{R} \times \cos 2\varphi = \frac{3 \times 12860}{3.05^2} \times \left(-0.0042\right) \times \cos 2\varphi$$
$$= -17.2 \times \cos 2\varphi \text{ kNm/m}$$

Final internal parameters and deformations

$$N = N_0 + N_2 = -849.7 - 160.1 \times \cos 2\varphi \text{ kN/m}$$

$$Q = Q_2 = 11.3 \times \sin 2\varphi \text{ kN/m}$$

$$M = M_2 = -17.2 \times \cos 2\varphi \text{ kNm/m}$$

Radial displacements:

$$w = \quad w_2 = -0.0042 \times \cos 2\varphi \text{ m}$$

F) STRESS ANALYSIS

Using Ahrens *et al.* (1982) section 2.3.4.2 and the following equation,

$$\sigma = \frac{N}{A} \pm \frac{M}{W}$$

The crown stresses are:

Extrados:

$$\sigma_e = -\frac{1010}{0.175} + \frac{17.23}{0.0051} = |-2.4| \text{ MPa} < 11.3 \text{ MPa}$$

Intrados:

$$\sigma_i = -\frac{1010}{0.175} - \frac{17.23}{0.0051} = |-9.15| \text{ MPa} < 11.3 \text{ MPa}$$

The bench stresses are:

Extrados:

$$\sigma_e = -\frac{690}{0.175} - \frac{17.23}{0.0051} = |-7.32| \text{ MPa} < 11.3 \text{ MPa}$$

Intrados:

$$\sigma_i = -\frac{690}{0.175} + \frac{17.23}{0.0051} = |-0.56| \text{ MPa} < 11.3 \text{ MPa}$$

The invert stresses are:

Extrados:

$$\sigma_e = -\frac{1010}{0.175} + \frac{17.23}{0.0051} = |-2.4| \text{ MPa} < 11.3 \text{ MPa}$$

Intrados:

$$\sigma_i = -\frac{1010}{0.175} - \frac{17.23}{0.0051} = |-9.15| \text{ MPa} < 11.3 \text{ MPa}$$

This shows that all the stresses are within the allowable stresses for the sprayed concrete lining.

G) PRESENTATION OF THE INTERNAL FORCES AND THE DEFORMATION

The internal forces are shown in Figure B.5.

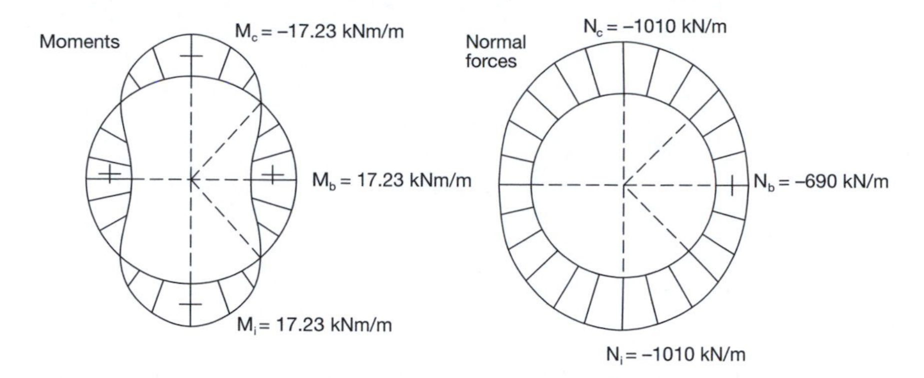

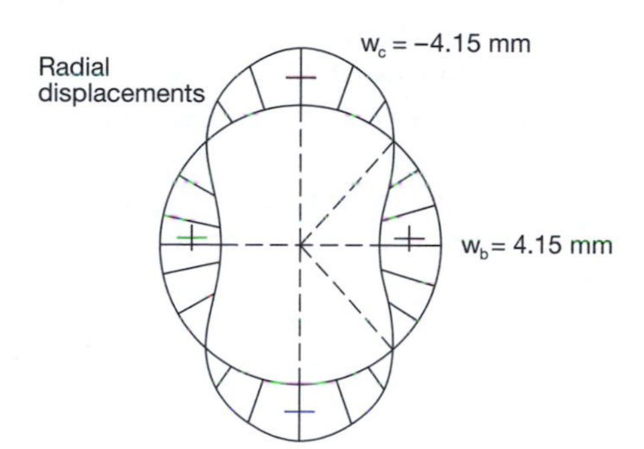

Figure B.5
Moments, normal forces and radial displacements

References and bibliography

References

Addenbrooke T.I. and Potts D.M. (2001). 'Twin tunnel interaction surface and subsurface effects'. *International Journal of Geomechanics*, 1(2): 249–71.

Ahrens H., Lindner E. and Lux K.H. (1982). 'Zur Dimensionierung von Tunnelausbauten nach den "Empfehlungen zur Berechnungen von Tunneln im Lockergestein (1980)"'. *Die Bautechnik*, Heft 8.

Allen R. (2006). 'Ground investigation – procedures and practice'. Course Notes. British Tunnelling Society, Notes on Tunnel Design and Construction. University of Surrey, Guildford, UK.

Allenby D. and Ropkins J.W.T. (2004). 'The use of jacked-box tunnelling under a live motorway'. *Proceedings of the Institution of Civil Engineers, Geotechnical Engineering*, 157, GE4, Thomas Telford, London, pp. 229–38.

Allenby D. and Ropkins J.W.T. (2007). 'Jacked box tunnelling: using the Ropkins System™, a non-intrusive tunnelling technique for constructing new underbridges beneath existing traffic arteries'. Lecture presented at the Institution of Mechanical Engineers on Wednesday 17 October 2007, London.

Anagnostou G. and Kovari K. (1996). 'Face stability in slurry and EPB shield tunnelling'. *Proceedings of the International Symposium on Geotechnical Aspects of Underground Construction in Soft Ground* (City University, London) (eds R.J. Mair and R.N. Taylor), Balkema, Rotterdam, The Netherladnds, pp. 453–8.

Anderson N., Croxton N., Hoover R. and Sirles P. (2008). 'Geophysical methods commonly employed for geotechnical site characterization'. *Transportation Research Circular E-C130*, Transportation Research Board, Washington, DC.

Anon (1977). 'The description of rock masses for engineering purposes. Engineering Group of the Geological Society Working Party report'. *Quarterly Journal of Engineering Geology*, 10: 355–88.

Anon (2000). *The Collapse of NATM Tunnels at Heathrow Airport*, Crown Copyright, Norwich, UK.

Appleby M. and Lamont D. (2009). 'Health and Safety Legislation'. Chapter in *Construction Law Handbook* (eds V. Ramsey, A. Minoque and M. O'Reilly), Institution of Civil Engineers and Thomas Telford, London.

Ata A.A. (1996). 'Ground settlements induced by slurry shield tunnelling in stratified soils'. *Proceedings of the North American Tunnelling 1996 Conference* (ed. L. Ozdemir), Vol. 1, pp. 43–50. Publisher unknown.

Atahan C., Leca E. and Guilloux A. (1996). 'Performance of a shield driven sewer tunnel in the Val-de-Marne, France'. *Proceedings of the International Symposium on Geotechnical Aspects of Underground Construction in Soft Ground* (City University, London) (eds R.J. Mair and R.N. Taylor), Balkema, Rotterdam, The Netherlands, pp. 641–6.

Atkinson J.H. and Mair R.J. (1981). 'Soil mechanics aspects of soft ground tunnelling'. *Ground Engineering*, 14(5): 20–8.

Attewell P.B. (1978). 'Ground movements caused by tunnelling in soil'. *Proceedings of the International Conference on Large Movements and Structures* (ed. J.D. Geddes), Pentech Press, London, pp. 812–948.

Attewell P.B. (1995). *Tunnelling Contracts and Site Investigation*, E & FN Spon, London.

Attewell P.B. and Woodman J.P. (1982). 'Predicting the dynamics of ground settlement and its derivatives caused by tunnelling in soil'. *Ground Engineering*, November: 13–36.

Attewell P.B., Yeates J. and Selby A.R. (1986) *Soil Movements Induced by Tunnelling and Their Effects on Pipelines and Structures*, Blackie, Glasgow, Scotland, p. 325.

Baker W.H. (1982). 'Planning and performing structural chemical grouting'. *Grouting in Geotechnical Engineering* (ed. W.H. Baker), American Society of Civil Engineers, Reston, VA, pp. 515–39.

Barton N. (1999). 'TBM performance estimation in rock using Q_{TBM}'. *Tunnels and Tunnelling International*, September: 30–4.

Barton N. (2000). *TBM Tunnelling in Jointed and Faulted Rock*, Balkema, Rotterdam, The Netherlands, p. 172.

Barton N. (2002). 'Some new Q-value correlations to assist in site characterisation and tunnel design'. *International Journal of Rock Mechanics & Mining Sciences*, 39: 185–216.

Barton, N., Lien, R. and Lunde, J. (1974). 'Engineering classification of rock masses for the design of tunnel support'. *Rock Mechanics*, 6: 189–236. Springer-Verlag, New York.

BASF (2009a). 'MEYCO® ABR 5'. Product Literature, BASF Construction Chemicals.

BASF (2009b). 'MEYCO® FIX TSG 6'. Product Literature, BASF Construction Chemicals.

Bassett R.H., Kimmance J.P. and Rasmussen R.C. (1999). 'An automated electrolevel deformation monitoring system for tunnels'. *Proceedings of the Institution of Civil Engineering, Geotechnical Engineering*, 137(3): 117–26.

Bayer H.J. (ed.) (2005). *HDD Practice Handbook*, Vulkan-Verlag GmbH, Essen, Germany.

Beamish R. (1862). *Memoir of the Life of Sir Marc Isambard Brunel*, 1st edition. Longman & Co., London.

Beaver P. (1973). *A History of Tunnels*, The Citadel Press, Secaucus, NJ.

Beth M. and Obre X. (2005). 'Ground movement monitoring at King's Cross Station London'. *Proceedings of the Institution of Civil Engineers, Geotechnical Engineering*, 158, GE3: 125–33.

BetonKalender (2005). *Fertigteile und Tunnelbauwerke*. Vols 1 and 2, Ernst & Sohn, Berlin.

Bieniawski Z.T. (1984). *Rock Mechanics Design in Mining and Tunneling*, Balkema, Rotterdam, The Netherlands.

Bieniawski Z.T. (1989). *Engineering Rock Mass Classifications*, John Wiley & Sons, New York, p. 251.

Bloodworth A.G. (2002). 'Three-dimensional analysis of tunnelling effects on structures to develop design methods'. Unpublished Ph.D. thesis, University of Oxford, Oxford, UK.

Boscardin M.D. and Cording E.J. (1989). 'Building response to excavation induced settlement'. *Journal of Geotechnical Engineering*, American Society of Civil Engineers, 115(1): 1–21.

Bracegirdle A., Mair R.J., Nyren R.J. and Taylor R.N. (1996). 'A simple methodology for evaluating the potential damage to buried cast iron pipes from ground movements'. *Proceedings of the International Symposium on Geotechnical Aspects of Underground Construction in Soft Ground* (City University, London) (eds R.J. Mair and R.N. Taylor), Balkema, Rotterdam, The Netherlands, pp. 659–64.

Broms B.B. and Bennermark H. (1967). 'Stability of clay at vertical opening'. *Journal of Soil Mechanics and Foundations*, American Society of Civil Engineers, 93 (SM1): 71–94.

Brown D.A. (2004). 'Hull wastewater flow transfer tunnel: recovery of tunnel collapse by ground freezing'. *Proceedings of the Institution of Civil Engineers, Geotechnical Engineering*, 157 (GE2): 77–83.

BSI (1986). *Code of Practice for Foundations*, BS 8004: 1986, British Standards Institution, London.

BSI (1989). *Code of Practice for Ground Anchorages*, BS 8081: 1989, British Standards Institution, London.

BSI (1990). *Soil Testing – Shear Strength Tests*, BS 1377–8: 1990, British Standards Institution, London.

BSI (1997). *Safety of Unshielded Tunnel Boring Machines and Rodless Shaft Boring Machines for Rock*, BS EN 815:1996 + A2:2008, British Standards Institution, London.

BSI (1999). *Code of Practice for Site Investigations*, BS 5930: 1999, British Standards Institution, London.

BSI (2000). *Eurocode 7: Execution of Special Geotechnical Work: Ground Anchors*, BS EN 1537: 2000, British Standards Institution, London.

BSI (2001a). *Execution of Special Geotechnical Works – Jet Grouting*, BS EN ISO 12716: 2001, British Standards Institution, London.

BSI (2001b). *Code of Practice for Safety in Tunnelling in the Construction Industry*, BS 6164: 2001, British Standards Institution, London.

BSI (2002a). *Geotechnical Investigations and Tests – Identification and Description*, BS EN ISO 14685–1: 2002, British Standards Institution, London.

BSI (2002b). *Tunnelling Machines. Roadheaders, Continuous Miners and Impact Rippers. Safety Requirements*, BS EN 12111:2002 + A1:2009, British Standards Institution, London.

BSI (2002c). *Tunnelling Machines. Air Locks. Safety Requirements*, BS EN 12110:2002 + A1:2008, British Standards Institution, London.

BSI (2003). *Geotechnical Investigations and Tests, Identification and Classification of Rocks – Identification and Description*, BS EN ISO 14689–1: 2003, British Standards Institution, London.

BSI (2004a). *Geotechnical Investigations and Tests – Classification*, BS EN ISO 14685–2: 2004, British Standards Institution, London.

BSI (2004b). *Eurocode 2: Design of Concrete Structures*, BS EN 1992–01–02: 2004, British Standards Institution, London.

BSI (2004c). *Geotechnical Investigation and Testing – Identification and Classification of Soil – Part 2: Principles for a Classification*, BS EN ISO 14688–2: 2004, British Standards Institution, London.

BSI (2005). *Tunnelling Machines. Shield Machine, Thrust Boring Machines, Auger Boring Machines, Lining Erection Equipment. Safety Requirements*, BS EN 12336:2005 + A1:2008, British Standards Institution, London.

BSI (2007). *Eurocode 7: Geotechnical Design – Ground Investigation and Testing*, BS EN 1997–2: 2007, British Standards Institution, London.

BSI (2008a). *Execution of Special Geotechnical Work: Diaphragm Walls*, BS EN 1538 (standard number 08/30192726 DC), British Standards Institution, London.

BSI (2008b). *Execution of Special Geotechnical Work: Bored Piles*, BS EN 1536 (standard number 08/30192723 DC), British Standards Institution, London.

BTS (2008). *Occupational Exposure to Nitrogen Monoxide in a Tunnel Environment: Best Practice Guide*, British Tunnelling Society, London.

BTS/ICE (2004). *Tunnel Lining Design Guide*, Thomas Telford, London.

BTS/ICE (2005). *Closed-face Tunnelling Machines and Ground Stability: A Guideline for Best Practice*, Thomas Telford, London.

Büchi E., Mathier J.F. and Wyss C. (1995). 'Gesteinsabrasivität – ein bedeutender Kostenfaktor beim mechanischen Abbau von Fest- und Lockergestein'. *Tunnel*, 5: 38–44.

Bundesgesetzblatt Teil I, Nr. 110 (1972, last updated 2008). 'Verordnung über Arbeit in Druckluft (Druckluftverordnung) vom 4. Oktober 1972, 1909–1928'. Last update: *Bundesgesetzblatt Teil I* from 18 December 2008, p. 2768, Bundesanzeiger Verlag, Bonn, Germany.

Burd H.J., Houlsby G.T., Augarde C.E., Liu G. (2000). 'Modelling tunnelling-induced settlement of masonry buildings'. *Proceedings of the Institution of Civil Engineers, Geotechnical Engineering*, 143: 17–30.

Burger W. and Wehrmeyer G. (2008a). 'The making of the Mixshield – Part 1'. *Tunnels and Tunnelling International*, May: 35–40.

Burger W. and Wehrmeyer G. (2008b). 'Lifting the lid on Mixshield performance'. *Tunnels and Tunnelling International*, June: 31–6.

Burgess M. and Davies H. (2007). 'Channel Tunnel Rail Link Section 2: Thames tunnel'. *Proceedings of the Institution of Civil Engineers, Civil Engineering*, 160, November: 14–18.

Burland J.B. (1995). 'Assessment of risk of damage due to tunnelling and excavations'. Invited Special Lecture to the *1st International Conference on Earthquake Geotechnical Engineering*, International Symposium, Tokyo, Japan.

Burland J.B., Broms B.B. and de Mello V.F. (1977). 'Behaviour of foundations and structures'. *Proceedings of the 9th International Conference on Soil Mechanics and Foundation Engineering, Tokyo. State-of-the-art Report*, Vol. 2, pp. 495–546.

Burland J.B., Standing J.R. and Jardine F.M. (2001a). *Building Response to Tunnelling, Case Studies from Construction of the Jubilee Line Extension, London, Volume 1: Projects and Methods*, Construction Industry Research and Information Association (CIRIA) Special Publication 200, CIRIA and Thomas Telford, London.

Burland J.B., Standing J.R. and Jardine F.M. (2001b). *Building Response to Tunnelling, Case Studies from Construction of the Jubilee Line Extension, London,*

Volume 2: Case Studies, Construction Industry Research and Information Association (CIRIA) Special Publication 200, CIRIA and Thomas Telford, London.

Burland J.B. and Wroth C.P (1975). 'Settlements on buildings and associated damage'. *Proceedings of the Conference on Settlement of Structures*, British Tunnelling Society, London, pp. 611–54.

Caiden D. (2008). 'Minimizing risk and protecting investment'. Presentation given at the Tunnelling 2008 Symposium, October 2008, Earls Court, London.

Cashman P.M. and Preene M. (2001). *Groundwater Lowering in Construction: A Practical Guide*, Spon, London.

Chang C., Sun C., Duann S.W. and Hwang R.N. (2001). 'Response of a Taipei Rapid Transit System (TRTS) tunnel to adjacent excavation'. *Tunnelling and Underground Space Technology*, 16: 151–58.

Chapman D.N. (1999). 'A graphical method for predicting ground movements form pipe jacking'. *Proceedings of the Institution of Civil Engineering, Geotechnical Engineering*, 137(2): 87–96.

Chapman D.N., Ahn S.K. and Hunt D.V. (2007). 'Investigating ground movements caused by the construction of multiple tunnels in soft ground using laboratory model tests'. *Canadian Geotechnical Journal*, 44(6), June: 631–43.

Clarke R.P.J. and Mackenzie C.N.P. (1994). 'Overcoming ground difficulties at Tooting Bec'. *Proceedings of the Institution of Civil Engineers, Civil Engineering, Thames Water Ring Main*, 102, Special Issue 2: 60–75.

Clarkson T.E. and Ropkins J.W.T. (1977). 'Pipe jacking applied to large structures'. *Proceedings of the Institution of Civil Engineers*, 62(1): 539–61.

Clayton C.R.I. (1995). *The Standard Penetration Test: SPT*, Construction Industry Research and Information Association (CIRIA), London.

Clayton C.R.I. (2001). *Managing Geotechnical Risk – Improving Productivity in UK Building and Construction*, Institution of Civil Engineers, London.

Clayton C.R.I., Hope V.S., Heymann G., van der Berg J.P. and Bica A.V.D. (2000). 'Instrumentation for monitoring sprayed concrete lined soft ground tunnels'. *Proceedings of the Institution of Civil Engineering, Geotechnical Engineering*, 143(3): 119–30.

Clayton C.R.I., Mathews M.C. and Simons N.E. (1995). *Site Investigation*, 2nd edition, Blackwell Science, Oxford (not in print, but can be downloaded free from www.geotechnique.info).

Clayton P., Pitkin F. and Allenby D. (2001). 'The use of hydraulic roadheader equipment in mining and civil engineering'. *Proceedings of the International Conference Underground Construction 2001*, London Docklands, London. Publisher unknown.

Cooper M.L. (2002). 'Tunnel-induced ground movements and their effects on existing tunnels and tunnel linings'. Unpublished Ph.D. thesis, University of Birmingham, Birmingham, UK.

Cooper M.L., Chapman D.N., Rogers C.D.F. and Chan A.H.C. (2002). 'Movements in the Piccadilly Line tunnels due to the Heathrow Express construction'. *Géotechnique*, 52(4): 243–57.

Cording E.J. (1991). 'Control of ground movements around tunnels in soil'. *General Report 9th Pan-American Conference on Soil Mechanics and Foundation Engineering, Chile*, pp. 2195–2244. Publisher unknown.

Cording E.J. and Hansmire W.H. (1975). 'Displacements around soft ground tunnels – General Report'. *Proceedings of the 5th Pan-American Conference on Soil*

Mechanics and Foundation Engineering, Buenos Aires, Session IV, pp. 571–632. Publisher unknown.

Court D.J. (2006). 'Soft ground tunnel boring machines'. Course Notes. British Tunnelling Society, Course on Tunnel Design and Construction, University of Surrey, Guildford, UK.

Davis A. (2006). 'Rock mass classifications'. Course Notes. British Tunnelling Society, Course on Tunnel Design and Construction, University of Surrey, Guildford, UK.

Deere D.U. and Deere, D.W. (1989). *Rock Quality Designation (RQD) After 20 Years*, US Army Corps Engineers Contract Report GL-89-1 (Contract No. DACW39-86-M-4273), Waterways Experimental Station, Vicksburg, MS.

Deere D.U., Hendron, A.J., Patton, F.D. and Cording, E.J. (1967). 'Design of surface and near surface construction in rock'. *Proceedings of the 8th US Symposium on Rock Mechanics 'Failure and breakage in rock'* (Minnesota) (ed. C. Fairhurst), September 15–17, 1966, Society of Mining Engineers, American Institute of Mining, Metallurgy and Petroleum Engineers, New York, pp. 237–302.

Devriendt M. (2006). 'Tunnel induced ground movements and control instrumentation'. Course Notes. British Tunnelling Society, Course on Tunnel Design and Construction, University of Surrey, Guildford, UK.

DGGT (1980). 'Empfehlungen zur Berechnung von Tunnel im Lockergestein'. Deutsche Gesellschaft Für Geotechnik e.V. (previously Deutsche Gesellschaft für Erd- und Grundbau). *Die Bautechnik*, Wilhelm Ernst & Sohn, Berlin.

Dimmock P.S. and Mair R.J. (2007a). 'Volume loss experienced on open face London Clay tunnels'. *Proceedings of the Institution of Civil Engineers, Geotechnical Engineering*, 160(GE1): 3–11.

Dimmock P.S. and Mair R.J. (2007b). 'Estimating volume loss for open-face tunnels in London Clay'. *Proceedings of the Institution of Civil Engineers, Geotechnical Engineering*, 160(GE1): 13–22.

Dimmock P.S. and Mair R.J. (2008). 'Effect of building stiffness on tunnelling-induced ground movement'. *Tunnelling and Underground Space Technology*, 23: 438–50.

DIN (2005). *Sprayed Concrete: Production and Inspection*, DIN 18551 Deutsches Institut für Normung, e.V., Berlin.

DIN (2006). *Earthworks and Foundations: Soil Classifications for Civil Engineering Purposes*, DIN 18196 Deutsches Institut für Normung, e.V., Berlin.

Duddeck H. and Erdman J. (1985). 'On structural design models for tunnels in soft ground'. *Underground Space*, 9: 245–59.

Dumpleton M.J. and West G. (1976). *A Guide to Site Investigation Procedures for Tunnels*, Transport and Road Research Laboratory Digest LR740, Department of the Environment, Crown Copyright, London.

Dunnicliff J. and Green G.E. (1993). *Geotechnical Instrumentation for Monitoring Field Performance*, John Wiley & Sons, New York.

Eddie C. and Neumann C. (2003). 'LaserShell™ leads the way for SCL tunnels'. *Tunnels and Tunnelling*, June 2003: 38–42.

Eddie C. and Neumann C. (2004). 'Development of LaserShell™ Method of Tunneling'. *Proceedings of the North American Tunneling Conference 2004* (Atlanta, GA), Balkema, Rotterdam, The Netherlands.

EFNARC (1996). *European Specification for Sprayed Concrete*, European Federation of Producers and Applicators of Specialist Products for Structures, Farnham, UK.

Einstein H., Chiaverio F. and Koppel U. (1994). 'Risk analysis for the Adler Tunnel'. *Tunnels and Tunnelling*, 26(11): 28–30.

Eskesen S.D., Tengborg P., Kampmann J. and Veicherts T.H. (2004). 'Guidelines for tunnelling risk management', International Tunnelling Association Working Group 2. *Tunnelling and Underground Space Technology*, 19: 217–37.

Essex R.J. (ed.) (1997). 'Geotechnical Baseline Reports for underground construction'. *The Technical Committee on Geotechnical Reports of the Underground Technology Research Council*, ASCE, Reston, VA, p. 40.

Essler R.D. (2009). 'Use of grout treatment'. Course Notes. British Tunnelling Society, Course on Tunnel Design and Construction. Brunel University, Uxbridge, UK.

Fang Y.S., Lin S.J. and Lin J.S. (1993). 'Time and settlement in EPB shield tunneling'. *Tunnels and Tunnelling*, 25(11), November: 27–28.

Fecker E. and Reik G. (1987). *Baugeologie*, Ferdinand Enke, Stuttgart, Germany.

Forschungsgesellschaft für das Straßenwesen (Hg.) (1980). 'Merkblatt über Felsgruppenbeschreibung für bautechnische Zwecke im Straßenbau'. *Taschenbuch für den Tunnelbau 1981*. Glückauf GmbH, Essen, Germany.

FSTT (2004). *Microtunnelling and Horizontal Directional Drilling: Recommendations*, The French Society for Trenchless Technology, Hermes Science Publishing, London.

Franzen T., Garshol K.F. and Tomisawa N. (2001). 'Sprayed concrete for final linings', ITA working group report, *Tunnelling and Underground Space Technology*, 16(4): 295–309.

Franzius J.N., Potts D.M. and Burland J.B. (2006). 'The response of surface structures to tunnel construction'. *Proceedings of the Institution of Civil Engineers, Geotechnical Engineering*, 157(GE1): 3–17.

Golser J. and Burger D. (2001). 'Bolu Tunnel – Tunnelbau im nordanatolischen Erdbebengebiet'. *Felsbau*, 19(5): 179–85.

Grant R.J. and Taylor R.N. (1996). 'Centrifuge modelling of ground movements due to tunnelling in layered ground'. *Proceedings of the International Symposium on Geotechnical Aspects of Underground Construction in Soft Ground* (City University, London) (eds R.J. Mair and R.N. Taylor), Balkema, Rotterdam, The Netherlands, pp. 507–12.

Grantz W.C. (2001). 'Immersed tunnel settlements part 2: case histories'. *Tunnelling and Underground Space Technology*, 16: 203–10.

Grimstad E. and Barton N. (1993). 'Updating the Q-system for NMT'. *Proceedings of the International Symposium on Sprayed Concrete – Modern Use of Wet Mix Sprayed Concrete for Underground Support* (Fagernes, Norway) (eds Kompen, Opsahl and Berg), Norwegian Concrete Association, Oslo, pp. 46–66.

Hansmire W.H. (2007). Personal communication with the author.

Harris F.C. (1983). *Ground Engineering Equipment and Methods*, Granada Technical Books, London.

Harris J.S. (1995). *Ground Freezing in Practice*, Thomas Telford, London.

Head K.H. (1997). *Manual of Soil Laboratory Testing: Effective Stress Tests, Pt. 3*, 2nd edition, John Wiley & Sons, New York.

Head K.H. (2008). *Manual of Soil Laboratory Testing: Soil Classification and Compaction Tests, Pt. 1*, 3rd revised edition, Whittles Publishing, Caithness, UK.

Head K.H. and Keeton P. (2009). *Manual of Soil Laboratory Testing: Permeability, Shear Strength and Compressibility Tests, Pt. 2*, 3rd revised edition, Whittles Publishing, Caithness, UK.

Heissenberger R., Lackner J. and Koch D. (2008). 'Implementierung des Sicherheitsmanagements in die Planung und Bauüberwachung am Beispiel des Lainzer Tunnels Baulos LT31 Maxing'. *Geomechanik und Tunnelbau 1*, Heft 3: 172–81.

Herrenknecht M. and Bäppler K. (2003). 'Segmental concrete lining design and installation'. Soft Ground and Hard Rock Mechanical Tunneling Technology Seminar, Colorado School of Mines (CSM), Colorado, DN.

Herrenknecht M. and Rehm U. (2003). 'Earth pressure balance shield technology'. Internal lecture. Excavation, Engineering and Earth Mechanics Institute, Colorado School of Mines (CSM), Colorado, DN.

Hilar M., Thomas A. and Falkner L. (2005). 'The latest innovation in sprayed concrete lining – the LaserShell™ method'. *Tunel* 4/2005 (Czech edition): 11–19.

Hoek E. and Brown E.T. (1997). 'Practical estimates of rock mass strength'. *International Journal of Rock Mechanics and Mining Sciences*, 34(8): 1165–86.

Hoek E. and Diederichs M.S. (2006). 'Empirical estimation of rock mass modulus'. *International Journal of Rock Mechanics and Mining Sciences*, 43: 203–15.

Holden J.T. (1997). 'Improved thermal calculations for artificial frozen shaft excavations'. *Journal of Geotechnical and Geoenvironmental Engineering*, American Society of Civil Engineers, 123(8): 696–701.

Holmberg R. (ed.) (2000). 'Explosives and blasting technique'. *Proceedings of the European Federation of Explosives Engineers 1st World Conference on Explosives and Blasting Technique*, September 2000, Balkema, Rotterdam, The Netherlands.

Holmberg R. (ed.) (2003). 'Explosives and blasting technique'. *Proceedings of the European Federation of Explosives Engineers 2nd World Conference on Explosives and Blasting Technique*, September 2003, Balkema, Rotterdam, The Netherlands.

Hopler R.B. (ed.) (1998). *Blaster's Handbook™*, 17th edition, International Society of Explosives Engineers, Cleveland, OH.

HSE (1996a). *Safety of New Austrian Tunnelling Method (NATM) Tunnels: A Review of Sprayed Concrete Lined Tunnels with Particular Reference to London Clay*, HSE Books, Sudbury, UK.

HSE (1996b). *A Guide to the Work in Compressed Air Regulations 1996*, HSE Books, Sudbury, UK.

HSE (2000). *The Collapse of NATM Tunnels at Heathrow Airport*, HSE Books, Sudbury, UK.

HSE (2006). *Tunnelling and Pipejacking: Guidance for Designers. Internal Dimensions for Pipejacks and Tunnels Below 3m Diameter and Indicative Drive Lengths*, The Health and Safety Executive and the Pipe Jacking Association. www.hse.gov.uk/construction/information.htm#other.

Huang M., Zhang C. and Li Z. (2009). 'A simplified analysis method for the influence of tunneling on grouped piles'. *Tunnelling and Underground Space Technology*, 24: 410–22.

Hunt D.V.L. (2005). 'Predicting the Ground Movements Above Twin Tunnels Constructed in London Clay'. Unpublished Ph.D. thesis, University of Birmingham, Birmingham, UK.

Ingerslev L.C.F. (1990). 'Concrete immersed tunnels the design process'. *Proceedings of the Conference on Immersed Tunnel Techniques, April 1989*, Institution of Civil Engineers, Thomas Telford, London, pp. 221–33.

ICE (1996). *Sprayed Concrete Linings (NATM) for Tunnels in Soft Ground*, ICE Design and Practice Guides, Thomas Telford, London.

ISRM (1985). 'Suggested method for determining point load strength'. *Journal of Rock Mechanics and Mining Sciences & Geomechanics Abstracts*, 22(2): 51–60.

ITA (1988). 'Guidelines for the design of tunnels'. *Tunnelling and Underground Space Technology*, 3(3): 237–49.

ITA (1997). 'State of the art report in immersed and floating tunnels'. *Tunnelling and Underground Space Technology*, 12(2): 83–355.

ITA (2000). 'Guidelines for the design of shield tunnel lining'. *Tunnelling and Underground Space Technology*, 15(3): 303–31.

ITA/AITES (2007). 'Settlements induced by tunnelling in soft ground'. *Tunnelling and Underground Space Technology*, 22(2): 119–49.

ITIG (2006). *A Code of Practice for Risk Management of Tunnel Works*, The International Tunnelling Insurance Group (in association with the Munich Re Group). www.munichre.com/publications/tunnel_code_of_practice_en.pdf.

Jäger J. and Stärk A. (2007). 'Deformation prediction for tunnels at PiccEx Junction in London Heathrow – An engineering approach. Underground Space – the 4th Dimension of Metropolises'. *Proceedings of the 33rd ITA-AITES World Tunnel Congress* (Prague), Balkema, Rotterdam, The Netherlands, pp. 463–68.

Jacobsz S.W. (2002). 'The Effects of Tunnelling on Piled Foundations'. Unpublished Ph.D. thesis, University of Cambridge, Cambridge, UK.

Jacobsz S.W., Standing J.R. Mair R.J., Hagiwara T. and Sugitama T. (2004). 'Centrifuge modelling of tunnelling near driven piles'. *Soils and Foundations*, Japanese Geotechnical Society, 44(1): 49–56.

Jancsecz S. and Steiner W. (1994). 'Face support for a large mixshield in heterogeneous ground condition'. *Proceedings of Tunnelling 1994 Conference*, The Institution of Mining and Metallurgy, London, pp. 531–50.

Jewell P.J. (2002). 'Geotechnical interpretation for tunnelling schemes'. Course Notes. British Tunnelling Society Course on Soft and Hard Ground Tunnelling, Royal Holloway, University of London, Surrey, UK.

John M., Wogrin J. and Heißel G. (1987). 'Analyse des Verbruches im Landrückentunnel, Baulos Mitte'. *Felsbau*, 5(2).

Jones B.D. (2007). 'Stresses in Sprayed Concrete Lining Junctions'. Unpublished Eng.D. thesis, University of Southampton, Southampton, UK.

Jones B.D., Thomas A.H., Yu S.H. and Hilar M. (2008). 'Evaluation of innovative sprayed-concrete-lined tunnelling'. *Proceedings of the Institution of Civil Engineers, Geotechnical Engineering*, 161, GE3: 137–49.

Karakus M. (2007). 'Appraising the methods accounting for 3D tunnelling effects in 2D plane strain FE analysis'. *Tunnelling and Underground Space Technology*, 22(1): 47–56.

Karol R.H. (1990). *Chemical Grouting*, 2nd edition, Marcel Dekker, New York.

Kavvadas M.J. (2005). 'Monitoring ground deformation in tunnelling: current practice in transportation tunnels'. *Engineering Geology*, 79: 93–113.

King M. (2006). 'Segmental lining design'. Course Notes. British Tunnelling Society, Course on Tunnel Design and Construction, University of Surrey, Guildford, UK.

Klar A., Vorster T.E.B., Soga K. and Mair R.J. (2005). 'Soil–pipe interaction due to tunnelling: comparison between Winkler and elastic continuum solutions'. *Géotechnique*, 55(6): 461–6.

Klengel J. and Wagenbreth O. (1987). *Ingenieurgeologie für Bauingenieure*, Bauverlag GmbH, Wiesbaden and Berlin, Germany.

Kuesel T.R. and King E.H. (eds) (1996). *Tunnel Engineering Handbook*, 2nd edition, Chapman and Hall, Boca Raton, FL.

Lamont D.R. (2006). 'Occupational health and safety risk management in tunnel works'. Keynote Lecture and Open Session, *ITA-AITES 2006 World Tunnel Congress*, Korean Tunnelling Association, Seoul, Korea.

Lamont D.R. (2007). 'Decompression Illness and Its Regulation in Contemporary UK Tunnelling – An Engineering Perspective'. Unpublished Ph.D. thesis, Aston University, Birmingham, UK.

Lauffer H. (1988). 'Zur Gebirgklassifizierung bei Fräsvortrieben'. *Felsbau*, 6(3): 137–49.

Lee G.T.K. and Ng C.W.W. (2005). 'Effects of advancing open face tunnelling on an existing loaded pile'. *Journal of Geotechnical and Geoenvironmental Engineering*. ASCE, 131(2): 193–201.

Lee Y.J. and Bassett R.H. (2007). 'Influence zones for 2D pile-soil-tunnelling interaction based on model test and numerical analysis'. *Tunnelling and Underground Space Technology*, 22: 325–42.

Legge N. (2006). 'Tunnel lining design – hard ground'. Course Notes. British Tunnelling Society, Course on Tunnel Design and Construction, University of Surrey, Guildford, UK.

Loganathan N., Poulos H.G. and Xu K.J. (2001). 'Ground and pile-group responses due to tunnelling'. *Soils and Foundations*, Japanese Geotechnical Society, 41(1): 57–67.

Ma B. and Najafi M. (2008). 'Development and applications of trenchless technology in China'. *Tunnelling and Underground Space Technology*, 23: 476–80.

Macklin S.R. (1999). 'The prediction of volume loss due to tunnelling in overconsolidated clay based on heading geometry and stability number'. *Ground Engineering*, April: 30–3.

Macnab A. (2002). *Earth Retaining Systems Handbook*, McGraw-Hill Professional, New York.

Maidl B., Schmid L., Ritz W. and Herrenknecht M. (2008). *Hardrock Tunnel Boring Machines*, Ernst & Sohn, Berlin, Germany.

Mair R.J. and Taylor R.N. (1997). 'Theme lecture: Bored tunnelling in the urban environment'. *Proceedings of the 14th International Conference on Soil Mechanics and Foundation Engineering* (Hamburg), 4, pp. 2353–85.

Mair R.J., Taylor R.N. and Bracegirdle A. (1993). 'Subsurface settlement profiles above tunnels in clays'. *Géotechnique*, 43(2): 315–20.

Mair R.J., Taylor R.N. and Burland J.B. (1996). 'Prediction of ground movements and assessment of building damage due to bored tunnelling'. *International Symposium on Geotechnical Aspects of Underground Construction in Soft Ground* (eds R.J. Mair, R.N. Taylor), Balkema, Rotterdam, The Netherlands, pp. 713–18.

Marshall A.M. and Klar A. (2008). 'Shell versus beam representation of pipes in the evaluation of tunneling effects on pipelines'. *Tunnelling and Underground Space Technology*, 23: 431–7.

Mayne P.W. (2007). *Cone Penetration Testing: A Synthesis of Highway Practice*, Synthesis 368, National Cooperative Highway Research Program, NCHRP, Transportation Research Board, National Research Council, Washington, DC.

McDowell P.W. (2002). *Geophysics in Engineering Investigations*, Construction Industry Research and Information Association (CIRIA), London.

Megaw T.M. and Bartlett J.V. (1981). *Tunnels: Planning, Design and Construction*. Vol. 1, Ellis Horwood, Chichester, UK.

Megaw T.M. and Bartlett J.V. (1982). *Tunnels: Planning, Design and Construction*. Vol. 2, Ellis Horwood, Chichester, UK.

Meigh A.C. (1987). 'Cone penetration testing: methods and interpretation'. Ground Engineering Report, *in situ* testing, CIRIA, Butterworths, London.

Metje N., Chapman D.N., Rogers C.D.F., Henderson P. and Beth M. (2008). 'An optical fibre sensor system for remote displacement monitoring of structures – prototype tests in the laboratory'. *Structural Health Monitoring Journal*, 7(1): 51–63.

Milligan G.W.E. and Norris P. (1999). 'Pipe-soil interaction during pipe jacking'. *Proceedings of the Institution of Civil Engineering, Geotechnical Engineering*, 134(1): 27–44.

Mohamad H., Bennett P.J., Soga K., Mair R.J., Lim C.S., Knight-Hassell C.K. and Ow, C.N. (2007). 'Monitoring tunnel deformation induced by close-proximity bored tunneling using distributed optical fiber strain measurements'. *Proceedings of the Seventh International Symposium on Field Measurements in Geomechanics* (FMGM2007), ASCE Geotechnical Special Publication, Boston, MA, p. 175.

Morgan S.R. (2006). 'Geotechnical interpretation of tunnel schemes'. Course Notes. British Tunnelling Society, Course on Tunnel Design and Construction, University of Surrey, Guildford, UK.

Moritz B., Matt R., Graf F. and Brandtner M. (2008). 'Advanced Observation Techniques for Sophisticated Shallow Tunnel projects – Experience Gained Using Innovative Monitoring Methods at the Lainzer Tunnel LT31'. *Geomechanik und Tunnelbau 1*, Heft 5: 466–76.

Morrison P.R.J., McNamara A.M. and Roberts T.O.L. (2004). 'Design and construction of a deep shaft for Crossrail'. *Proceedings of the Institution of Civil Engineers, Geotechnical Engineering*, 157, GE4: 173–82.

Moseley M.P. and Kirsch K. (eds) (2004). *Ground Improvement*, 2nd edition, Spon Press, London.

Moss N.A. and Bowers K.H. (2006). 'The effect of tunnel construction under existing metro tunnels'. *Proceedings of the 5th International Conference of TC28 of the ISSMGE, Geotechnical Aspects of Underground Construction in Soft Ground*, June, Taylor & Francis, London, pp. 151–7.

Mroueh H. and Shahrour I. (2002). 'Three-dimensional finite element analysis of the interaction between tunneling and pile foundations'. *International Journal for Numerical and Analytical Methods in Geomechanics*, 26: 217–30.

Müller, L (1978). *Der Felsbau: Dritter Band Tunnelbau*, Ferdinand Enke Verlag, Stuttgart, Germany.

Müller, L. and Fecker, E. (1978). 'Grundgedanken und Grundsätze der Neuen Österreichischen Tunnelbauweise'. *Trans Tech Publications*: 247–2.

Muir Wood A. (2000). *Tunnelling: Management by Design*, Spon, London and New York.

Najafi M. and Gokhale S.B. (2004). *Trenchless Technology: Pipeline and Utility Design, Construction and Renewal*, McGraw-Hill, New York.

New B.M. (1990). 'Ground vibration caused by civil engineering work'. *Tunnelling and Underground Space Technology*, 5(3): 179–90.

New B.M. and Bowers K.H. (1994). 'Ground movement model validation at the Heathrow Express trial tunnel'. *Tunnelling 1994, Proceedings of the 7th International Symposium of the Institute of Mining and Metallurgy and the British Tunnelling Society*, Chapman and Hall, London, pp. 310–29.

New B.M. and O'Reilly M.P. (1991). 'Tunnelling induced ground movements predicting their magnitude and effects'. *Proceedings of the 4th International Conference on Ground Movements and Structures, Cardiff*, invited review paper, Pentech Press, London, pp. 671–97.

Ng C.W.W., Simons N. and Menzies B. (2004). *A Short Course in Soil-structure Engineering of Deep Foundations, Excavations and Tunnels*, Thomas Telford, London.

Nicholson D., Tse C., Penny C., O'Hana S. and Dimmock R. (1999). *The Observational Method in Ground Engineering: Principles and Applications*, Report 185, Construction Industry Research and Information Association (CIRIA), London.

ÖBV (1999). *Sprayed Concrete Guideline: Application and Testing*, Österreichischer Beton Verein, Austria (current English language edition published 2006).

O'Reilly M.P. and New B.M. (1982). 'Settlements above tunnels in the United Kingdom – their magnitude and prediction'. *Proceedings of the Tunnelling 1982 Symposium*, London, pp. 173–81.

Österreichisches Nationalkomitee 'Hohlraumbau' (1980). 'Definition der "Neuen Österreichischen Tunnelbaumethode"'. *Schriftenreihe der Forschungsgesellschaft für das Straßenwesen im Österreichischen Ingenieur- und Architektenverein*, Heft 74, Selbstverlag.

Palmström A. (1982). 'The volumetric joint count – a useful and simple measure of the degree of jointing'. *Proceedings of the IVth International, Congress by the International Association of Engineering Geology* (New Delhi), pp. v.221–8. Publisher unknown.

Palmström A. and Broch E. (2006). 'Use and misuse of rock classification systems with particular reference to the Q system'. *Tunnelling and Underground Space Technology*, 21: 575–593.

Pang C.H., Yong K.Y, Chow Y.K. and Wang J. (2006). 'The response of pile foundations subjected to shield tunnelling'. *Proceedings of the 5th International Conference of TC28 of the ISSMGE, Geotechnical Aspects of Underground Construction in Soft Ground*, June, Taylor & Francis, London, pp. 737–43.

Peck R.B. (1969a). 'Advantages and limitations of the Observational Method in applied soil mechanics'. *Géotechnique*, 19(2): 171–87.

Peck R.B. (1969b). 'Deep excavations and tunnelling in soft ground'. *Proceedings of 7th International Conference on Soil Mechanics and Foundation Engineering* (Mexico), State-of-the-art Volume, pp. 225–90. Publisher unknown.

Pelizza S. and Piela D. (2005). 'TBM tunnelling in rock: ground probing and treatments'. *Proceedings of the International Conference on World Long Tunnels 2005* (Taipei, Taiwan), November. Publisher unknown.

Persson P., Holmberg R. and Lee J. (1994). *Rock Blasting and Explosives Engineering*, CRC Press, Danvers, MA.

PJA (1995). *Guide to Best Practice for the Installation of Pipe Jacks and Microtunnels*, Pipe Jacking Association, London.

Potts D.M. and Addenbrooke T.I. (1997). 'A structure's influence on tunnelling-induced ground movements'. *Proceedings of the Institution of Civil Engineers, Geotechnical Engineering*, 125, April: 109–25.

Potts D.M. and Zdavković L. (1999). *Finite Element Analysis in Geotechnical Engineering: Theory*, Thomas Telford, London.

Potts D.M. and Zdavković L. (2001). *Finite Element Analysis in Geotechnical Engineering: Application*, Thomas Telford, London.

Powderham A.J. (1994). 'An overview of the observational method: development in cut and cover and bored tunnelling projects'. *Géotechnique*, 44(4): 619–36.

Powderham A.J. (2002). 'The observational method – learning from projects'. *Proceedings of the Institution of Civil Engineering, Geotechnical Engineering*, 155(1): 59–69.

Powers J.P., Corwin A.B., Schmall P.C. and Kaeck W.E. (2007). *Construction Dewatering and Groundwater Control: News Methods and Applications*, 3rd edition, Wiley-Interscience, Hoboken, NJ.

Preene M., Roberts T.O.L., Powrie W. and Dyer, M.D. (2000). *Groundwater Control: Design and Practice*, Construction Industry Research and Information Association (CIRIA) C515, London.

Prinz H. and Strauß R. (2006). *Abriss der Ingenieurgeologie*, Elsevier GmbH, Spektrum Akademischer Verlag, Munich, Germany.

Quick H., Meißner S., Michael J. and Arslan U. (2001). 'Schuldwald-Tunnel: Vergleich von Ergebnissen numerischer Berechnungen mit in-situ Messungen während des Vortriebs'. *Proceedings Stuva-Conference 2001*, Forschung and Praxis 39, pp. 178–84, Bertelsmann Fachzeitschriften, Gütersloh, Munich, Germany.

Rabcewicz, L.V. (1963). 'Bemessung von Hohlraumbauten. Die "Neue Österreichische Bauweise" und ihr Einfluß auf Gebirgsdruckwirkungen und Dimensionierung'. *Felsmechanik und Ingenieurgeologie*, Vol. I/3–4, Springer Verlag, Vienna.

Rankin W.J. (1988). 'Ground movements from urban tunnelling: prediction and effects' (eds F.G. Bell, M.G. Culshaw, J.C. Cripps and M.A. Lovell), *Engineering Geology of Underground Movements*, Geological Society UK, Special Publication No. 5: 79–92.

Reuter F. (1992). *Ingenieurgeologie*, Grundstoffindustrie GmbH, Leipzig, Stuttgart, Germany.

Riggall T.J. (2008). 'Analysis of steering data from horizontal directional drilling using downhole motors'. Unpublished M.Sc. thesis, University of Birmingham, Birmingham, UK.

Rokahr R.B. (1995). 'Wie sicher ist die NÖT?'. *Felsbau*, 13(6): 334–40.

Rokahr R.B. and Mussger K. (2001). 'Der Heathrow Verbruch – Fakten und Hypothesen zur Wahrheitsfindung'. *Proceedings Stuva-Conference 2001*, Forschung and Praxis 39, Bertelsmann Fachzeitschriften, Gütersloh, Munich, Germany, pp. 185–91.

Rokahr R.B., Stärk A. and Zachow R. (2002). 'On the art of interpreting measurement results'. *Felsbau*, 20(2): 16–21.

Rokahr, R.B. and Zachow R. (1997). 'Ein neues Verfahren zur täglichen Kontrolle der Auslastung einer Spritzbetonschale'. *Felsbau*, 15(6): 430–4.

Rokahr R.B. and Zachow R. (2009). *Vergleich der Wiener Methode mit dem Verfahren IUB zur Bestimmung des Auslastungsgrades in der Spritzbetonschale am Beispiel des Lainzer Tunnels – Baulos LT31*, Leibniz Universität Hannover, Institut für Unterirdisches Bauen, Hanover, Germany.

Ropkins J.W.T. and Allenby D. (2000). 'Jacked box tunnelling'. *Proceedings of the Railway Technology Conference at Railtex 2000* (NEC, Birmingham, UK), November.

Rowe R.K., Lo K.Y. and Kack G.J. (1983). 'A method of estimating surface settlement above tunnels constructed in soft ground'. *Canadian Geotechnical Journal*, 20: 11–22.

Rubinstein R.Y. and Kroese D.P. (2007). *Simulation and the Monte Carlo Method*, 2nd edition, Wiley-Interscience, Hoboken, NJ.

Sala A. (2001). 'Geological/hydrogeological factors and operational requirements for tunnel linings'. Course Notes. Requirements for Greater Durability, Training Course for Practitioners by TSB, Switzerland.

Sandström G.E. (1963). *The History of Tunnelling: Underground Workings Through the Ages*, Barrie and Rockliff, London.

Sapigni M., Bert M., Bethaz E., Busillo A. and Cardone G. (2002). 'TBM performance estimation using rock mass classifications'. *Rock Mechanics and Mining Sciences*, 39: 771–88.

Schmidt B. (1969). 'Settlements and ground movements associated with tunnelling in soil'. Unpublished Ph.D. thesis, University of Illinois, Urbana-Champaign, IL.

Schubert, W. (1999). 'Perspektiven der NÖT. Neue Entwicklungen in der Geotechnik'. *4. Stuttgarter Geotechnik Symposium*, Stuttgart, Germany, October, pp. 35–45.

Scott J.S. (1984). *The Penguin Dictionary of Civil Engineering*, 3rd edition, Penguin Books, London.

Selby A.R. (1988). 'Surface movements caused by tunnelling in two layered soil'. *Engineering Geology of Underground Movements* (eds F.G. Bell, M.G. Culshaw, J.C. Cripps and M.A. Lovell), Geological Society Engineering Geology Special Publication No. 5, pp. 71–7.

Selemetas D., Standing J.R. and Mair R.J. (2006). 'The response of full-scale piles to tunnelling'. *Proceedings of the 5th International Conference of TC28 of the ISSMGE, Geotechnical Aspects of Underground Construction in Soft Ground*, Taylor & Francis, London, pp. 763–769.

Simons N., Menzies B. and Matthews M. (2002). *A Short Course in Geotechnical Site Investigation*, Thomas Telford, London.

SISG (1993a). *Without Site Investigation Ground is a Hazard*, Site Investigation in Construction, Part 1, Thomas Telford, London.

SISG (1993b). *Planning Procurement and Quality Management*, Site Investigation in Construction, Part 2, Thomas Telford, London.

SISG (1993c). *Specification for Ground Investigation*, Site Investigation in Construction, Part 3, Thomas Telford, London.

Standing J.R., Nyren R.J., Burland J.B. and Longworth T.I. (1996). 'The measurement of ground movements due to tunnelling at two control sites along the Jubilee Line Extension'. *Proceedings of the International Conference of TC28 of the ISSMGE, Geotechnical Aspects of Underground Construction in Soft Ground*, Balkema, Rotterdam, The Netherlands, pp. 751–6.

Stärk A. (2002). 'Standsicherheitsanalyse von Sohlsicherungen aus Spritzbeton'. Unpublished Ph.D. thesis, *Forschungsergebnisse aus dem Tunnel- und Kavernenbau*, Heft 22, Leibniz Universität Hannover, Institut für Unterirdisches Bauen, Hanover, Germany.

Stärk A. (2009). 'NATM – SCL, Displacement monitoring and quality control'. Course Notes. British Tunnelling Society, Course on Tunnel Design and Construction, Brunel University, Uxbridge, UK.

Stärk A., Rokahr R.B. and Zachow R. (2001). 'A new method for a daily monitoring of the stress intensity index of a sprayed concrete lining'. *Proceedings of the AITES-ITA 2001 World Tunnel Congress, Progress in Tunnelling after 2000*, June, Pàtron Editore, Bologna, Italy.

Stein D. (2005). *Trenchless Technology for Installation of Cables and Pipelines*, Stein and Partners GmbH, Germany.

Stone P.A., Lunniss R.C. and Shah S.J. (1990). 'The Conwy tunnel – detailed design'. *Proceedings of the Conference on Immersed Tunnel Techniques*, Institution of Civil Engineers, Thomas Telford, London, pp. 277–300.

Stroud M.A. (1989). 'The standard penetration test: its application and interpretation'. *Penetration Testing in the UK*, Thomas Telford, London, pp. 29–49.

Swoboda G. (1979). 'Finite element analysis of the New Austrian Tunnelling Method (NATM)'. *Proceedings of the 3rd International Conference on Numerical*

Methods in Geomechanics (Aachen, Germany), Vol. 2, A.A. Balkema, Rotterdam, The Netherlands, pp. 581–6.

Taylor R.N. (1995a). 'Buried structures and underground excavations'. *Geotechnical Centrifuge Technology* (ed. R. N. Taylor), Taylor & Francis, London.

Taylor R.N. (1995b). 'Tunnelling in soft ground in the UK'. *Proceedings of the International Conference on Underground Construction in Soft Ground* (eds K. Fujita and O. Kusakabe), Balkema, Rotterdam, The Netherlands, pp. 123–6.

Terzaghi K. (1950). 'Geologic aspects of soft ground tunnelling'. *Applied Sedimentation* (ed. P.D. Trask), John Wiley & Sons, New York, pp. 193–209.

TGN25 (2005). *Geotechnical Risk Management for Tunnel Works*, GEO Technical Guidance Note No. 25 (TGN25), Geotechnical Engineering Office, Civil Engineering and Development Department, The Government of Hong Kong Special Administrative Region.

Thomas A.H. (2006). 'Tunnel lining design – sprayed concrete linings'. Course Notes. British Tunnelling Society, Course on Tunnel Design and Construction, University of Surrey, Guildford, UK.

Thomas A.H. (2009a). *Sprayed Concrete Lined Tunnels*, Taylor & Francis, London.

Thomas A.H. (2009b). 'Numerical modelling'. Course Notes. British Tunnelling Society, Course on Tunnel Design and Construction, Brunel University, Uxbridge, UK.

Thomson J. (1995). *Pipejacking and Microtunnelling*, Blackie Academic and Professional, London.

Thuro K. and Plinninger R.J. (2003). 'Klassifizierung und Prognose von Leistungs- und Verschleißparametern im Tunnelbau'. *Taschenbuch für den Tunnelbau*, Verlag Glückauf Essen, pp. 62–126.

UK Government (1996). *The Work in Compressed Air Regulations 1996*, Statutory Instrument 1996 No. 1656.

Van der Poel J.T., Gastine E. and Kaalberg F.J. (2006). 'Monitoring for construction of the North/South metro line in Amsterdam, The Netherlands'. *Proceedings of the 5th International Conference of TC28 of the ISSMGE, Geotechnical Aspects of Underground Construction in Soft Ground*, Taylor & Francis, London, pp. 745–9.

Van Hasselt D.R.S., Hentschel V., Hutteman M., Kaalberg F.J., van Liebergen J.C.G., Netzel H., Snel A.J M., Teunissen E.A.H., and de Wit J.C.W.M. (1999). 'Amsterdam North/South Metroline'. *Tunnelling and Underground Space Technology*, 14(2): 191–210.

Vorster T.E.B., Mair R.J, Soga K., Klar A. and Bennett P.J. (2006). 'Using BOTDR fiber optic sensors to monitor pipeline behavior during tunnelling'. *Proceedings of the 3rd European Workshop on Structural Health Monitoring* (ed. A. Guemes), Destech Publications, Granada, Spain, pp. 930–7.

Wahlstrom E.E. (1973). *Tunnelling in Rock*, Developments in Geotechnical Engineering 3, Elsevier, Amsterdam, The Netherlands.

Waltham T. (2002). *Foundations of Engineering Geology*, Spon Press, London.

West G. (1988). *Innovation at the Rise of the Tunnelling Industry*, Cambridge University Press, Cambridge, UK.

Whittaker B.N. and Frith R.C. (1990). *Tunnelling: Design Stability and Construction*, Institution of Mining and Metallurgy, London.

Whitten D.G.A. and Brooks J.R.V. (1972). *Dictionary of Geology*, Penguin Books, London.

Wickham G.E., Tiedemann H.R. and Skinner E.H. (1972). 'Support determination based on geologic predictions'. *Proceedings of the Rapid Excavation and Tunnelling Conference*, ASCE-AIME, New York, pp. 43–64.

Wild H.W. (1984). *Sprengtechnik im Bergbau, Tunnel- und Stollenbau sowie in Tagebauen und Steinbrüchen*, Verlag Glückauf GmbH, Essen, Germany.

Williams I. (2008). 'Heathrow Terminal 5: tunnelled underground infrastructure'. *Proceedings of the Institution of Civil Engineers, Civil Engineering*, 161, May, pp. 30–7.

Williams I., Neumann C., Jäger J. and Falkner L. (2004). 'Innovativer Spritzbetontunnelbau für den neuen Flughafenterminal T5 in London'. *Proc. Österreichischer Tunneltag*, Salzburg, Austrian Committee of the International Tunnelling Association, Vienna, Austria, pp. 41–61.

Winter A. (2006). 'Lining types and where used'. Course Notes. British Tunnelling Society, Course on Tunnel Design and Construction, University of Surrey, Guildford, UK.

Wongsaroj J., Soga K. & Mair R.J. (2007). 'Modelling of long-term ground response to tunnelling under St James's Park, London'. *Géotechnique*, 57(1): 75–90.

Woods E., Battye G., Bowers K. and Mimnagh F. (2007). 'Channel Tunnel Rail Link section 2: London Tunnels'. *Proceedings of the Institution of Civil Engineers, Civil Engineering*, 160, Special Issue 2, pp. 24–8.

Woodward J. (2005). *An Introduction to Geotechnical Processes*, Spon Press, London.

Xanthakos P.P., Abramson L.W. and Bruce D.A. (1994). *Ground Control and Improvement*. Wiley-Interscience, Weinheim, Germany.

Yeates J. (1985). 'The response of buried pipelines to ground movements caused by tunnelling in soil'. *Ground Movements and Structures* (ed. Geddes, J.D.), Pentech Press, Plymouth, UK, pp. 129–44.

Zachow R. and Vavrovsky G.M. (1995). 'Abschätzung der Auslastung des Spritzbetons mit Hilfe geotechnischer Messungen'. *Felsbau*, 13(6): 382–5.

Bibliography

Augarde C.E. (1997). 'Numerical modelling of tunnelling processes for assessment of damage to buildings'. Unpublished Ph.D. thesis, University of Oxford, Oxford, UK.

Barla G. (2001). 'Tunneling under squeezing rock conditions'. *Eurosummer-School in Tunnel Mechanics*, Innsbruck, Austria.

Barla M., Ferrero S. and Barla G. (2003). 'A new approach for predicting the swelling behaviour of expansive clays in tunnelling'. *International Conference on New Developments in Soil Mechanics and Geotechnical Engineering ZM2003, Lefkosa*, Near East University Press, Turkish Republic of Northern Cyprus, pp. 29–31.

Barton N. and Grimstad E. (1994). 'The Q-System following twenty years of application in NMT support selection'. *Felsbau* 12(6): 428–36.

Deane A.P. and Bassett R.H. (1995). 'The Heathrow Express Trial Tunnel'. *Proceedings of the Institution of Civil Engineers, Geotechnical Engineering*, 113, July: 144–56.

Hoeks' Corner – useful information on rock engineering www.rocscience.com/hoek/Hoek.asp (accessed 9/09).

HSE website – www.hse.gov.uk (accessed 9/09).

Jones B.D. (2007). 'Stresses in sprayed concrete tunnel junctions'. Unpublished Ph.D. thesis, Faculty of Engineering, Science and Mathematics, School of Civil Engineering and the Environment, University of Southampton, Southampton, UK.

Kastner R., Kjekstad O. and Standing J. (2003). *Avoiding Damage Caused by Soil-interaction: Lessons Learnt from Case Histories*, Thomas Telford, London.

Maidl B., Herrenknecht M. and Anheuser L. (1996). *Mechanical Shield Tunnelling*, Ernst & Sohn, Berlin, Germany.

Mair R.J. (2008). 'Tunnelling and geotechnics: new horizons'. *Géotechnique*, 58(9): 695–736.

Matthews M., Simons N. and Menzies B. (2008). *A Short Course in Geology for Civil Engineers*, Thomas Telford, London.

Mitchell R. (2003). *Jubilee Line Extension: From Concept to Completion*, Thomas Telford, London.

Möller S.C. and Vermeer P.A. (2008). 'On numerical simulation of tunnel installation'. *Tunnelling and Underground Space Technology*, 23(4): 461–75.

Rokahr R.B. (2000). *Tunnelling: The Last Eldorado for Civil Engineers*, Institut für Unterirdisches Bauen, Leibniz Universität Hannover, Hanover, Germany.

Standing J.R. and Potts D.M. (2008). 'Contributions to Géotechnique 1948–2008: Tunnelling'. *Géotechnique*, 58(5): 391–8.

Stipek W. and Galler R. (eds) (2008). *The Austrian Art of Tunnelling in Construction Consulting and Research*, Austrian National Committee of ITA – ITA Austria, Wilhelm Ernst & Sohn, Berlin.

Széchy K. (1973). *The Art of Tunnelling*, Akadémiai Kiadó, Budapest.

Tatiya R. (2005a). *Civil Excavations and Tunnelling*, Thomas Telford, London.

Tatiya R. (2005b). *Surface and Underground Excavation*, Taylor & Francis, London.

Thomas A.H. (2004). 'The numerical modelling of sprayed concrete tunnel linings'. Unpublished Ph.D. thesis, University of Southampton, Southampton, UK.

US Army Corps of Engineers (1997). *Tunnels and Shafts in Rock*, Engineering Design Manual, EM 1110–2–2901.

Index

abrasiveness test (CERHAR) 35
accuracy 281–2, 286–7, 292, 308
advance length 67 165–6, 175, 177, 179–81, 301, 302
advance rate 57, 67, 134, 137, 289, 298, 302, 354, 355
age-dependent elastic models 83
age-dependent nonlinear models 83
analytical methods 73–8, 356; *see also* Bedded Beam Spring Method, Continuum Method, Tunnel Support Resistance Method
anchoring 143, 183, 185, 188, 302–3, 314
anchors 96, 141, 165, 183–5, 187, 189, 196–9, 206, 301–3, 337; *see also* bolts, dowels
anti-drag system (ADS) xix, 216–21, 223–4, 228, 230
Archimedean screw 158–9
Atterberg limits 33; *see also* liquid limit, plastic limit
automated total stations 308, 310

Bassett Convergence system™ 309–10
backfill 190–1, 193, 203–5, 211, 215, 286, 303, 312–15, 318
Bedded Beam Spring Method 71, 74
bench 7, 56, 180, 186–8, 190–2, 285–6, 290, 315, 324, 331, 333, 336–7, 339, 342, 366
bentonite 95, 119, 154, 160, 198–200, 220, 224, 232, 235
blasting 51, 56, 61, 149, 165, 166–9, 172–80, 186, 352; *see also* drill and blast
blowout 5, 105, 106
bolts 56, 96–8, 116, 119, 189, 285, 287; *see also* anchors, dowels
boom-in-shield tunnelling machine 135–7
bored piles 193, 196, 199, 201
borehole geophysical logging 17
boreholes 11, 16–18, 21–3, 28–9, 86, 90, 165, 172, 178–9, 284, 301, 305, 350
bottom-up method 193
Brunel's shield (Brunel's Thames tunnel shield) 4, 134, 218
building damage (classification of) 279
buoyancy 196
buried services (buried utilities)12, 272
burn cut 178–9

cable percussion boring (shell and auger) 14, 18–21, 23–4
caisson 5, 106, 122, 123
case studies *see* tunnel examples
centrifuge modelling 69, 275
Channel Tunnel 5, 126, 158, 245, 251, *see also* fire
Channel Tunnel Rail Link xix, 118, 158, 162, 269, 275
chemical grouts 95
classification 7, 12, 35, 43, 49, 50–1, 53–4, 59, 62, 258, 272, 276, 279, 346, 347, 351; *see also* rock mass classification
charging 165, 168
coarse grained soils xxii, 14–15, 25, 40, 67, 69–70, 86, 100, 267–8
coefficient of lateral earth pressure (K_0) xxii, xxiii, 15, 30, 64–5, 70, 72, 74, 361
coefficient of volume compressibility xxiii, 34
cohesion (apparent) xxii, 15, 39, 40, 47–8, 70, 76, 302, 347
cohesionless soil 14–15, 153
compaction grouting 93–4
compensation grouting 78, 102, 104–5, 263, 277, 281, 284, 309
compressed air 5, 70, 105, 107, 111, 133–4, 153, 155, 159–60, 163, 188, 245–6, 249, 252–4, 268, 271
compression seals 119
compressive strength xxiii, 34–5, 45–7, 53, 58, 83, 118, 142, 337, 345, 348
concrete linings 110, 114, 116, 123, 125–6, 189, 297, 311
cone penetration test (CPT) xix, 26–8
consistency index xxii, 14, 33–4, 160
constitutive models 81–2
contiguous pile wall 201
contingency measures 298–300, 303, 312, 314; *see also* advance rate, anchoring, backfill, divided face, elephant's feet, face support, footing piles, forepoling, grouting, post shotcreting, sealing the ground, sheet

piling, support core, temporary invert, tree trunks
Continuum Method 60, 73–6, 356
conditioning 158, 160
convergence–confinement method xxi, 80
convergence gauges 305
corrosion 96, 108, 118, 273
crack control 207
cross-hole seismic techniques 17
crown xxii, xxiii, 7, 31, 36, 51, 56, 69, 73–6, 79–80, 85–6, 89, 98, 100–1, 104, 137, 180, 186–8, 191, 285–6, 289–93, 295, 297, 303–4, 308, 315–19, 324, 327, 331, 333, 335–7, 339–40, 353, 365
cut types 174–180; *see also* burn cut, fan cut, parallel drill hole cut, pre-splitting, smooth cut, wedge cut
cut-and-cover 6, 62, 63, 90, 193–5, 202, 206, 215–16, 225, 280; *see also* bottom-up method, contiguous pile wall, diaphragm wall, ground anchors, king piles, bored piles; secant pile wall, shoring systems, sheet piling, top-down method
cutterhead 119, 129–33, 135, 137–8, 140–1, 143, 145, 147–50, 155–60, 244, 250, 253
cutting tools (dressing) 35, 131, 138, 148, 150, 155, 239; *see also* discs, drag bits, round shank cutters, scraping tools

decompression 106–8, 246, 253–4
decompression illness 106, 246, 253–4
deep well 90–1
deflectometer 305
deformation 28, 30, 36, 39–40, 74, 76–8, 80, 94, 105, 108–9, 114, 184, 272, 274, 276, 286, 288, 290, 293–4, 296–8, 305, 325, 353, 356, 358–9, 361, 363, 365, 366; critical deformation 77; deformation characteristics 30, 39, 46; deformation measurement 77, 305; deformation modulus xxii, 15, 28, 31, 38, 352, 353; deformation of the lining 109, 275; ground deformation 262, 272, 274, 277, 306
Demec™ gauge 309
desk study 10–13
detonation 168–73, 175–7, 179–81, 188, 251
detonator (blasting) cord 173–4
detonators 172–4, 176–7, 179; *see also* detonator (blasting) cord, electronic detonator, millisecond detonator, non-electrical detonator
diaphragm wall (or slurry wall) 193, 198–200
dilatometer 23, 28–30; *see also* pressuremeter
dip (dipping) 22, 41–3, 71, 346–7, 350
disc(s) 145, 147–8, 150, 153, 161
displacement xxii, xxiii, 2, 36, 41, 84, 102, 105, 110, 152, 183–4, 210, 225, 231, 262–73, 276–7, 280, 282–3, 286–96, 298–9, 305, 308, 310–19, 325, 327–9, 337, 339, 340–4, 358, 363–6; development of 287–8; displacement curve 287, 288; horizontal xxii, 269, 272, 293–5, 305, 316, 327, 329, 341; radial 363, 365, 366; tangential 364; vertical xxiii, 265, 269, 289–90, 292–4, 310, 316–19, 327
divided face 302
dowels 92, 95–8, 100–3, 116, 189; *see also* anchors, bolts, face dowels
double load plate test 30, 31
double shell lining 113
dressing *see* cutting tools
drag bits 150
drill and blast 3, 53, 61, 92, 127, 149, 164–5, 172, 175, 182, 184, 249, 251, 311, 331; *see also* charging, detonation, detonators, dust, explosive, mucking, stemming, ventilation
drill rig 166, 249
drilling xix, 18–23, 44. 52, 56, 89, 92, 94, 101, 165, 167–8, 176–7, 180, 197, 235, 239, 240, 251
drilling carriage 165; *see also* jumbo
drilling mud 235, 240
dry mix 111
durability 15, 95, 108, 110, 116, 148, 256
dust 111, 131, 148, 169, 170, 174, 181–2, 251–2, 260, 280, 286

earth pressure balance machine (EPB and EPBM) xix, 67, 88, 116–17, 153, 156–64, 230, 253, 262, 265, 269, 275; *see also* Archimedean screw, conditioning, screw conveyor
effective stress xxii, 15, 66, 195, 306, 352–3
effects of tunnelling 80, 271, 274–5; on buried utilities 272; on existing tunnels 84, 105, 272, 274–5, 308, 310; on piled foundations 272, 275; on surface and subsurface structures 164, 271
Eggemouse 312–13, 315, 319; *see also* invert control
electrolevel 305, 307, 309–10
electromagnetic 18
electronic detonator 174
elephant's feet 300–1; *see also* contingency measures
emulsion 168, 171–2
excavation chamber 154, 156–7, 163; *see also* plenum
excavation sequence 183, 184, 189–90, 192, 290, 334, 337
existing tunnels 84, 105, 272, 274–5, 308, 310
explosive 16, 147, 168–75, 177–82, 250, 251; *see also* emulsion, gelatin-dynamite, powder
extensometer 286–8, 305–7, 309–10; borehole magnet 305; rod (or invar tape) 305, 307; strain gauge 306, 309; tape 286–7, 304–5, 309–10

face dowels 92, 100–1
face support xxi, 67, 69, 105, 132–3, 138, 150, 152–3, 162, 216–19, 265, 301
face stability 62, 68, 70, 129, 137, 153, 192, 302
falsework 123
fan cut 177–8
Fenner-Pacher curve 76–7
fibre 57–9, 96, 110, 116–18, 126, 189, 232, 301, 309, 322–4
fibre optics 309
fibre reinforced 116–18, 232
field investigation 13, 23
fine grained soils 15, 25, 33, 40, 67, 69, 86, 95, 150, 198, 262, 267–8, 270
fire 84, 85, 108, 125–6, 244–5, 247, 250–2, 254
fire resistance 125–6
footing piles 301; *see also* contingency measures
forepoling 56, 98, 100–1, 103, 301, 337; forepoling plates (or sheets) 98, 100; *see also* dowels
formwork 123, 125, 206, 333
full face excavation 132, 138, 184, 192

'Gap' method 79
gelatin-dynamite 170–1
geophysical methods 16, 18; *see also* borehole geophysical logging, cross-hole seismic techniques, electromagnetic, magnetic methods, resistivity/conductivity, seismic reflection, seismic refraction
geotechnical baseline (report) xix, 61
geotechnical factual (report) xix, 60
geotechnical interpretive (report) xix, 60
Gina gasket 210
girders 100, 114–15, 184, 187, 301, 311, 322, 333–4
gripper TBM 138, 140–5; gripper shoes 140, 142–5
ground: hard 6, 62, 115, 127, 262; soft xx, 6–8, 14–15, 18–21, 23, 25, 29–30, 35, 47, 49, 53–4, 62, 64–6, 73–5, 77, 82, 84, 90, 98, 102, 115, 127–8, 132, 145, 150–2, 157, 189, 192, 197, 199, 239, 244, 249, 253, 262–7, 271, 274, 284, 293, 298, 302, 322, 330, 331–3, 336, 363; *see also* hard rock
ground conditions 2, 6, 9, 20–2, 57, 60–1, 78, 88, 97, 109–10, 113, 115, 127, 129, 138, 146, 149–50, 159, 161, 163, 164, 177, 185–6, 188, 192, 216–18, 221–2, 227, 232, 261, 262, 268, 284–5, 321
ground (rock mass) classification *see* rock mass classification
ground anchors 196–9
ground freezing 85–9, 101, 227, 281
ground improvement 1, 64, 67, 84, 94, 127, 209; *see also* ground treatment
ground investigation 9–13, 35
ground movement 8, 45, 69, 164, 227, 262–3, 265, 268–71, 274–77, 280; *see also* horizontal displacement, long term settlement, multiple tunnels, surface settlement, trough width parameter, volume loss
ground reinforcement 95; *see also* anchors, bolts, dowels
ground risk 246, 248–9
ground treatment 78, 94, 102, 222, 284; *see also* ground improvement
groundwater xix, xxi, 11–12, 16–17, 23, 43, 53, 62–3, 65–6, 69–70, 72, 85–6, 88–91, 95, 105, 113, 119, 122, 139, 145, 152, 162, 194, 196, 253, 262, 281, 302, 306, 333, 337, 350, 361
groundwater table xxi, xxiii, 16, 65, 69–70, 89, 105, 119, 152, 194, 196; lowering of 89–90, 196, 333; *see also* deep well, wellpoints
grouting 5, 78, 86, 90, 92–5, 97, 102, 104–5, 109, 115, 120, 198, 221–2, 224, 263, 270, 277, 280–1, 284, 302, 309; *see also* chemical grouts, compaction grouting, compensation grouting, jet grouting, permeation grouting, suspension grouts

hard rock 14–15, 36, 47, 53–4, 57, 96–7, 109, 114, 138, 140, 145, 147–50, 152, 164, 169, 239, 244, 311, 330–3; *see also* ground
hazard 12, 41, 174, 238, 244–6, 249–52, 254–58, 260–1
health and safety *see* safety; *see also* safety, compressed air, ground risk, hazards, occupational health, risk management
horizontal directional drilling (HDD) xvii, xix, 21–2, 235, 239–43; *see also* drilling mud, pilot tunnel, pre-reaming, pullback
horizontal displacement xxii, 269, 272, 293–5, 305, 316, 327, 329, 341
hydraulic fracturing 29, 30
hydrophilic seals 119
hydroshield 153
Hypothetical Modulus of Elasticity (HME) xix, 83

immersed tube tunnels 62, 201–3, 205–9, 211–13; *see also* crack control, water tightness
inclinometer 305, 307
in situ concrete linings 123
in situ testing 10, 20, 23; *see also* cone penetration test, dilatometer, double load plate test, hydraulic fracturing, pressuremeter, standard penetration test
instrumentation 8, 21, 105, 189, 280–3, 304, 306–7
interjacks *see* intermediate jacking station
intermediate jacking station 227, 231–3
internal friction angle xxii, 26, 40, 47, 69, 73

in-tunnel monitoring 110, 262, 285, 304, 338
invert closure 184, 189, 192
invert control 313; *see also* Eggemouse

jacked box tunnelling 216, 226, 230; *see also* anti-drag system (ADS), jacking base, jacking rig
jacking base 216–18, 220–1, 223–5
jacking pit 202, 216–7, 220, 225, 227, 231–3
jacking rig 217–18, 220, 224–5, 231–2, 217
jacks 5, 115, 117, 122–3, 132–4, 136–7, 140, 145, 154, 156, 198, 203, 210, 217, 219–20, 227, 231–2, 238, 315–16; *see also* rams
jet grouting 93, 94
joints xxii, 41, 43, 48, 53, 97, 106, 119, 179, 189, 200–1, 207, 209–11, 232, 251, 253, 273–4, 284, 289, 323, 325–6, 329, 336, 341, 351–3
jumbo 165–6, 249

king piles 196, 199
knee 7

laboratory tests 23, 28, 31, 35; *see also* abrasiveness test (CERHAR), Atterburg limits, point load index, triaxial test, uniaxial test
LaserShell™ 189, 192, 319, 322–3
lattice girder 100, 114, 187, 191, 311
layering 2, 12, 21–2, 41–2, 44–5, 71, 175; *see also* stratum
lining design 6, 108, 113
liquid limit xxiii, 14, 33
liquidity index xxii, 14, 33
London Clay 6, 73, 105, 128–9, 137, 151, 162, 192, 269, 275, 307, 321–2, 325–6, 361–2
long term settlement 270–1, 312

measuring profile 285–6
magnetic methods 18
mesh reinforcement 187, 311
microtunnelling 230–6, 250; *see also* pipe jacking
millisecond detonator 179
Mixshield 138, 151–2, 166–8
modulus xix, xxii, 12, 15, 28, 31, 34–9, 41, 44–5, 58–9, 62, 74–6, 83, 296, 352–3, 362–3
modulus ratio 38
monitoring xviii, 21, 110, 113, 165, 183–5, 189, 223, 225, 247, 249, 251, 259–62, 269, 274–5, 280–90, 292–3, 295, 298–9, 304–11; *see also* contingency measures, in-tunnel monitoring, observational method, stress-intensity-index, trigger values
monitoring targets 285
mucking 131, 149, 160, 165, 182, 187, 189, 331, 333
multi-mode TBM 145, 161–3
multiple tunnels 78, 271

nails 96 (*see also* dowels)
NATM (*see* New Austrian Tunnelling Method)
New Austrian Tunnelling Method (NATM) xix. 79–80, 82, 110, 183–90, 192, 201, 249, 262–3, 280, 283–5, 298
non-electrical detonator 173–4
numerical modelling 7, 78, 81–2, 275; *see also* age-dependent elastic models 83, age dependent nonlinear models 83, constitutive models, convergence–confinement method, 'Gap' method, progressive softening method, volume loss control method

observational method 8, 49, 113, 249, 283–4, 311
occupational health 238, 245–6, 252–3
Omega seal 210
one pass lining 113; *see also* single shell lining
open face 62, 67, 88, 128, 133, 149–50, 152–3, 161, 218, 223, 262, 265

parallel drill hole cut 177
parallel tunnels 160, 271
partial excavation 133, 135, 138
partial face boring machine 129–31, 135, 186; *see also* roadheader
particle size distribution 14, 32, 163
percussive boring 18, 92; cable percussion (shell and auger) 14, 18–21, 23–4
permeability (hydraulic conductivity) xxii, 12, 15, 32, 34, 43, 46, 48, 62, 64, 67, 84, 86, 88, 90, 95, 117, 164, 207, 235, 270
permeation grouting 93, 94
piezometer 88, 306, 307, 350; pneumatic 306, 307; standpipe 306; vibrating wire 306
piled foundations 272, 275
pilot tunnel (pilot drilling or pilot bore) 57, 129, 190, 235, 239
pipe jacking xix, 106, 122, 134, 220, 227, 230–1, 235–6, 238; *see also* intermediate jacking station, jacking pit, jacking rig, microtunnelling, reception pit, thrust wall
plastic limit xxiii, 14, 33
plasticity index xxii, 14, 25, 26, 33, 50–1
plenum 119, 153–4, 158–60
plumb-lines 308, 309
point load index xxii, xxiii, 15, 34
Poisson's ratio xxi, 15, 31, 37–8, 71, 361, 363
poling boards 196, 198
pore water pressure 28, 40, 66–7, 263, 270–1, 306
portal 5, 21, 57, 72, 88–9, 124, 206, 216–17, 247, 262, 311, 321, 351
post shotcreting 303
powder 168–9, 172; gunpowder 3–4
precise liquid level settlement gauges 308
precision 174, 179, 281
pre-reaming 239
pre-splitting 180

pressure cell 28, 159, 304, 306–7
pressuremeter 23, 28–30; *see also* dilatometer
pressurized tunnelling 85, 105; *see also* compressed air, blowout, decompression
primary lining 113
primary stresses 7, 30, 64–6, 72, 356
progressive softening method 80
pullback 235, 239, 242–3

rams 117, 135, 328; *see also* jacks
reaction frame 139
reception pit 216–17, 231
resistivity/conductivity 17
ribbed systems 114; *see also* lattice girder
risk 9, 12, 68, 92, 94, 108, 116, 145, 160, 173, 193, 205, 211, 238, 244–51, 253–61, 273–5, 277–8, 283; assessment 116, 245, 248, 256, 260, 261, 264, 275, 277–8, 338; management 8, 9, 244–5, 246, 255–7; mitigation 238, 247–8; *see also* hazard
roadheader 62, 100, 127, 129–31, 135, 249–50; *see also* partial face boring machine
rock mass classification 47–9, 54, 57–60, 345, 350, 354; *see also* Rock Mass Rating System (RMR), Rock Mass Quality Rating (Q-method), Rock Quality Designation (RQD)
Rock Mass Rating System (RMR) xix, 53, 345–7
Rock Mass Quality Rating (Q-method) xxiii, 49, 54, 350
Rock Quality Designation (RQD) xx, xxii, 14, 22, 31, 49, 52–4, 58–9, 97, 345–7, 348–9, 351–4
roof pipe umbrella 101, 103, 331, 333
rotary drilling 18–21, 92
round shank cutters 150

safety 1, 4, 54, 84, 108, 171–2, 176, 189, 203, 215, 244–54, 258, 278, 291, 319, 336, 338–9, 343, 363; health and safety xviii, xix, 3, 8, 11, 108, 110–11, 121, 127, 134, 192–3, 235, 238, 244–7, 254–5, 322–3; legislation 246
sampling 10–11, 18–21, 23–5, 31–2, 44
satellite geodesy 305
scaling 180
scraping tools (flat bits or cutting teeth) 150
screw conveyor 158–60, 163–4
sealing the ground 302
secant pile wall 201–2
secondary lining 113
segmental lining 5, 74, 81, 115–20, 123, 128, 132, 137, 139, 141–2, 185, 284, 358–9; *see also* fibre reinforced, spheroidal graphite (cast) iron (SGI)
seismic xx, xxiii, 16–17, 28, 40, 57, 157, 205, 210–11, 352–3; seismic loading 211; seismic velocity xxiii, 57, 352, 353
seismic reflection 16, 157
seismic refraction 16, 17
settlement trough xxiii, 264,–8, 270–2, 277
shafts 12, 21, 63, 88, 92, 104–5, 110, 113, 120, 122–3, 125, 133, 158, 182, 193, 230–1, 235, 237, 247, 276, 287, 319, 331–4, 336, 339; *see also* caisson, underpinning
sheet piling (sheet piles) 194, 196–7, 301
shield tail seal 132, 139, 140–1
shield; double-shield TBM (or telescopic shield) 145–6, 149; *see also* full face excavation, partial excavation, single-shield TBM, *see also* boom-in-shield tunnelling machine, tunnel boring machines
shotcrete 58–9, 98, 109, 143, 188, 190–1, 290; *see also* sprayed concrete
shoring systems 196
shoulder 7, 295
shrinkage 33, 83, 110, 206, 270, 299
side wall drift 112, 286, 290–1, 311, 330–1, 334–43
single shell lining 113; *see also* one pass lining
single-shield TBM 145–6
site investigation reports 60; *see also* geotechnical baseline (report), geotechnical factual (report), geotechnical interpretive (report)
site investigation xx, 2–3, 7, 9–13, 41, 44, 57, 60–1, 90, 149, 165, 216, 218, 248, 283; *see also* desk study, field investigation, ground investigation, site reconnaissance
site reconnaissance 10–12
sleeve port tube 93 ; *see also* tube-a-manchette
slip form 123, 125
slump test 158
slurry tunnelling machines (STM) xx, 70, 153, 155–6, 162–4, 230
slurry walls *see* diaphragm walls
smooth cut 179–80
solid core recovery xx, 14, 31
spiles 100; *see also* forepoling
spheroidal graphite (cast) iron (SGI) xx, 116, 118, 325
sprayed concrete 2, 56, 57, 81, 111–12, 114, 123, 137, 142, 143, 149, 191–2, 260–1, 286, 289, 293, 301–2, 304, 321–4, 333, 337, 344, 362; *see also* age-dependent elastic models, age-dependent nonlinear models, double shell lining, dry mix, Hypothetical Modulus of Elasticity (HME), shotcrete, single shell lining, wet mix
sprayed concrete lining (SCL) 67, 77, 80, 83, 98–9, 101, 109–10, 113, 128–9, 183–7, 189–90, 248–9, 284–5, 290, 296–300, 303, 311–15, 319, 325, 336, 339–40, 343, 356, 366; *see also* New Austrian Tunnelling Method
springline 7, 85
stability ratio xxiii, 67–8, 269
stand-up time 38, 40, 48, 53–5, 61–2, 67, 109, 123, 127–8, 132, 149–50, 186, 211, 347

standard penetration test (SPT) xx, xxiii, 20, 23–6, 94
stemming 165, 169–70, 174
stratum (strata) xxiii, 12, 18–20, 21–2, 32, 42–3, 66, 72, 84, 88, 90, 97, 134, 162, 222–4
stress-intensity-index xxi, 2, 296, 297, 298, 316, 317, 318, 319, 342, 343
stress-strain 37, 83, 296
strike 42, 346–7
support core 301
surface settlement xxiii, 197, 263–8, 275, 299, 305, 307, 337
suspension grouts 94–5
swelling 12, 33–4, 45–6, 50, 352, 353

TBM *see* tunnel boring machine (TBM)
Tell Tale™ 299–300, 309
temporary invert 300, 315, 319
thrust wall 220, 231, 233
timber heading 128, 236, 337
top-down method 193
total core recovery xx, 14, 31
total stress xxi, 66
tree trunks 303–4, 314
tremie 198–200, 210
trenchless technology (trenchless technologies) xix, 230; *see also* horizontal directional drilling (HDD), pipe jacking
triaxial test xxi, 23, 35, 38–40
trigger values 277, 282, 293
trough width parameter xxii, 264–8, 271
tube-a-manchette (TAM) xx, 92–3, 104; *see also* sleeve port tube
tunnel boring machine (TBM) xx, xxiii, 3, 54–5, 57, 61–2, 67, 70, 80, 84, 88–9, 92, 100, 115–17, 119, 125, 127, 132–3, 135, 138–59, 161–4, 185, 201–2, 230, 249–53, 262, 269, 280, 319, 328, 354–5
tunnel design 1–3, 12–15, 30, 46, 57, 61, 161, 247
tunnel examples: 4th Elbe Tunnel (Germany) 157–8; Airside Road Tunnel (UK) 160–1; Angel Islington Station (escalator tunnel, UK) 129 ; Channel Tunnel Rail Link (CTRL) xix, 118, 157, 160–2, 269, 275; Docklands Light Railway (UK) 219; Dublin Port tunnel (Ireland) 137, 138; Eggetunnel (Germany) xviii, 286, 292, 304, 311–12, 315, 339; Glendoe (UK) 144–5; Gotthard Base Tunnel (Switzerland) 23, 115, 143; Heathrow Express Extension (UK) 136, 137; Heidkopf Tunnel (Germany) 98, 123–5, 175; I-90 Highway Extension (Boston, Massachusetts, US) 225–6; Jubilee Line Extension (UK) 194, 271, 306, 308; Katschberg Tunnel 114, 165, 167–8, 174, 187, 189; King's Cross Station (UK) 128, 308, 361; Lainzer Tunnel LT31 (Austria) xviii, 90, 98–9, 102, 103, 112, 290, 313, 320; Limerick immersed tube tunnel (Ireland) 213; Owen Street (UK) 202; Piccadilly Line Extension (UK) xx, 135, 136, 319, 321; Spillvatten Tunnel (Sweden) 167–8; Storm Water Outfall Tunnel (SWOT) (UK) xx, 151; Ted Williams Tunnel (USA) 212, 214–15; Vehicular under-bridge, M1 motorway, J15A (UK) 221
tunnel lining systems 108
Tunnel Support Resistance Method 73, 76
TunnelBeamer™ 323, 329
twin tunnels 137, 222, 271

unconfined compressive strength xxiii, 34–5, 58
underpinning 88, 120
uniaxial test xxii, 35, 38, 40

ventilation 131, 143, 165, 180–2, 188, 194, 206, 238, 247, 250–1, 286, 319
vibrating wire strain gauges 306, 309
vibration 121, 131, 134, 164, 172, 174, 238, 249, 251–2, 282
volume loss xxiii, 79–80, 81, 104, 268, 269–71, 276
volume loss control method 80

waterproofing 15, 110, 113, 119, 125, 205; membrane 110, 113, 123, 184, 205–6, 357; *see also* compression seals, hydrophilic seals, Gina gasket, Omega seal
water tightness 207
weathering 13, 43, 47, 72, 347, 350
wedge cut 175–76
wellpoints 90–1
wet mix 111
working platform 4, 134, 137

Young's modulus xxii, 12, 30, 37, 41, 44, 296, 362